Handbook of AI-Driven Threat Detection and Prevention

In today's digital age, the risks to data and infrastructure have increased in both range and complexity. As a result, companies need to adopt cutting-edge artificial intelligence (AI) solutions to effectively detect and counter potential threats. This handbook fills the existing knowledge gap by bringing together a team of experts to discuss the latest advancements in security systems powered by AI. The handbook offers valuable insights on proactive strategies, threat mitigation techniques, and comprehensive tactics for safeguarding sensitive data.

Handbook of AI-Driven Threat Detection and Prevention: A Holistic Approach to Security explores AI-driven threat detection and prevention, and covers a wide array of topics such as machine learning algorithms, deep learning, natural language processing, and so on. The holistic view offers a deep understanding of the subject matter as it brings together insights and contributions from experts from around the world and various disciplines including computer science, cybersecurity, data science, and ethics. This comprehensive resource provides a well-rounded perspective on the topic and includes real-world applications of AI in threat detection and prevention emphasized through case studies and practical examples that showcase how AI technologies are currently being utilized to enhance security measures. Ethical considerations in AI-driven security are highlighted, addressing important questions related to privacy, bias, and the responsible use of AI in a security context. The investigation of emerging trends and future possibilities in AI-driven security offers insights into the potential impact of technologies like quantum computing and blockchain on threat detection and prevention.

This handbook serves as a valuable resource for security professionals, researchers, policymakers, and individuals interested in understanding the intersection of AI and security. It equips readers with the knowledge and expertise to navigate the complex world of AI-driven threat detection and prevention. This is accomplished by synthesizing current research, insights, and real-world experiences.

Handbook of AI-Driven Threat Detection and Prevention

A Holistic Approach to Security

Edited by
Pankaj Bhambri and A. Jose Anand

CRC Press
Taylor & Francis Group
Boca Raton London New York

CRC Press is an imprint of the
Taylor & Francis Group, an **Informa** business

Designed cover image: Shutterstock

First edition published 2025
by CRC Press
2385 NW Executive Center Drive, Suite 320, Boca Raton FL 33431

and by CRC Press
4 Park Square, Milton Park, Abingdon, Oxon, OX14 4RN

CRC Press is an imprint of Taylor & Francis Group, LLC

ISBN: 9781032859743 (hbk)
ISBN: 9781032860435 (pbk)
ISBN:9781003521020 (ebk)

DOI: 10.1201/9781003521020

Typeset in Times
by KnowledgeWorks Global Ltd.

Contents

Preface

The significance of security in the contemporary digital era cannot be overemphasized. The continuous progression of technology has presented remarkable prospects for both enterprises and individuals; but, it has also rendered us susceptible to a constantly changing array of risks and weaknesses. The increasing prevalence of cyberattacks targeting vital infrastructure and the theft of personal data has underscored the imperative for the implementation of resilient and flexible security solutions.

The primary objective of this edited volume is to critically examine the urgent concerns pertaining to security within the context of the era of artificial intelligence (AI). AI has swiftly evolved as a potent instrument in the possession of both security experts and malevolent entities. Therefore, it is crucial to comprehend, utilize, and mitigate the potential security ramifications of AI. A broad assemblage of specialists and prominent figures from academia, industry, and government has been convened to delve into the intricate realm of AI-driven threat identification and prevention. The objective of this book is to offer a complete examination of the impact of AI on contemporary security procedures, encompassing both the obstacles and opportunities that arise from its use.

The authors of this compilation have explored a diverse array of subjects, encompassing the utilization of machine learning and deep learning in the identification of potential risks, the ethical implications of AI in the realm of security, the implementation of AI in responding to incidents, and the influence of AI on the formulation of forthcoming security plans. The chapters in this book delve into both the technical components of security driven by AI, and the wider socio-political and ethical considerations associated with it.

In our capacity as the editors of this publication, we express our aspirations for this book to function as a significant and beneficial asset for security professionals, researchers, policymakers, and individuals with a vested interest in comprehending the convergence of AI and security. Our objective is to provide readers with the essential information and skills to effectively navigate the dynamic and intricate realm of AI-driven threat identification and prevention. This will be achieved by consolidating current research, insights, and practical experiences.

Author Bios

Dr. Pankaj Bhambri is affiliated with the Department of Information Technology at Guru Nanak Dev Engineering College in Ludhiana. He fulfills the role of the Convener for his Departmental Board of Studies. He possesses nearly two decades of teaching experience. He is an active member of IE India, ISTE New Delhi, IIIE Navi Mumbai, IETE New Delhi, and CSI Mumbai. He has contributed to the various research activities while publishing articles in the renowned *SCIE* and *Scopus* journals and conference proceedings. He has also published several international patents. Dr. Bhambri has garnered extensive experience in the realm of academic publishing, having served as an editor/author for a multitude of books in collaboration with esteemed publishing houses. Dr. Bhambri has been honored with several prestigious accolades, including the ISTE Best Teacher Award in 2023 and 2022, the I2OR National Award in 2020, the Green ThinkerZ Top 100 International Distinguished Educators award in 2020, the I2OR Outstanding Educator Award in 2019, the SAA Distinguished Alumni Award in 2012, the CIPS Rashtriya Rattan Award in 2008, the LCHC Best Teacher Award in 2007, and numerous other commendations from various government and non-profit organizations. He has provided guidance and oversight for numerous research projects and dissertations at the postgraduate and Ph.D. levels. He has successfully organized a diverse range of educational programs, securing financial backing from esteemed institutions such as the AICTE, the TEQIP, among others. Dr. Bhambri's areas of interest encompass machine learning, bioinformatics, wireless sensor networks, and network security.

Dr. A. Jose Anand is currently associated with Halifax Regional Center for Education, Halifax, Nova Scotia, Canada. Earlier, he had worked as a professor at the Department of Electronics and Communication Engineering, KCG College of Technology, Chennai, Tamil Nadu, India. He has one year of industrial experience, 24 years of teaching experience, and one year of experience as an assistant for Halifax Regional Center for Education at Saint Mary's Elementary School, Halifax, Nova Scotia, Canada. He has presented several papers at national and international conferences. He has published several papers in national and international journals and has published books for polytechnic and engineering subjects. He is a member of CSI, IEI, IET, IETE, ISTE, INS, QCFI, and EWB. His current research interests are wireless sensor networks, embedded systems, Internet of Things, machine learning, and image processing.

List of Contributors

A. Jose Anand
KCG College of Technology
Chennai, Tamil Nadu, India

Sepideh Bazzaz Abkenar
Islamic Azad University
Tehran, Iran

Shaista Alvi
Amity University
United Arab Emirates

Halil Arslan
Sivas Cumhuriyet University
Merkez, Türkiye

S. Benjamin Arul
Jeppiaar University
Chennai, Tamil Nadu, India

Gayathri B.
Bishop Heber College (Autonomous)
Trichy, Tamil Nadu, India

Paula Bajdor
Częstochowa University of Technology
Częstochowa, Poland

Pankaj Bhambri
Guru Nanak Dev Engineering College
Ludhiana, Punjab, India

Özkan Canay
Sakarya University of Applied Sciences
Sakarya, Türkiye

Abeny Ramadan Chadar
Kenyatta University
Nairobi, Kenya

Utpal Ghosh
Sarojini Naidu College for Women
Kolkata, West Bengal, India

Gowtham H.
Dalhousie University
Halifax, Nova Scotia, Canada

Ahmed Hamad
Benha University
Benha, Egypt

Ashok J.
CHRIST (Deemed to be University)
Bengaluru, Karnataka, India

Nandha Gopal J.
Velammal Institute of Technology
Chennai, Tamil Nadu, India

Mostafa Haghi Kashani
Islamic Azad University
Tehran, Iran

Karthick K.
R.M.K. Engineering College
Kavaraipettai, Tamil Nadu, India

Praveen Kumar
Datta Meghe Institute of Higher Education and Research
Wardha, Maharashtra, India

A. Vijaya Lakshmi
Vardhaman College of Engineering
Hyderabad, Telangana, India

Julien Michel
EPITA, Université de Strasbourg
Strasbourg, France

Uttam Kr. Mondal
Vidyasagar University
Midnapore, West Bengal, India

M. Narender
TKR College of Engineering and Technology
Hyderabad, Telangana, India

Mohammad Nikravan
Islamic Azad University
Tehran, Iran

Vedat Önal
Detaysoft, Türkiye

Özgür Önday
Anadolu University
Tepebaşı/Eskişehir, Türkiye

Pierre Parren
EPITA, Université de Strasbourg
Strasbourg, France

Ilona Pawełoszek
Częstochowa University of Technology
Częstochowa, Poland

J. S. Prasath
Sapthagiri NPS University
Bengaluru, Karnataka, India

Rachna Rana
Ludhiana Group of Colleges
Ludhiana, Punjab, India

David Raymond
Datta Meghe Institute of Higher Education and Research
Wardha, Maharashtra, India

Mayakannan S.
Rathinam Technical Campus
Coimbatore, Tamil Nadu, India

Arunmozhi S.A.
Saranathan College of Engineering
Trichy, Tamil Nadu, India

Mansi Sharma
Datta Meghe Institute of Higher Education and Research
Wardha, Maharashtra, India

Marta Starostka-Patyk
Częstochowa University of Technology
Częstochowa, Poland

Satya Subrahmanyam
Holy Spirit University of Kaslik
Jounieh, Lebanon

Jayapratha T.
Sri Eshwar College of Engineering
Coimbatore, Tamil Nadu, India

Induni Weerarathna
Datta Meghe Institute of Higher Education and Research
Wardha, Maharashtra, India

1 Understanding AI and Machine Learning in Security

Pankaj Bhambri

1.1 INTRODUCTION

Artificial intelligence (AI) and machine learning (ML) are transforming the cybersecurity domain by providing sophisticated functionalities to promptly identify, assess, and counteract cyber threats in real time [1]. These technologies augment conventional security measures by utilizing extensive data to detect trends and abnormalities that could suggest hostile behavior. AI and ML methods, including supervised, unsupervised, and reinforcement learning, facilitate the automation of threat identification and prevention. This leads to a substantial decrease in the amount of time and effort needed for manual monitoring and reaction [2]. AI and ML offer agile and resilient security solutions that may proactively identify and mitigate possible breaches, hence preventing substantial damage by consistently acquiring knowledge and adjusting to emerging threats [3]. Given the growing complexity of cyber threats, it is crucial to incorporate AI and ML into security frameworks to ensure the preservation of the integrity, confidentiality, and availability of digital assets.

1.1.1 Importance of AI and ML in Modern Cybersecurity

The significance of AI and ML in contemporary cybersecurity cannot be exaggerated, as they offer crucial improvements to conventional security methods in response to progressively sophisticated cyber threats. AI and ML provide the examination of extensive volumes of data at unparalleled velocities, enabling the detection of patterns and irregularities that could indicate possible breaches of security [4]. The ability to detect threats in real time is crucial for taking preemptive measures to prevent damage. AI and ML streamline several elements of identifying and addressing threats, lessening the workload on human analysts, and enabling faster and more effective countermeasures against attacks [5]. Through the process of continuously acquiring knowledge from fresh data, these technologies adjust to changing environments of potential harm, guaranteeing that security systems maintain their ability to withstand rising attacks. The incorporation of AI and ML into cybersecurity frameworks is crucial in safeguarding sensitive data, preserving system integrity, and assuring the resilience of organizations, as cyber-attacks become increasingly intricate and frequent.

DOI: 10.1201/9781003521020-1

1.1.2 Overview of the Chapter

This chapter presents a thorough examination of the influential impact of AI and ML in the realm of cybersecurity. Due to the increasing complexity and volume of cyber threats, conventional security methods frequently prove inadequate in effectively reducing these risks. This chapter aims to bridge this gap by exploring how AI and ML technologies enhance security protocols and create more resilient digital defenses.

The chapter is started by laying a foundational understanding of AI and ML, highlighting their distinct and synergistic contributions to security measures. This section delves into the basic principles and historical evolution of these technologies, setting the stage for more advanced discussions. The chapter classifies several AI and ML strategies used in danger identification and prevention, including supervised, unsupervised, and reinforcement learning. Each category is explained with an emphasis on its unique capabilities and applications in identifying and neutralizing cyber threats. Through detailed case studies and practical examples, real-world applications of AI and ML in cybersecurity are demonstrated. These examples demonstrate how these technologies can detect patterns, predict future security breaches, and respond to threats in real time.

In addition to technical applications, the chapter also covers the integration of AI and ML in different security domains, including network security, endpoint protection, and data security. Advanced topics such as anomaly detection, behavioral analysis, and the use of neural networks in identifying malicious activities are discussed in depth. Ethical and legal implications of AI-powered security solutions are examined, with a focus on transparency, accountability, and privacy protection. Challenges and obstacles in implementing these technologies are also addressed, including technical, organizational, and ethical concerns. Finally, the chapter explores future trends in AI and ML for cybersecurity, highlighting emerging technologies and innovations that are expected to shape the field. Predictive analytics and the potential for forecasting future security breaches are discussed, providing insights into the prospective trends that will influence the discipline.

By the end of this chapter, readers will have a thorough understanding of the impact of AI and ML on cybersecurity, the challenges involved in their application, and the future directions of these technologies. This knowledge will equip security professionals, researchers, and policymakers with the necessary insights to leverage AI and ML for building robust, adaptive, and resilient security frameworks.

1.2 FUNDAMENTALS OF ARTIFICIAL INTELLIGENCE AND MACHINE LEARNING

AI and ML are crucial technologies in contemporary cybersecurity, with each offering distinct and complimentary capacities to strengthen safety precautions [6]. AI involves the creation of systems capable of executing activities that traditionally necessitate human intelligence, including thinking, learning, and problem-solving. ML, a branch of AI, is concerned with creating algorithms that enable computers to learn from data and make predictions or judgments [7–9]. AI and ML depend on

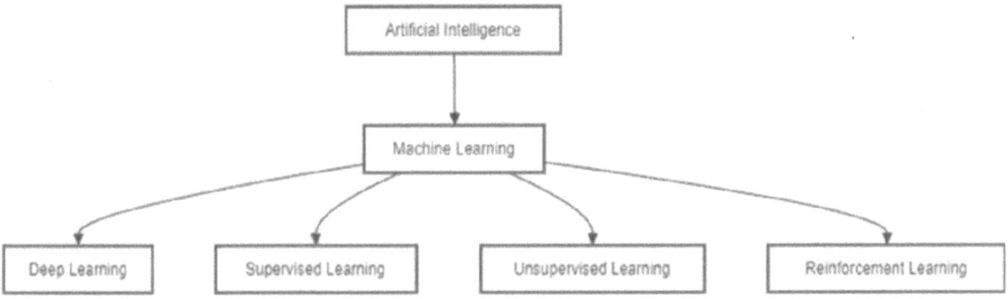

FIGURE 1.1 Artificial intelligence and machine learning.

extensive data and advanced algorithms to spot patterns and abnormalities, allowing for proactive identification and reaction to threats. Supervised learning algorithms, trained on labeled data, can classify and predict threats; while unsupervised learning can detect previously unknown threats by identifying outliers. Reinforcement learning, which learns optimal actions through trial and error, can enhance adaptive security measures [10]. Together, these technologies provide powerful tools for automating threat detection, analyzing vast datasets in real time, and continuously improving security protocols in the face of evolving cyber threats. Figure 1.1 depicts the relation of AI with ML.

1.2.1 Basic Principles of AI and ML

Some of the basic principles of AI [11] are as follows:

- *Automation*: AI seeks to automate processes that traditionally necessitate human intelligence, such as logical thinking, acquiring knowledge, and resolving problems.
- *Data-driven*: AI systems depend on huge amounts of data to acquire knowledge of patterns and render judgments. Accurate data is essential for optimal AI performance.
- *Intelligence*: AI systems aim to replicate human cognitive functions, such as sensing, reasoning, and learning via experience.
- *Adaptability:* AI systems have the ability to adjust to new sources and enhance their capabilities over time, frequently by means of self-learning methods.
- *Interactivity*: Many AI systems are designed to interact with users in a natural manner, understanding and processing human language or behaviors.

And basic principles of ML [12] are as follows:

- *Learning from data*: ML algorithms acquire knowledge from past data to recognize patterns and provide predictions or judgments without the need for explicit programming.

- *Types of learning:*
  - *Supervised learning*: The algorithm acquires knowledge from labeled data by utilizing input-output pairs to make predictions about future outcomes.
  - *Unsupervised learning*: The algorithm acquires knowledge from data that lacks labels, discerning patterns or clusters within the data.
  - *Reinforcement learning*: The algorithm acquires knowledge by its interaction with an environment, where it is provided with rewards or punishments depending on its behaviors.
- *Generalization:* One of the main objectives in ML is to extrapolate from the learning data in order to generate precise predictions on data that has not been previously encountered.
- *Model evaluation*: ML models undergo evaluation using metrics such as precision, recall, and F1 score to ascertain their performance on new data.
- *Feature selection*: Selecting relevant features (attributes) from the data is crucial for improving model performance and reducing complexity.

These principles serve as the basis for creating and executing AI and ML systems in several fields, including healthcare, finance, entertainment, and autonomous systems.

1.2.2 Key Differences and Complementary Roles

Key differences between AI and ML are as follows [13]:

- *Scope:*
  - AI: It is a wide-ranging discipline that aims to develop systems capable of doing activities that usually necessitate human intelligence, like logical thinking, comprehending natural language, and seeing the environment.
  - ML: It is a branch of AI that is dedicated to developing methods that allow systems to learn through data and enhance their performance without explicit programming.
- *Functionality:*
  - AI: It includes a range of methods, such as rule-based systems, trained systems, along with neural networks, and others.
  - ML: It utilizes statistical methods and algorithms to facilitate machine learning and data-driven predictions.
- *Data requirement:*
  - AI: It can operate with predefined rules and logic, sometimes requiring less data.
  - ML: It heavily relies on data; the performance of ML models improves with more high-quality data.
- *Applications:*
  - AI: It has broader applications including robotics, natural language processing (NLP), computer vision, and more.
  - ML: It has more focused applications, often used within AI systems for tasks like recommendation systems, fraud detection, and image recognition.

The complementary roles of AI and ML are as follows [14]:

- *Enhancement of AI:* ML enhances AI capabilities by providing the tools to learn from data, making AI systems more adaptive and intelligent.
- *Automation*: While AI provides the framework for intelligent behavior, ML automates decision-making processes by learning from data [15].
- *Problem-solving*: AI can integrate various techniques (including ML) to address complex problems across diverse domains, from healthcare to finance.
- *Innovation*: The combination of AI and ML drives innovation, enabling new applications and solutions that were not possible with traditional programming methods.

1.2.3 Historical Context and Evolution

The historical origins of AI and ML may be traced back to the second half of the twentieth century, specifically to the Dartmouth Conference in 1956 where the word "AI" was first introduced. This conference played a significant role in establishing the foundation for future study in the field of AI [16]. Initially, AI was primarily concerned with symbolic thinking and problem-solving. This was demonstrated by programs such as the Logic Theorist and General Problem Solver [17]. The advent of neural networks in the 1980s was a significant milestone, but enthusiasm diminished due to constraints in computing capabilities and data availability. The reemergence of AI in the twenty-first century, propelled by advancements in computer power, the availability of large datasets, and improved algorithms, led to the rapid development of ML, specifically deep learning. This has brought about significant transformations in areas such as image and speech recognition [18]. The advancement of AI and ML technologies has resulted in their extensive use across several industries, revolutionizing our interaction with machines and the way we handle information [19].

1.3 CATEGORIES OF AI AND ML TECHNIQUES IN SECURITY

The application of AI and ML techniques in the field of security can be classified into various important domains: Anomaly detection refers to the employment of algorithms to find atypical patterns in web traffic and user behavior, with the purpose of detecting potential dangers. Intrusion detection systems (IDS) employ ML models to identify and promptly react to harmful actions. Fraud detection involves the use of ML models to analyze transaction patterns and identify fraudulent transactions in real time. User authentication utilizes biometric recognition as well as behavioral analysis to improve security measures. Threat intelligence employs AI to analyze large amounts of data from different sources in order to predict and mitigate possible threats to security [20]. These categories demonstrate the growing integration of AI and ML techniques into security structures in order to improve the capabilities of detecting, preventing, and responding to threats.

1.3.1 Supervised Learning

Supervised learning, a fundamental subset of ML in the wider field of AI, is often used in security situations to improve the identification and response to threat processes. This method entails instructing algorithms using labeled datasets, wherein input data is matched with appropriate output labels. This enables the model to discern patterns linked to certain security risks, such as spyware, phishing scams, or attempts to gain access [21]. Supervised learning is often employed to categorize emails as either legitimate or spam by utilizing past data, thus enhancing the precision of spam-filtering systems. Organizations can enhance their defenses against progressively complex assaults by consistently providing the model with fresh data and retraining it to react to evolving security threats. Supervised learning is essential in automating and enhancing processes for making decisions in cybersecurity, hence increasing the responsiveness and effectiveness of systems [22].

1.3.2 Unsupervised Learning

Unsupervised learning is an essential subset of AI and ML methods that are employed in security applications, namely for the purpose of detecting anomalies and identifying threats. Unsupervised learning algorithms differ from supervised learning algorithms in that they do not require labeled data. Instead, they evaluate datasets without labels to discover concealed structures or patterns [23]. These techniques in cybersecurity can detect abnormal activity in network traffic by differentiating normal patterns from aberrant ones, hence identifying probable security breaches or attacks [24]. Unsupervised learning aids in the categorization of similar occurrences, enabling security teams to identify patterns and determine the order of importance for their replies. Organizations can improve their ability to detect threats and reduce risks by utilizing techniques such as clustering (e.g., k-means) and dimensionality reduction (e.g., principal component analysis [PCA]). This can be done without relying on large amounts of labeled datasets.

1.3.3 Reinforcement Learning

Reinforcement learning (RL) is a potent subset of AI and ML methodologies that concentrates on instructing agents to make decisions by repeatedly attempting different actions and receiving feedback in the form of incentives or penalties. In the context of security, RL can be utilized to enhance threat detection, response systems, and adaptive security protocols [25]. For example, RL algorithms can learn optimal strategies for identifying anomalies in network traffic or predicting potential breaches by continuously interacting with the environment and adjusting their policies based on evolving threats. This dynamic learning process allows security systems to adapt to new attack patterns and vulnerabilities in real time, significantly improving their efficacy in safeguarding sensitive data and infrastructure [26]. By leveraging RL, organizations can develop proactive security measures that evolve alongside emerging cyber threats, thereby enhancing overall resilience.

1.4 APPLICATIONS OF AI AND ML IN THREAT DETECTION AND PREVENTION

Nowadays, AI and ML are being used more and more to detect and prevent threats in several areas, especially in the field of cybersecurity. These technologies let enterprises to examine large quantities of data in actual time, detecting trends and irregularities that could suggest possible dangers. Algorithms for ML can be trained using past attack data to identify the distinctive characteristics of malware and phishing efforts. This enables the automated identification of such threats in real time. AI-driven systems can enhance incident response by predicting potential vulnerabilities and recommending preventive measures. By utilizing NLP, AI can also sift through security logs and alerts, prioritizing those that require immediate attention. The integration of AI and ML into threat detection frameworks significantly enhances an organization's ability to proactively defend itself against cyber threats and minimize response times.

1.4.1 Network Security

Network security utilizes AI and ML to improve the detection and prevention of threats. This is achieved by allowing systems to promptly recognize and address abnormalities and potential attacks as they occur. By analyzing extensive volumes of network data related to traffic, AI systems can identify patterns that suggest malicious activity, such as abnormal access requests or attempts to steal data. ML models, specifically anomaly detection methods, are trained using past network data to identify normal patterns, enabling them to detect anomalies that could indicate breaches of security [27]. AI-powered technologies have the ability to automatically carry out response activities, such as isolating systems that have been affected or blocking suspicious traffic. This results in faster response times and reduces the negative effects of possible attacks. The incorporation of AI and ML into the security of networks not only increases the precision of identifying threats but also boosts the effectiveness of responding to incidents, ultimately resulting in stronger security measures against developing cyber threats.

1.4.2 Endpoint Protection

Endpoint protection utilizes AI and ML to improve the identification and prevention of threats. It is achieved by constantly tracking and evaluating data from endpoint devices, including PCs and mobile phones [28]. Through the utilization of ML algorithms, these systems have the capability to recognize regular behavioral patterns and promptly identify irregularities that could suggest possible risks, such as malware or unwanted attempts to get access. Endpoint protection systems powered by AI can dynamically acquire knowledge from emerging threats, enhancing their precision as time progresses and reducing the occurrence of incorrect detections. These solutions have the capability to automatically respond to identified threats by isolating infected endpoints and launching remediation operations. This results in a reduction of the time and resources needed for human intervention [29]. By incorporating

AI and ML into endpoint protection, a business can enhance its security position by gaining immediate and actionable knowledge about emerging cyber threats and implementing proactive defense measures.

1.4.3 Data Security

Data security utilizes AI and ML technologies to improve the identification and prevention of threats in different digital settings [30]. Through the analysis of extensive data in real time, AI and ML algorithms have the capability to detect trends and abnormalities that may indicate possible breaches of security, such as atypical access attempts or anomalous network traffic. These systems have the ability to acquire knowledge from past data, consistently enhancing their precision in differentiating between harmless and harmful behaviors. For instance, ML models can classify email content to detect phishing attempts, while anomaly detection algorithms can flag deviations in user behavior that might suggest compromised accounts. By automating the threat detection process, AI and ML not only reduce the response time to incidents but also allow security teams to focus on higher-priority tasks, thereby significantly strengthening overall data security posture [31].

1.5 ADVANCED AI AND ML TECHNIQUES IN CYBERSECURITY

The application of advanced AI and ML methods is revolutionizing the field of cybersecurity by improving the ability to detect, respond to, and avoid threats. Neural networks and deep learning are used to evaluate intricate patterns in extensive datasets, enabling more precise detection of sophisticated threats, such as zero-day attacks with advanced persistent threats (APTs). NLP is employed to scrutinize textual data, such as safety logs and user communications, in order to detect indications of phishing or threats from insiders [32]. RL is applied to develop adaptive security systems that evolve in response to new threats, optimizing incident response strategies over time [33]. These advanced methodologies enable organizations to not only detect and mitigate threats more effectively but also predict potential vulnerabilities, thereby fostering a proactive cybersecurity posture that evolves with the ever-changing landscape of cyber threats.

1.5.1 Anomaly Detection

Anomaly detection is an advanced use of AI and ML in the field of cybersecurity. Its main objective is to find abnormal patterns or departures from the expected behavior within extensive datasets. Anomaly detection models can utilize algorithms like supervised, unsupervised, or semi-supervised learning to analyze historical data and identify abnormal actions that could indicate future security breaches or attacks [34]. The utilization of this strategy is of utmost importance in the field of cybersecurity as it allows for the timely identification of new methods of attack and internal dangers that may get overlooked by conventional rule-based systems. Anomaly detection models continually adjust to changing threats by upgrading their comprehension of what defines normal behavior, thereby offering a proactive protection mechanism toward sophisticated cyberattacks. By incorporating the identification of anomalies into cybersecurity frameworks, the overall resilience is improved as it allows for quick response and mitigation steps to address possible hazards before they worsen.

1.5.2 Behavioral Analysis

Behavioral analysis is an advanced technique in cybersecurity that employs AI and ML to monitor and analyze the actions of users and entities. Its purpose is to detect and identify potential security problems [35]. Through the utilization of ML algorithms, such systems are able to identify abnormal activity by establishing a reference point of typical behaviors. The deviations can serve as indicators of potential hostile actions such as insider threats or hacked accounts. For instance, if a user who usually accesses files during regular working hours suddenly starts downloading substantial volumes of data at unusual hours, the behavior analysis system can identify this deviation and highlight it for further examination. By adopting a proactive approach, companies can promptly address possible dangers, hence minimizing the chances of successful assaults [36]. Behavioral analysis can enhance user awareness and compliance by providing insights into risky behaviors, ultimately contributing to a more robust cybersecurity framework.

1.5.3 Neural Networks in Detecting Hostile Actions

Neural networks represent an advanced AI and ML technique that is increasingly used in cybersecurity for detecting hostile actions and threats. Deep learning architectures, including other models, are highly proficient in analyzing intricate patterns within extensive datasets. Consequently, they are exceptionally suitable for detecting advanced cyber threats such as malware, phishing attempts, and insider threats. By training on extensive historical data, neural networks can learn to recognize subtle indicators of malicious behavior, even those that may be imperceptible to traditional rule-based systems [37]. For instance, convolutional neural networks (CNNs) can be employed to analyze traffic patterns and classify network anomalies, while recurrent neural networks (RNNs) are adept at modeling sequential data, enabling them to detect unusual user activity over time. The ability to acquire knowledge and adjust accordingly improves the precision and swiftness of identifying potential dangers, enabling enterprises to react more efficiently to emerging cyber threats and strengthen their overall security measures. Figure 1.2 shows the neural networks in detecting hostile actions.

1.6 CASE STUDIES AND PRACTICAL EXAMPLES

1.6.1 Real-World Applications

AI and ML have numerous real-world applications in the domain of security, significantly enhancing threat detection, response, and prevention mechanisms across various sectors [38]. These are discussed below:

- *Intrusion detection systems (IDS)*: AI and ML algorithms are utilized to monitor network activity in real time with the purpose of detecting abnormal patterns that could potentially signify a security breach. Through the

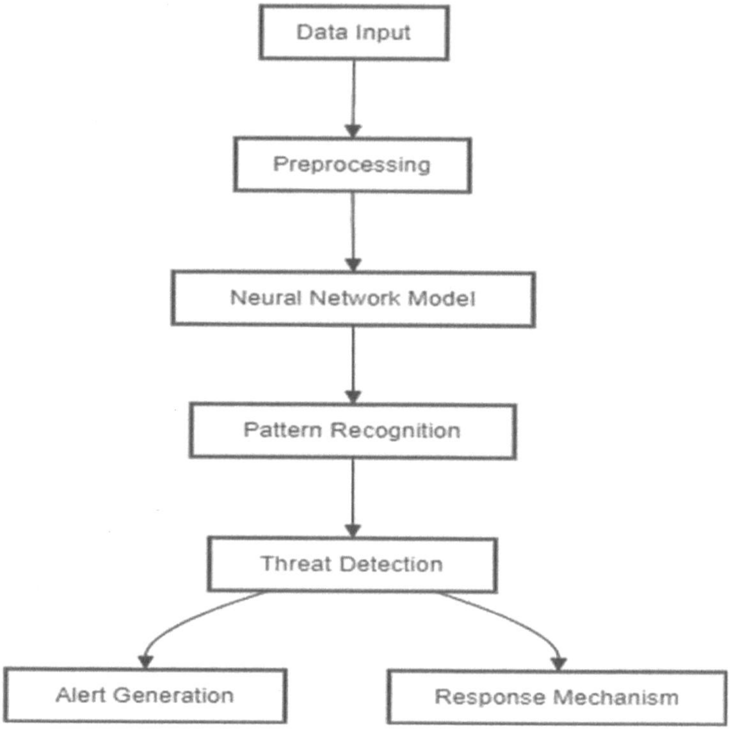

FIGURE 1.2 Neural networks in detecting hostile actions.

process of analyzing previous data, these systems are able to adjust to changing risks and reduce the occurrence of incorrect identifications.

- *Fraud detection*: ML models are utilized in the banking industry to identify fraudulent transactions through the examination of patterns and abnormalities in user behavior. These systems can flag suspicious activities, such as unusual purchase amounts or locations, for further investigation.
- *Identity and access management*: AI-driven solutions enhance identity verification processes by using biometric data, such as facial recognition or fingerprint scanning, to ensure secure access to systems and sensitive information [39].
- *Phishing detection*: ML algorithms can scan emails and web pages to identify phishing attempts by analyzing textual and visual content for suspicious indicators, thereby protecting users from potential scams.
- *Predictive analytics for threat intelligence*: AI systems have the capability to evaluate extensive quantities of data from diverse sources, including threat feeds as well as social media. This analysis enables the systems to forecast upcoming security threats, empowering businesses to take proactive measures to enhance their defenses.
- *Automated incident response*: AI has the capability to optimize incident response procedures by automatically executing activities according

to pre-established criteria. This allows for quicker resolution of security threats and alleviates the burden on security personnel.

1.6.2 Detailed Case Studies

Here are some detailed case studies highlighting the application of AI and ML in security [40]:

- *Darktrace*: Darktrace, a prominent company in the field of AI-powered cybersecurity, employs advanced ML techniques to promptly detect and counteract cyber threats. Darktrace uses unsupervised learning to examine network traffic patterns and build a baseline of usual conduct for users and devices. Darktrace's system is designed to detect and respond to deviations from the normal baseline activity in digital environments. These deviations could include abnormal data transfers or unauthorized access attempts. When such deviations are detected, the system provides alerts and can take automatic actions to reduce possible risks. In this way, Darktrace's system acts as a digital "immune system."
- *IBM QRadar*: IBM's QRadar Security Information and Event Management (SIEM) software integrates AI and ML to improve the identification and handling of security threats. QRadar utilizes algorithms for identifying anomalies to monitor log data as well as network flows, detecting abnormal behaviors that could potentially suggest security breaches. The technology employs sophisticated analytics to rank warnings according to their risk levels, enabling security personnel to concentrate on the most crucial threats and decrease response times.
- *Cisco's Security Solutions*: Cisco integrates ML in its security products, particularly in the Cisco Talos Intelligence Group, which focuses on threat intelligence. By leveraging ML algorithms, Cisco can rapidly analyze and classify threats, including malware and phishing attacks, based on vast datasets of known threats. This proactive approach enables faster updates to security protocols and real-time threat intelligence dissemination, enhancing overall security posture [41].
- *Google's Cloud Security*: Google employs AI and ML to detect threats in its cloud services. The Google Cloud Security system utilizes ML algorithms to examine user patterns and identify anomalies, such as unwanted access attempts or atypical data movements. This feature enables enterprises to promptly detect potential breaches and implement proactive measures to protect sensitive data preserved in the cloud.

1.6.3 Practical Implementation Examples

Here are some practical implementation examples of AI and ML in security domain [42]:

- *User behavior analytics (UBA)*: ML is employed to monitor and analyze user behavior within an organization, establishing baselines for normal activity. Alerts may be triggered for possible threats from insiders or compromised accounts if there are any variations from these trends, such as unexpected login locations or irregular data access.

- *Security automation and response*: AI-driven security orchestration tools automate incident response by analyzing alerts and determining appropriate actions based on predefined criteria. For example, platforms like Splunk can integrate ML to prioritize alerts and automate responses, significantly reducing the time to mitigate threats.

1.7 ETHICAL AND LEGAL IMPLICATIONS

Organizations must negotiate the ethical and legal considerations that arise from deploying AI and ML in security.

- *Importance of transparency and accountability*: Transparency in AI algorithms is crucial, as stakeholders need to understand how decisions are made, particularly in sensitive areas like threat detection and response. Ensuring accountability involves establishing clear lines of responsibility for the outcomes produced by AI systems, especially when false positives or negatives can lead to substantial repercussions, such as wrongful accusations or missed threats [43].
- *Safeguarding privacy*: Utilizing AI in security frequently entails the analysis of extensive quantities of private information, which may give rise to problems regarding privacy [44]. Organizations must establish strong data protection protocols to preserve the privacy rights of individuals, while also ensuring adherence to rules such as the General Data Protection Regulation (GDPR). This entails limiting data collection to what is necessary, anonymizing personal data when possible, and securing explicit consent from users where applicable.
- *Legal ramifications*: Legal frameworks surrounding AI technologies are still evolving, but there are implications regarding liability and compliance. For example, if an AI-driven security solution incorrectly flags an individual as a threat, questions arise about liability for damages caused [45]. Organizations must be up-to-date with current and upcoming rules to minimize legal risks and guarantee that their AI platforms are created with ethical concerns.

Overall, addressing these ethical and legal implications is not only crucial for compliance but also for fostering public trust in AI-powered security solutions. Ensuring that these systems are transparent, accountable, and respectful of privacy rights will help build confidence among users and stakeholders.

1.8 CHALLENGES AND OBSTACLES IN IMPLEMENTING AI AND ML IN SECURITY

Implementing AI and ML in the field of security faces several challenges and obstacles across various domains that are discussed as follows:

- Technical challenges:

- *Data quality and quantity*: AI and ML models necessitate substantial quantities of meticulously curated data for the purpose of training. In the field of security, data may exhibit noise, incompleteness, or bias, hence resulting in the development of inefficient models.
- *Complexity of attacks*: Cyber threats are continuously evolving, making it difficult for static models to keep up. Adaptive techniques are needed, but these can be complex to develop and deploy [46].
- *False positives and negatives*: Excessive occurrences of false positives might overpower security staff, while false negatives can result in overlooked dangers. Striking a balance between sensitivity and specificity poses a notable difficulty [47].
- *Integration with existing systems*: Integrating AI and ML solutions with legacy systems can be difficult, requiring substantial resources and technical expertise.

- Organizational and operational challenges:
    - *Skills gap*: There is often a shortage of skilled personnel who understand both cybersecurity and AI/ML, making it hard to implement and maintain these technologies effectively.
    - *Resistance to change*: Organizational culture can hinder the adoption of AI and ML. Some employees may exhibit resistance toward adopting new technologies or harbor concerns about potential loss of employment [48].
    - *Resource allocation*: Organizations with restricted funds may face challenges in implementing modern AI and ML solutions due to the substantial investment required in technology and training.
- Addressing bias and ethical concerns:
    - *Bias in algorithms*: AI and ML models have the potential to perpetuate or magnify biases that exist in the data used for training. This can result in unfair or discriminating outcomes, particularly in automated decision-making systems [49].
    - *Transparency and explainability*: Several AI systems function as "opaque entities," posing challenges in comprehending the decision-making process. The absence of openness can give rise to ethical concerns and make it more difficult to adhere to regulations.
    - *Privacy concerns*: The use of personal data for training models poses privacy risks, requiring organizations to navigate data protection laws and ethical standards [50].

To tackle these difficulties, a strategic procedure is necessary. This approach involves investing in highly talented individuals, promoting a culture of creativity, guaranteeing the accuracy of data, and giving priority to ethical considerations while developing and implementing AI and ML tools in security.

1.9 FUTURE TRENDS IN AI AND ML FOR CYBERSECURITY

The future of AI and ML in the field of cybersecurity is expected to experience substantial progress due to the emergence of new technology and innovative approaches.

- *Emerging technologies and innovations:* An important trend is the merging of AI with blockchain computing to improve the reliability and safety of data, guaranteeing that records and transactions cannot be altered or tampered with. The progress in quantum computing has the potential to completely transform encryption techniques, leading to the creation of algorithms that are resistant to quantum attacks. These algorithms will incorporate AI to carry out security evaluations in real time [51]. Furthermore, the rise of edge computing will enable more localized data processing, allowing AI-driven security solutions to operate with lower latency and increased efficiency, especially in Internet of Things (IoT) environments.
- *Predictive analytics and forecasting future security breaches*: AI and ML will increasingly utilize predictive analytics to anticipate and identify future breaches of security before they actually happen. ML algorithms can utilize extensive datasets from several sources, such as security alert feeds and previous attack patterns, to detect weaknesses and forecast emerging attacks [52]. By adopting this proactive strategy, businesses can strengthen their protections and implement preventative measures, thereby moving their focus from reactionary to anticipatory security tactics.
- *Prospective trends shaping the discipline*: The cybersecurity landscape will be shaped by the growing adoption of AI-driven automated response systems that can rapidly mitigate threats without human intervention, enhancing response times significantly. Additionally, as organizations face an increasing volume of cyberattacks, there will be a greater emphasis on explainable AI (XAI) to ensure transparency in decision-making processes, allowing security teams to understand and trust AI-generated insights [53]. Moreover, the convergence of cybersecurity with other domains such as privacy protection and compliance will drive the development of integrated AI solutions that address a broader range of security concerns.

1.10 CONCLUSION

1.10.1 Summary of Key Points

This chapter has examined the profound influence of AI and ML in the field of cybersecurity. Key takeaways include:

- The foundational principles of AI and ML, highlighting their complementary roles in enhancing security measures.
- Comprehensive examination of different AI and ML methods, including supervised, unsupervised, and reinforcement learning. These approaches are crucial for ensuring efficient identification and mitigation of potential threats.
- Practical implementations of these technologies in several fields, such as network security, safeguarding endpoints, and data security.
- Insights into advanced techniques such as anomaly detection and neural networks, showcasing their effectiveness in identifying and responding to cyber threats.

- An analysis of the legal and moral consequences associated with the integration of AI in security, with a particular focus on the importance of openness and accountability.
- Challenges faced in the adoption of AI and ML, particularly concerning bias, technical hurdles, and organizational readiness.

1.10.2 Final Thoughts on the Role of AI and ML in Building Robust Security Frameworks

Given the increasing complexity and sophistication of cyber threats, the use of AI and ML in cybersecurity measures is not just advantageous, but absolutely necessary. These technologies offer the flexibility and ability to handle increased demands in order to predict, identify, and react to potential risks immediately, hence strengthening the durability of security systems. Nevertheless, it is crucial that the deployment of AI and ML is undertaken with meticulous regard for ethical norms and legal ramifications. By cultivating a culture characterized by accountability and openness, security experts may effectively utilize the complete capabilities of AI and ML to develop strong and flexible security systems which not only safeguard against existing risks but also proactively predict forthcoming issues. As we look ahead, continuous innovation and collaboration among stakeholders will be crucial in shaping a secure digital environment that safeguards individuals and organizations alike.

REFERENCES

1. Brown, T. (2022). Ethical implications of AI in cybersecurity. Cybersecurity Ethics Review, 8(1), 50–65.
2. Kumar, A., & Singh, S. (2022). Machine learning for malware detection: A comparative study. Journal of Digital Security, 10(4), 220–235.
3. Cooper, D. (2023). Automated response systems powered by AI. Security Automation Journal, 3(1), 15–29.
4. Martinez, E., & Johnson, B. (2023). Neural networks in intrusion detection systems. Journal of Network Security, 34(1), 30–45.
5. Turner, M. (2023). Exploring the role of AI in proactive cybersecurity measures. Journal of Information Security and Applications, 28, 100–115.
6. Moore, G. (2024). Future-proofing cybersecurity: AI and beyond. Future Security Journal, 15(1), 125–140.
7. Bhambri, P. (2014a). Effectiveness of performance management system in IT industries: Empirical approach. In S. Kumar & B. Anjum (Eds.), Management of globalised business: Plethora of new opportunities (pp. 114–121). Bharti Publications.
8. Bhambri, P. (2014b). Estimation and non-response in clustering sampling design. In S. Kumar & B. Anjum (Eds.), Management of globalised business: Plethora of new opportunities (pp. 111–113). Bharti Publications.
9. Bhambri, P. (2014c). Consumer rights awareness. In S. Kumar & B. Anjum (Eds.), Management of globalised business: Plethora of new opportunities (pp. 93–110). Bharti Publications.
10. Smith, P. (2024). Legal ramifications of AI in cybersecurity. Journal of Cyber Law, 7(1), 45–60.

11. Foster, R. (2024). Challenges in AI deployment for cybersecurity. Security Challenges Review, 6(2), 110–125.
12. Reed, V. (2024). Future directions for AI in security technology. Journal of Emerging Technologies in Security, 8(1), 15–30.
13. Garcia, E. (2024). Exploring the impact of AI on cybersecurity workforce dynamics. Cyber Workforce Journal, 2(1), 90–105.
14. Wood, J. (2024). The influence of AI on incident response strategies. Incident Response Review, 3(2), 75–90.
15. Parker, S. (2024). Integrating AI and machine learning into existing security frameworks. Cybersecurity Framework Journal, 16(1), 110–125.
16. Bhambri, P., & Paika, V. (2013). Image recognition using neuro-fuzzy techniques: Developing a Mamdani's fuzzy inference system in MATLAB using fuzzy logic toolbox (Vol. 1). Lap Lambert Academic Publishing. ISBN: 9783659460838.
17. Bhambri, P., & Goyal, F. (2013). Development of phylogenetic tree based on Kimura's method: Based on un-weighted pair group method with arithmetic mean (UPGMA) and neighbor joining (NJ) scoring techniques (Vol. 1). Lap Lambert Academic Publishing. ISBN: 9783659336539.
18. Bhambri, P., Kaur, R., & Kaur, S. (2015). Hosiery management system: An automation software (Vol. 1). Lap Lambert Academic Publishing. ISBN: 9783659675782.
19. Bhambri, P., & Bansal, P. (2013). Secondary structure prediction of amino acids using GOR method: On different input formats (Vol. 1). Lap Lambert Academic Publishing. ISBN: 9783659107306.
20. Evans, T., & Smith, C. (2024). Case studies in AI-driven threat detection. Cybersecurity Case Studies, 2(1), 20–35.
21. Doe, J. (2022). Advances in AI for cybersecurity: Techniques and applications. Journal of Cybersecurity Research, 15(3), 234–250.
22. Smith, A., & Johnson, L. (2022). Machine learning in network security: Challenges and solutions. International Journal of Information Security, 21(4), 456–472.
23. Green, P., & Zhao, Q. (2022). AI-driven anomaly detection in cloud computing environments. Cloud Security Journal, 9(3), 90–107.
24. Bhambri, P. (2011). Data mining model for protein sequence alignment: Bioinformatics (Vol. 1). Lap Lambert Academic Publishing. ISBN: 9783844321531.
25. Price, T. (2024). Machine learning applications in financial cybersecurity. Financial Cybersecurity Journal, 5(1), 35–50.
26. Nguyen, H. (2024). AI and machine learning for predictive cybersecurity analytics. Predictive Security Journal, 4(2), 72–88.
27. Scott, H. (2024). Implementing machine learning in cybersecurity operations. Operational Security Journal, 20(3), 60–75.
28. Bhambri, P., & Kaur, A. (2013). Novel technique for robust image segmentation: New technique of segmentation in digital image processing (Vol. 1). Lap Lambert Academic Publishing. ISBN: 9783659331831.
29. Bell, C. (2024). AI and machine learning for securing IoT devices. IoT Security Journal, 11(1), 30–45.
30. Bhambri, P., & Garg, D. (2012). Enhanced model for fusion of multi-modality images: Discrete wavelet transformation using region based fusion rules (Vol. 1). Lap Lambert Academic Publishing. ISBN: 9783659208089.
31. King, P. (2024). Collaborative AI approaches to cybersecurity challenges. Journal of Collaborative Cybersecurity, 4(1), 15–29.
32. Robinson, L. (2023). Future trends in machine learning for cybersecurity. Cybersecurity Horizons, 19(2), 120–135.
33. Bhambri, P., & Bedi, S. (2013). Consumer's perception for the two wheelers: Analytical and comparative study with the inclusion of different locomotives (Vol. 1). Lap Lambert Academic Publishing. ISBN: 9783659499890.

34. Grant, L., & Cooper, J. (2024). AI algorithms for cyber threat hunting. Cyber Threat Journal, 10(2), 40–55.
35. Bhambri, P. (2013). Roadmaps to e-business: Translating e-business strategy into action (Vol. 1). Lap Lambert Academic Publishing. ISBN: 9783659347870.
36. Black, S. (2024). The role of AI in mitigating cyber risks. Journal of Risk Management in Cybersecurity, 14(3), 70–85.
37. Wang, Y., & Lee, C. (2022). Reinforcement learning for threat detection in IoT networks. Journal of Internet of Things, 7(2), 118–135.
38. Patel, R. (2022). The future of AI in cybersecurity: A review of recent advancements. Security Technology Insights, 12(5), 300–315.
39. Williams, K. (2023). Deep learning applications in cybersecurity: An overview. International Journal of Artificial Intelligence, 5(1), 88–104.
40. Adams, R. (2023). Using machine learning for real-time threat intelligence. Cyber Defense Review, 14(1), 45–60.
41. Walker, J., & Harris, T. (2023). The intersection of AI and cybersecurity regulations. Journal of Law and Cyber Policy, 6(3), 77–94.
42. Chen, L., & Gupta, N. (2023). Privacy concerns in AI-powered security solutions. Journal of Cyber Ethics, 11(2), 112–130.
43. Lee, A., & Kim, J. (2023). Behavioral analysis using AI: Enhancing cybersecurity protocols. Cyber Behavior Journal, 22(4), 200–215.
44. Bhambri, P., & Singh, S. (2013). Temporal weather prediction using genetic algorithm: Utilizing the techniques of back propagation algorithms (Vol. 1). Lap Lambert Academic Publishing. ISBN: 9783659401237.
45. Thompson, R. (2024). AI in endpoint security: Opportunities and challenges. Endpoint Protection Review, 18(1), 90–105.
46. Brooks, A. (2024). AI frameworks for enhancing cybersecurity resilience. Cyber Resilience Journal, 1(1), 100–115.
47. Carter, B. (2024). The role of AI in enhancing data security practices. Journal of Data Protection, 9(3), 150–165.
48. Wilson, J. (2024). The importance of transparency in AI security solutions. Journal of AI Transparency, 11(1), 50–65.
49. Taylor, M. (2024). AI's role in adaptive security measures. Adaptive Security Review, 3(2), 85–100.
50. Simmons, R. (2024). Trends in AI applications for cyber threat intelligence. Cyber Intelligence Journal, 3(1), 20–35.
51. Sanchez, D. (2024). AI in security operations centers: Enhancing efficiency. Journal of Security Operations, 7(1), 20–35.
52. Lewis, F. (2024). Ethical AI usage in cybersecurity. Cyber Ethics and Policy Journal, 9(2), 55–70.
53. Murphy, T. (2024). The evolution of threat detection: AI's impact on cybersecurity tools. Journal of Security Technology, 5(2), 80–95.

2 Data Collection and Preprocessing for Security

Satya Subrahmanyam

2.1 INTRODUCTION

Data has become an invaluable resource in this era of digital transformation, as it is used to power innovations and improve the performance of many systems, especially those involving security systems powered by artificial intelligence (AI). Data plays a crucial role in these systems since it is used to train machine learning (ML) models and algorithms that identify, assess, and counteract security risks. Data gathering and preparation are emphasized as critical steps in threat identification and prevention in this chapter.

2.1.1 Overview of the Importance of Data in AI-Driven Security Systems

Developing new security solutions that use AI has become necessary due to the growth of cyber threats and the rising complexity of assaults. Extensive datasets are crucial to the operation of AI-driven security solutions. Many types of information are included in this data, such as recordings of system activities, analytics of user behavior, and logs of network traffic. To what extent AI models are able to detect and counteract risks depends on the amount, quality, and relevance of this data.

Effective AI-driven security systems require continuous data collection to maintain up-to-date insights into the evolving threat landscape. The data collected serves multiple purposes, such as training ML models, validating their accuracy, and enabling real-time threat detection. Without a robust and comprehensive data collection framework, these systems would lack the necessary inputs to recognize patterns, anomalies, and potential security breaches, rendering them ineffective in safeguarding critical assets [1].

2.1.2 Data Preprocessing and Collection's Function in Detecting and Preventing Threats

When building and launching security systems powered by AI, data collecting and preprocessing are crucial steps. These procedures guarantee that the data used to train AI models is precise, applicable, and organized in a way that allows for efficient analysis.

DOI: 10.1201/9781003521020-2

Collection of data: Data collection from several sources is the first stage in developing AI-driven security solutions. Included in this category are endpoint security solutions, intrusion detection systems (IDS), firewall logs, and network sensors. The objective is to gather a variety of data points that provide a complete picture of the user's actions and the network setting. In order to provide the most recent information for threat analysis, it is essential that data be gathered continually and in real time. This is what makes data gathering tactics effective.

Preprocessing of data: After data collection is complete, the data is processed to make it more suitable for AI applications and improve its quality. Data cleansing, normalization, and transformation are some of the preprocessing procedures. Imperfect data, including missing values, duplication, and noise, might hinder the efficacy of AI algorithms. Improving model performance is possible via normalization which entails scaling data to a consistent range. Data may also be transformed into forms that are suitable with ML algorithms.

To determine which qualities are most important for threat detection, preprocessing steps include feature extraction and selection. Preprocessing enhances the efficacy and precision of AI models by lowering the data complexity and zeroing down on essential characteristics. Take network security as an example. In order to spot suspicious behavior, it is common practice to extract and analyze data like packet size, connection length, and protocol type.

2.1.3 Chapter Objectives and Structure

An in-depth analysis of the steps required to gather and prepare data for security systems powered by AI is the major goal of this chapter. In order to effectively obtain and prepare data for threat detection and prevention, it seeks to clarify the methodology and best practices.

The chapter is structured as follows:

1. *Overview of data collection methods*: The purpose of this section is to examine the methods and resources available for security-related data collection from a variety of sources.
2. *Challenges in data collection*: This section will discuss common obstacles encountered during data collection, such as data privacy concerns, data volume, and the integration of heterogeneous data sources.
3. *Preprocessing techniques*: This section will provide an in-depth analysis of preprocessing methods, including data cleaning, normalization, transformation, and feature selection.
4. *Case studies and applications*: In order to demonstrate how data gathering and preprocessing might improve security systems, this section includes case studies and real-world examples.
5. *Future trends and developments*: The data gathering and preparation processes in AI-driven security are going through several changes and this section will go over some of those changes.

2.2 FUNDAMENTALS OF DATA COLLECTION

Data collection forms the backbone of AI-driven security systems, providing the essential inputs needed to identify, analyze, and mitigate threats. Understanding the fundamentals of data collection is critical for developing robust security solutions capable of responding to the complex and dynamic cyber threats. Data collection in security is defined and discussed in this chapter, along with the many kinds of data that are pertinent to security and where to get them.

2.2.1 Definition and Importance of Data Collection in Security

The term "data collection" describes methodical steps used to compile a complete dataset by measuring and acquiring information from a variety of sources. Gathering information is crucial for security purposes since it allows for the identification and mitigation of cyber threats. Security systems can keep an eye on things, spot strange occurrences, and foresee any security breaches, before any serious damage, if they gather data effectively [1].

Collecting data for security purposes is crucial. It helps enterprises stay ahead of potential risks by allowing continuous monitoring and evaluation of network environments. Security systems are able to identify trends that can be signs of malicious activity because they gather data from several sources and correlate it. This skill is crucial for finding advanced persistent threats (APTs), zero-day vulnerabilities, and other complex forms of attack that may bypass standard security protocols [2].

2.2.2 Types of Data Relevant to Security

When it comes to security, there are many kinds of data that may throw some light on various parts of how an organization handles security. Some important categories of security data are discussed in subsequent text.

Capturing and analyzing data packet flows over a network is what network traffic data is all about. By revealing trends in device-to-device communication, this data is useful for spotting outliers such sudden increases in traffic, intrusion attempts, and data theft. If you want to find and stop distributed denial of service (DDoS) assaults and other network breaches, you need statistics on network traffic [3].

Operating systems, apps, and network devices all keep recordings of what's happening in their own systems, which are called system logs. System faults, configuration changes, user logins, and other important events are documented in these logs. Security events, such as malware infections, illegal access, and insider threats, may be better identified by analyzing system logs. Forensic investigations and compliance reporting rely heavily on system logs [4].

Information on how people interact with a system or network is known as user behavior data. User actions, such as login times, file changes, and access patterns, are included in this data. It is possible to identify compromised accounts or insider threats by keeping an eye on user activity and noting any changes from the usual. With this information, user behavior analytics (UBA) may establish norms for user behavior and spot out-of-the-ordinary actions [5].

Indicators of compromise (IOCs), malware signatures, threat actor profiles, and other information on recognized threats make up threat intelligence data. Security companies, threat intelligence systems, and information-sharing forums are the places this data is collected from. To better identify and counter new threats, it is helpful to include threat intelligence data [6].

2.2.3 Sources of Security Data

Effective data collection relies on a variety of sources that provide comprehensive coverage of an organization's digital environment. Key sources of security data are discussed in subsequent text.

Firewalls are network security devices that control and filter data packets entering and exiting the system based on preexisting rules. They keep track of unusual traffic patterns, attempted port scanning, and authorized and denied connections in their records. The ability to detect and prevent network-based threats is greatly enhanced by firewall logs [7].

IDS are specifically designed to identify any harmful or unauthorized activity occurring inside a network. They look for indicators of possible danger in system activity and network traffic. IDSs collect data regarding intrusions, such as the kind and origin of the assault, and record it in alerts and logs. For incident response and real-time threat detection, this data is crucial [8].

The primary function of antivirus software is to identify, block, and eliminate malicious software from computer systems. It checks all data, including files and email attachments, for harmful code. Logs are created by antivirus software to record instances of malware detection, attempts at infection, and measures taken to remedy the situation. This information is useful for gauging the efficacy of security measures and comprehending the frequency of malware [9].

A security information and event management (SIEM) system collects and analyzes information from many sources, such as antivirus programs, firewalls, and IDSs. They provide a unified system for tracking security incidents in real time and analyzing them. SIEM systems provide proactive threat detection and incident response by generating comprehensive warnings and reports [10].

Building reliable AI-powered security systems requires a firm grasp of data collecting principles. Security systems may be built to identify and react to a broad variety of threats by using multiple kinds of data and sources. This improves an organization's overall security posture.

2.3 METHODS AND TECHNIQUES FOR COLLECTION OF DATA

Accurate data collection is crucial for the effectiveness of AI-powered security systems. This chapter delves into various data gathering methods and strategies, covering passive and active data collection approaches, automated data collection tools and frameworks, challenges and best practices in security data collection, and ensuring data integrity and authenticity.

2.3.1 Passive vs. Active Techniques

Data collection techniques can be broadly categorized into passive and active methods, each with its own advantages and challenges. Passive data collection involves monitoring and recording data without directly interacting with the data sources. This technique is often used to gather information unobtrusively, making it suitable for environments where continuous monitoring is essential. Examples of passive data collection include network sniffing, where tools like Wireshark capture network traffic, and log analysis, where system and application logs are reviewed for security-relevant information [11]. One benefit of passive data gathering is that it does not disrupt the system's or network's regular functioning, making it harder for attackers to notice. However, it may not always capture all pertinent data, especially if data encryption is in place or certain events aren't recorded.

Active data collection involves direct interaction with the data sources to gather information. This can include techniques such as port scanning, where tools like Nmap actively probe a network to identify open ports and services, and vulnerability scanning, where automated tools assess systems for known vulnerabilities [12]. Active data collection can provide more comprehensive data as it actively seeks out information that may not be readily available through passive methods. However, it can be more intrusive and may be detected by attackers, potentially alerting them about the presence of security measures.

2.3.2 Automated Data Collection Tools and Frameworks

The complexity and volume of data in modern networks necessitate the use of automated tools and frameworks for efficient data collection. These tools help streamline the process, ensuring that data is collected consistently and accurately. IDSs like Snort and Suricata are essential for automated data collection in security. They monitor network traffic and system activities for signs of potential threats, generating alerts and logs that can be analyzed for security incidents. IDS tools use predefined signatures and behavioral patterns to detect known and unknown threats, providing real-time data collection and analysis capabilities [8].

SIEM systems, such as ArcSight, IBM QRadar, and Splunk, aggregate data from several sources, including antivirus software, firewalls, and IDSs. These systems provide a unified platform for security event collection, correlation, and analysis, allowing for thorough threat identification and response. Automated data collection with SIEM systems enables real-time monitoring and eases the workload on security analysts [10].

Servers and workstations are examples of endpoints that endpoint detection and response (EDR) products like CrowdStrike Falcon and Carbon Black collect data from. They can detect attacks that can evade conventional network security by keeping tabs on endpoint activities. In order to help identify and mitigate complex attacks, EDR technologies provide comprehensive insight into endpoint actions [13].

Network traffic analysis (NTA) systems, such as Darktrace and Vectra Networks, use AI and ML to analyze network data in real time. They collect data on network flows and look for abnormalities and potential dangers when there are deviations from the norm. NTA tools excel at identifying covert attacks and lateral movements within a network [14].

2.3.3 Ensuring Data Integrity and Authenticity During Collection

Reliable security solutions must ensure that the data they gather is intact and legitimate. This may be accomplished in a number of ways. Data encryption safeguards information from prying eyes at every point in the data lifecycle, from preparation for transmission to storage and beyond. To protect sensitive information while it is in motion or stored, use an encryption protocol such as Transport Layer Security (TLS) or a sophisticated encryption standard such as Advanced Encryption Standard (AES) [15]. Digital signatures help confirm the authenticity and integrity of data. Security systems can ensure data has not been altered since it was signed by creating a unique cryptographic signature for each piece of data, which is particularly useful for validating logs and other security-related data.

One way to ensure data is intact is to utilize hash functions which take input data and produce a hash value of a specified size. Security systems can detect changes by comparing the hash value of newly acquired data with a previously stored hash. Common hash functions include MD5 and SHA-256, with SHA-256 being more secure [16]. Complete audit trails documenting data collection activities are essential for accountability and traceability. An audit trail should include data sources, collection methods, timestamps, and any changes to the data. This documentation is crucial for forensic investigations and compliance reporting [17].

2.3.4 Obstacles and Solutions in Security Data Collection

Data collection for security purposes must overcome several obstacles to ensure efficient threat detection and response. The sheer volume and variety of data generated by modern networks from numerous sources make it challenging to gather, process, and evaluate all pertinent information. Data filtering and prioritization strategies can help manage the volume and focus on the most critical data. Concerns around privacy and compliance with regulations like General Data Protection Regulation (GDPR) and California Consumer Privacy Act (CCPA) arise while collecting security data, as it often involves handling sensitive information. To safeguard user privacy and stay out of legal hot water, organizations should have strong data governance rules in place and make sure their data collecting methods are in line with applicable laws and regulations.

Ensuring high-quality data is crucial for effective threat detection. Security analytics may be rendered ineffective due to data quality concerns including noise, duplication, or missing information. Data cleaning and validation processes can help maintain data quality and reliability. Data collection and processing can be resource-intensive, requiring significant storage and computational power. Organizations can consider cloud-based technologies to enhance their data collection strategies, balancing thoroughness with resource efficiency. Security relies heavily on timeliness. Slower reaction times and lost detection possibilities might result from sluggish data collection, processing, and analysis. Frameworks for collecting and processing data in real time or near-real time are essential for the rapid detection and mitigation of risks.

2.3.5 Best Practices for Security Data Collection

Several recommended approaches can address these challenges effectively. Establishing a comprehensive data collection strategy that outlines what data to collect, where to collect it from, and how to use it is vital. The threat environment and organizational requirements are subject to change; thus, it is important to assess and update this plan on a regular basis. Utilizing state-of-the-art data collection tools, such as automated tools and frameworks like SIEM and EDR systems, can simplify data collection and analysis. These tools can reveal sophisticated threats in real time that automated systems might miss [8].

It is critical to encrypt the data, use digital signatures, and store it securely. The data can only be accessed by authorized workers with the use of access controls, and the data collecting procedures may be audited on a regular basis to make sure that security regulations are being followed. The precision and comprehensiveness of the acquired data may be guaranteed by directing attention toward data quality via the implementation of data cleansing and validation procedures. Keeping trustworthy security analytics requires routinely checking data quality metrics and fixing problems as soon as they arise.

Prioritizing critical data and using efficient data processing techniques can balance data thoroughness with available resources. Cloud-based solutions can be considered for scalable data collection and processing. Ensuring that data collection and processing frameworks operate in real time or near-real time is crucial for timely threat detection and response. Regularly testing and tuning system performance to minimize delays helps maintain efficiency.

2.4 DATA PREPROCESSING: AN OVERVIEW

Data preparation is a vital stage in AI-driven security systems, ensuring the usefulness and dependability of the data. This chapter provides a general outline of data preparation, covering its definition, importance, relevance to security, and the typical stages involved, such as cleaning, normalizing, and transforming data.

2.4.1 Data Preprocessing and Its Goals in Security Setting

Data preprocessing involves a series of procedures to prepare raw data for analysis and ML models. In the context of security, preprocessing aims to enhance data quality by ensuring accuracy, comprehensiveness, and appropriateness for identifying and mitigating risks. The primary goals of data preparation in security are to improve data quality by ensuring security data is accurate, comprehensive, and reliable. In addition, it involves enhancing consistency by standardizing data formats and structures, which facilitates easy analysis and integration. Noise reduction is another goal, where superfluous or irrelevant data that could interfere with detection and analysis is eliminated. Finally, the process prepares data for analysis by cleaning and formatting it to make it readable by analytical tools and ML algorithms.

2.4.2 The Value of Reliable and Consistent Data

The efficacy of security systems hinges on high-quality and consistent data. Poor data quality can lead to inaccurate threat detection, false positives, and missed security events, thereby weakening an organization's security posture. Consistent data enables the integration and coherent analysis of information from various sources, providing a comprehensive view of the security environment. Ensuring data quality involves checking for accuracy, completeness, and reliability. High-quality data supports precise predictions and decisions in security systems. Issues like missing values, duplicates, or incorrect inputs can significantly impair the effectiveness of analytical procedures and ML models. Consistent data, achieved through standard formats and organization, is crucial for integrating information from different security tools and systems, allowing for thorough analysis and event correlation. Inconsistent data can lead to misunderstandings and ineffective security measures.

2.4.3 Common Preprocessing Steps

Data preparation involves several steps that make raw data suitable for analysis, with essential preprocessing procedures including data cleansing, normalization, and transformation. Data cleaning identifies and corrects errors and inconsistencies. Steps involved in data cleaning include handling missing values using techniques such as imputation (estimating values to fill in missing ones), deletion (removing records with missing values), or algorithms designed to manage missing data. It also involves removing duplicates to ensure data accuracy and correcting errors by identifying and fixing inconsistencies such as typos or incorrect entries.

Data is made consistent and comparable by normalization, which involves converting information into a standard format. When dealing with numerical data, this is a very important step to do since it may be necessary to scale it to a common distribution or range. Z-score normalization uses the mean and standard deviation to normalize data, producing a distribution with a mean of 0 and a standard deviation of 1, and min-max scaling rescales data to a specific range, often between 0 and 1.

Data transformation is the process of transforming data so that it can be easily analyzed. The process involves transforming categorical data into numerical form by using techniques such as one-hot encoding or label encoding. Another part of it is feature engineering, which improves the predictive power of ML models by combining, deconstructing, or creating new features using domain expertise in order to extract more features from the data that already exists [18].

By lowering the number of features in the dataset and eliminating unwanted or duplicate information, dimensionality reduction enhances the efficiency and performance of ML models. Feature selection techniques and principal component analysis (PCA) are two common methodologies [19]. When working with security data, it is crucial to prepare it for analysis and ML. Businesses may improve the efficiency of their security systems, leading to improved threat detection and mitigation, by prioritizing data quality and consistency and employing relevant preprocessing techniques.

2.5 DATA PURIFICATION AND SCREENING

Before AI-driven security systems can analyze raw data, it must first undergo data cleansing and filtering, two crucial preprocessing procedures. For effective threat detection and mitigation, these methods guarantee that the data is accurate, consistent, and free of errors. Fundamental aspects of data cleansing and filtering include identifying and handling missing or incomplete data, discovering and removing duplicates, filtering out unnecessary or noisy data, and addressing outliers and anomalies.

2.5.1 Identifying and Handling Missing or Incomplete Data

Missing or incomplete data is a prevalent problem in data gathering which can significantly affect the dependability and quality of analysis. To maintain the dataset's integrity, it is essential to properly detect and handle missing data. Recognizing when data is missing is the first stage in addressing this issue. Several approaches can be employed for this purpose, such as visual inspection of data tables to spot gaps or blanks, using summary statistics to detect missing values by calculating the percentage of missing data per column, and utilizing data profiling tools that automatically identify missing values and provide reports on data completeness.

Missing data may be handled in a variety of ways after it has been detected. While erasing records with missing values (a process known as deletion) is suitable when the percentage of missing data is limited, doing so excessively might result in the loss of important information. The process of imputation entails using other available data to fill in missing values with approximated values. A few examples of common imputation methods are mean or median imputation, which uses the non-missing values as a replacement for missing ones, regression imputation, which creates multiple imputed datasets and combines their results to account for uncertainty in the imputations, and multiple imputation, which uses regression models as a prediction tool to fill in missing values.

2.5.2 Detecting and Removing Duplicates

Duplicates in data can lead to biased analysis and inaccurate results, making their detection and removal crucial for ensuring data quality. Duplicates can be detected through exact matching, which identifies records that are identical across all fields, and fuzzy matching, which uses algorithms to detect records that are similar but not identical due to typographical errors or variations in data entry. Once duplicates are detected, they can be removed using deduplication tools that efficiently identify and eliminate duplicates or through a manual review when automated tools are insufficient, ensuring accuracy [20].

2.5.3 Filtering Irrelevant or Noisy Data

Filtering irrelevant or noisy data is essential to ensure that the dataset contains only relevant information that can contribute to accurate analysis. Irrelevant data refers to information that does not contribute to the objectives of the analysis. Filtering out such data involves defining clear criteria for what constitutes relevant data based on the analysis objectives and implementing automated filters to exclude data that does not meet these criteria. Data that is noisy, meaning it includes mistakes, inconsistencies, or outliers that could skew analysis, can be handled by using noise detection algorithms to identify and eliminate the noise, or by utilizing smoothing methods such as exponential or moving averages to decrease the noise.

2.5.4 Techniques for Dealing with Outliers and Anomalies

The integrity of the dataset depends on the efficient management of outliers and anomalies since they may greatly impact the accuracy of security assessments and the performance of ML models. Data points that differ greatly from the average are called outliers. Methods for dealing with outliers include applying data transformation techniques, such as log transformation, to lessen the effect of outliers, using clustering algorithms to find and isolate outliers from normal data points, and utilizing statistical methods, such as the interquartile range (IQR) method, to identify outliers.

Isolation forest and one-class support vector machine (SVM) are two examples of anomaly detection algorithms that may be used to handle data points that are out of the ordinary and might potentially reveal security risks. Applying strong statistical approaches that are less vulnerable to outliers and anomalies further guarantees data integrity [21]. Domain-specific rules derived from domain expertise may aid in identifying and handling abnormalities.

2.6 STANDARDIZING AND TRANSFORMING DATA

When it comes to security-related ML applications, data standardization and transformation are two of the most important preparatory tasks. The data is prepared for analysis via these steps, thus improving the accuracy and performance of ML algorithms. Data normalization, data scaling, data transformation, and feature engineering and selection for improved security insights are all covered in this chapter.

2.6.1 Importance of Data Standardization for Machine Learning Models

The purpose of data standardization is to create a consistent distribution for all of the data's independent variables and characteristics. There are a number of reasons why normalization is so important when discussing ML. First, the model's performance is enhanced. A lot of ML techniques, such SVM and k-nearest neighbors, are very sensitive to data size since they use distance computations. Another benefit of normalization is that it ensures that features are of equal magnitude, which speeds up the convergence of algorithms that rely on gradient descent, such neural networks.

Because of this, the model is able to learn better and faster. Lastly, normalized data improves interpretability, which is critical in security scenarios where knowing the relative value of various characteristics is key. This makes it simpler to comprehend the findings of ML models.

2.6.2 Data Standardization and Scaling Methods

There are a number of methods for standardizing and scaling data, and each has its own set of benefits and uses. Data is transformed to fit inside a certain range, usually [0, 1], using min-max scaling, which is also called normalization. This method is great since it standardizes the scale for all features, which is especially helpful when their ranges and units are different [22].

The formula for min-max scaling is:

Xscaled=X−XminXmax−XminX_{\text{scaled}} = \frac{X - X_{\text{min}}} {X_{\text{max}} - X_{\text{min}}}Xscaled=Xmax−XminX−Xmin

Reducing the influence of outliers on the overall data distribution, Z-score normalization (or standardization) converts the data to have a mean of 0 and a standard deviation of 1, as described in reference [23].

The formula for z-score normalization is:

Xstandardized=X−μσX_{\text{standardized}} = \frac{X - \mu}{\sigma} Xstandardized=σX−μ, where μ\muμ is the mean and σ\sigmaσ is the standard deviation of the feature.

In decimal scaling, the greatest absolute value of a feature determines the number of places to relocate the decimal point in order to normalize the data. This approach isn't often used, although it works well for data that has established boundaries [24].

2.6.3 Data Transformation Methods

The process of data transformation entails altering data so it may be better analyzed. Among the most common transformation techniques are one-hot encoding and log transformation. For features with a large range of values, log transformation is particularly useful for reducing data skewness. This method is helpful for data that follows an exponential distribution since adding 1 to the value guarantees that the transformation is specified for zero values [18].

The formula for log transformation is:

Xlog=log(X+1)X_{\text{log}} =\log(X + 1)Xlog=log(X+1)

To convert numeric variables with several categories into a binary vector, one-hot encoding is used. The binary vectors are used to represent the categories; each vector has one high bit (1) and zero low bits (0). Algorithms for ML that work best with numerical data and struggle with categorical data naturally must use this technique [25].

The Box-Cox transformation is another tool for reducing outliers and bringing data closer to a normal distribution.

It is defined as:

y(λ)=(yλ−1)λ for λ≠0y(\lambda) =\frac{(y^\lambda - 1)}{\lambda}\text {for}\ lambda\neq 0y(λ)=λ(yλ−1) for λ=0 y(λ)=log(y) for λ=0y(\lambda) =\ log(y)\text {for}\lambda = 0y(λ)=log(y) for λ=0

The parameter λ\lambdaλ is estimated using maximum likelihood estimation. This transformation is particularly useful when the data does not conform to normality [26].

2.6.4 Feature Engineering and Selection for Enhanced Security Insights

Improving the security applications of ML models for prediction relies heavily on feature engineering and selection. In order to enhance the performance of a model, feature engineering is used to generate additional features from preexisting data. This might include integrating several characteristics to capture relationships between them, extracting temporal features to capture time-based patterns, or aggregating data to provide summary statistics in a security context [27]. Finding and choosing the best features for the model is what feature selection is all about. It helps with data dimensionality reduction, model performance, and interpretability. Some common feature selection methods are filter, wrapper, and embedded. Filter methods use statistical measures to evaluate features, wrapper methods use subsets of features to train models, and embedded methods use regularization and other techniques to penalize less important features during model training as part of feature selection [28].

2.7 HANDLING IMBALANCED DATA

Problems with data imbalance are prevalent in security datasets, when one class has a disproportionately large number of instances compared to other classes. The assessment and performance of ML models may be significantly affected by this mismatch. The effects of class imbalance on model performance and assessment, how to handle class imbalance, and the nature of unbalanced data in security situations are all covered in this chapter.

2.7.1 Understanding Imbalanced Data in Security Datasets

When the number of harmful activities is much lower than the number of regular activities, this results in unbalanced data in security applications. As an example, the number of attack instances is often much lower than the number of regular traffic instances in IDS. Multiple difficulties arise from this disparity. To start, the majority class is unfairly favored. ML models that are trained on data that is unbalanced are more likely to favor the majority class, which makes them bad at identifying instances of the minority class. The second issue is that performance measures are biased. When applied to datasets that are unbalanced, standard assessment criteria like accuracy might be deceptive. A high level of accuracy can only mean that the model is good at predicting the majority class and bad at spotting the minority.

2.7.2 Techniques to Address Class Imbalance

Resampling approaches and synthetic data creation are two of the ways that may be used to fix security datasets that have an imbalance in classes. The goal of resampling techniques is to get a more uniform distribution of classes in the training dataset. The method of oversampling is used to boost the representation of the minority group. Two methods that are used are random oversampling and the synthetic minority over-sampling technique (SMOTE). The former involves duplicating instances of the minority class, while the latter creates synthetic instances by interpolating between existing minority instances. To undersample, one must lower the percentage of the majority class. Although simpler approaches like random undersampling remove examples from the majority class at random, more complex techniques like Tomek links and cluster-based undersampling try to keep the most informative instances.

To achieve statistical parity, synthetic data creation methods generate new, fictitious members of the minority group. Earlier we discussed how SMOTE creates a more varied and balanced dataset by generating synthetic examples by interpolating between existing minority occurrences [29]. To address underrepresented minority groups and concentrate on challenging feature spaces, ADASYN (Adaptive Synthetic Sampling) builds on SMOTE by creating synthetic instances in such areas [30]. Several ML methods have been developed with the express purpose of dealing with data that is skewed in one direction or the other. A larger penalty is associated with misclassifying the minority class in cost-sensitive learning, which in turn encourages the model to pay more attention to occurrences of the minority class during training. Incorporating procedures to balance the distribution of classes within the ensemble is one way to adjust ensemble methods like boosting algorithms and Random Forests to tackle class imbalance.

2.7.3 How Data Inequality Affects Model Performance and Assessment

There are several ways in which data imbalances might affect how ML models function and are evaluated. To begin with, it has the potential to reduce the efficiency of the whole model. When it comes to security, the ability to identify harmful behaviors (the minority class) is of the utmost importance, and models trained on unbalanced data may show great accuracy overall but low recall for that class. Furthermore, it has the potential to cause assessment measures to be misleading. When applied to datasets that are unbalanced, traditional assessment criteria like accuracy might be deceptive. An area under the receiver operating characteristic curve (AUC-ROC), recall, F1-score, and accuracy are more instructive metrics in this setting. The third concern is the possibility of becoming "overfit" to the dominant group. When models are overfit to the dominant class, they are unable to generalize well to new data. To make sure that both the training and validation sets are balanced, methods like stratified sampling and cross-validation may be used to reduce the impact of this problem.

The development of successful ML models for security applications relies heavily on the handling of unbalanced data. Organizations may enhance the performance and reliability of their security systems, leading to improved threat identification and mitigation, by using appropriate strategies to resolve class imbalance and by applying relevant assessment metrics.

2.8 DATA ANNOTATION AND LABELING

Data annotation and labeling are crucial processes in preparing datasets for supervised learning, especially in the realm of security. The quality and accuracy of labeled data significantly influence the performance of ML models. This chapter discusses the importance of labeled data, various techniques for data annotation and labeling, the role of expert knowledge in ensuring accurate labeling, and the use of automated labeling tools and technologies.

2.8.1 Importance of Labeled Data for Supervised Learning in Security

Labeled data is essential for supervised learning algorithms to discover patterns and provide predictions. Labeled data is crucial in security applications for several reasons. To begin, training ML models need labelled data. It gives the algorithm the instances it needs to learn how input characteristics relate to the target variable. The second benefit is that the model is better able to detect risks like intrusions, malware, and fraudulent activities when the labels are correct and can discriminate between harmless and harmful actions. The evaluation of ML model performance relies heavily on labeled data. To evaluate the model's performance in detecting security risks, metrics including recall, accuracy, precision, and F1-score are calculated using labeled datasets.

2.8.2 Techniques for Data Annotation and Labeling

Several techniques can be employed for data annotation and labeling in security applications, each with its own set of advantages and challenges. Manual annotation involves human annotators reviewing and labeling data. This technique is often used for complex tasks that require domain expertise and nuanced understanding. Its advantages include high accuracy and reliability, as human annotators can apply their expertise and contextual knowledge. Nevertheless, it may be rather expensive, labor-intensive, and time-consuming, particularly when dealing with huge datasets. Combining automated technologies with human control is what semi-automatic annotation is all about. Automated algorithms provide initial labels, which are then reviewed and corrected by human annotators. This technique reduces the time and effort required for labeling while maintaining a reasonable level of accuracy. Still, it requires human intervention, and the quality of the initial automated labels can vary. Crowdsourcing involves distributing the annotation task to a large number of contributors via online platforms. It can quickly generate large volumes of labeled data at a lower cost compared to manual annotation. However, quality control can be challenging, and contributors may lack domain-specific expertise.

2.8.3 Leveraging Expert Knowledge for Accurate Labeling

In security applications, leveraging expert knowledge is crucial for accurate data labeling. While automated technologies may fail to notice small trends and abnormalities, cybersecurity experts have the expertise and knowledge to spot them. Expert-led annotation involves experts manually reviewing and labeling data, ensuring high accuracy and reliability. This method is particularly useful for complex security tasks, such as identifying sophisticated attacks or advanced persistent threats [1]. Developing detailed annotation guidelines can help standardize the labeling process and ensure consistency across different annotators. These guidelines should include definitions of various threat types, labeling criteria, and examples. Collaborating with domain experts can enhance the quality of labeled data. Experts can provide insights and feedback during the annotation process, helping to refine labeling criteria and improve accuracy.

2.8.4 Automated Labeling Tools and Technologies

A more efficient and scalable data annotation procedure is possible with the help of automated labeling tools and technologies. In ML-based labeling, algorithms are trained on labeled datasets that already exist to automatically classify new data. Automatic labeling systems may be much more effective when trained using active learning and transfer learning techniques [31]. In active learning, human annotators choose the most informative examples to label and use in training the model. The model's performance is enhanced with a smaller number of labelled examples via this iterative procedure. To decrease the quantity of labelled data needed for training, transfer learning makes use of pre-trained models on comparable tasks. This method shines in situations when there is a dearth of labelled data. Labelbox, Prodigy, and Amazon SageMaker Ground Truth are just a few examples of annotation systems that aim to make labeling easier by offering tools for collaborative annotation, quality control, and connection with ML frameworks [32]. Incident reports and log files are two examples of textual data that may be automatically labeled using natural language processing (NLP) techniques. To better detect and classify pertinent security events, methods like sentiment analysis and named entity recognition (NER) may be used [33].

Supervised learning in security applications relies heavily on data annotation and tagging. To guarantee the production of high-quality labeled datasets, companies may use a mix of manual, semi-automatic, and automated procedures, as well as specialist knowledge, cutting-edge tools, and technology. Better threat detection and mitigation are the results of improved ML model performance and dependability.

2.9 PRIVACY AND ETHICAL CONSIDERATIONS IN DATA COLLECTION

In the era of big data and AI-driven security solutions, ensuring user privacy and maintaining data confidentiality are paramount. This chapter delves into the ethical and legal implications of data collection in security contexts, strategies for anonymizing and protecting sensitive data, and compliance with data protection regulations like GDPR and CCPA.

2.9.1 Protecting the Privacy of Users and Their Data

Fundamental principles that are needed to govern any data gathering effort, especially in security, include user privacy and data secrecy. Significant privacy problems are raised by the collecting of massive volumes of data, which includes sensitive and personally identifiable information. To address these issues, it is essential to limit data collection to just what is needed for the current security objective. Reducing exposure and the hazards of data breaches is one goal of data reduction. Data security at rest and in transit requires the use of strong encryption methods. Encryption makes data unintelligible and safe even if it is intercepted or viewed without authority. To further guarantee that no unauthorized individuals have access to sensitive information, it is critical to establish stringent access restrictions. Two strong methods for protecting sensitive information are role-based access control and multiple factor authentication.

2.9.2 Legal and Ethical Implications of Data Collection in Security

Data gathering in security is fraught with ethical and legal complexities, since there are several rules and regulations that dictate how data should be used. A fundamental ethical concept is to get users' informed permission after explaining the data collection process, its intended purpose, and the individuals who will have access to their data. Transparency is promoted while user sovereignty is respected. Another important concept is purpose restriction, which states that data should only be acquired for certain, valid objectives and should not be used in a way that contradicts those goals. Organizations must also be honest about how they handle data and ensure that they are in compliance with all applicable laws and regulations. Assessing and auditing on a regular basis might assist in keeping people accountable.

2.9.3 Strategies for Anonymizing and Protecting Sensitive Data

Data anonymization and other privacy-preserving measures are essential for meeting regulatory standards and protecting users' personal information. To ensure that no one can be identified from datasets, data anonymization is used. This is achieved by obfuscating or deleting any personally identifying information (PII). Data masking, differential privacy, and k-anonymity are among methods that successfully anonymize data while keeping its analytical value [34]. While pseudonymization and anonymization may not provide the same degree of security, they can greatly lessen the likelihood of re-identification [35]. Data de-identification is the process of erasing or altering data components that may be used to identify persons, either directly or indirectly. This includes eliminating personal details like names and addresses and also masking or generalizing indirect identifiers [36].

2.9.4 Data Protection Regulation Compliance

Legal and ethical data collecting procedures need compliance with data protection legislation. In order to ensure data minimization, get express permission and provide users with the ability to view and erase their data. The GDPR imposes strict rules for data protection. This regulation applies to the European Union. Heavy penalties are levied for noncompliance [35]. People living in California have certain rights under the CCPA that deals with their personal data. These rights include being able to see what data is being gathered, having that data erased, and not having it sold. Companies need to make sure they are in compliance with CCPA regulations and provide transparent privacy notifications [37]. In addition, data protection regulations exist in different locations, for example, Singapore has the Personal Data Protection Act (PDPA) and Brazil has the Lei Geral de Proteção de Dados (LGPD). Depending on their operating area, organizations must guarantee compliance with all applicable rules [38, 39].

To successfully navigate the ethical and privacy challenges associated with security data collecting, one must strike a balance between the competing demands of strong security measures, user privacy protection, and legal compliance. Organizations may responsibly handle and safeguard data by following best practices for data minimization, encryption, access restrictions, anonymization, and compliance with regulatory requirements. This will create confidence and ensure ethical integrity.

2.10 DATA COLLECTION AND PREPROCESSING: WHAT's NEXT?

Emerging technologies and approaches are changing the way businesses manage data, especially when it comes to security, since the data gathering, and preparation environment keeps changing. In this chapter, we will look at what the future holds for data pretreatment and gathering, with a focus on how AI and ML will improve these processes, along with developments in predictive analytics and real-time data processing.

2.10.1 Emerging Technologies and Methodologies

Improving the efficiency, accuracy, and security of data gathering and preprocessing, a number of new technologies are on the horizon. Edge computing is one such technology; it eliminates the need for centralized data centers by processing data close to its point of origin. Data collection and preparation can be done more quickly and efficiently using this method since it decreases latency and bandwidth utilization. Applications in the security domain that need real-time analysis and rapid reactions greatly benefit from edge computing.

With the expansion of the Internet of Things (IoT), many devices, such as cameras, sensors, and smart appliances, are producing massive volumes of data. The integration of varied data kinds and the improvement of the comprehensiveness of security assessments are both made possible by advanced IoT frameworks, which permit smooth data gathering and preprocessing. Furthermore, blockchain technology provides a distributed and unchangeable record of data transfers. When it comes to security, blockchain technology can make sure that data is more genuine and less susceptible to tampering. For security model data to remain trustworthy, this is essential.

2.10.2 Data Preprocessing and the Importance of Artificial Intelligence and Machine Learning

With their advanced methods for dealing with complicated and big datasets, AI and ML are leading the way in data preparation approaches that are constantly expanding. Data cleansing tasks such as finding and fixing mistakes, filling in missing numbers, and eliminating duplicates may be automated with the use of AI-driven solutions. Without requiring a lot of human input, these technologies can analyze data patterns using sophisticated algorithms and guarantee great data quality.

Automating feature engineering, ML algorithms may sift through raw data for the most useful characteristics that improve prediction accuracy. This method improves the security-related performance of ML models by making their input data more high-quality. In addition, textual data pretreatment is being improved with the use of NLP methods. This includes security logs and incident reports. To facilitate the analysis of massive amounts of unstructured data, NLP may classify data, extract relevant information, and spot patterns and outliers.

2.10.3 Predictive Analytics and Real-Time Data Processing Advancements

Modern security systems rely on real-time data processing and predictive analytics to identify and respond to threats proactively. Modern innovations in real-time data processing make it possible to analyze newly received data in near-real time. In real time, systems may identify security concerns and react accordingly, reducing the likelihood of harm. Apache Kafka and Apache Flink are two of the most important frameworks for stream processing when dealing with data streams moving at high speeds.

The goal of predictive analytics is to foretell future outcomes by analyzing past data. When it comes to safety, this means seeing dangers in the air before they become real. Organizations may take preventative actions by training ML models on previous security data to anticipate attack trends. Furthermore, state-of-the-art anomaly detection methods use AI to spot out-of-the-ordinary actions that could be signs of security breaches. These approaches are becoming better and better at spotting intricate and subtle abnormalities that older ones could overlook.

Modern approaches and tools are molding the way security data gathering and preparation will be done in the future. More proactive and efficient security measures are made possible by developments in real-time data processing and predictive analytics, while AI and ML play crucial roles in automating and improving these processes. Improved data security and threat detection capabilities will be available when these technologies develop further.

2.11 CONCLUSION

This chapter has thoroughly explored the many aspects of security-related data collecting and preprocessing, illuminating their relevance, methods, and potential future paths. Through an examination of typical data kinds and sources—including user activity, system logs, and network traffic—this chapter has defined key concepts and shown their applicability to security. With the goal of guaranteeing the validity and integrity of the data, several methods of data collecting have been explored, including active and passive techniques as well as automated systems.

The importance of data preprocessing was emphasized, detailing steps like data cleaning, normalization, and transformation to ensure data quality and consistency. Addressing class imbalances through resampling methods and synthetic data generation was also covered, highlighting their impact on model performance. Data annotation and labeling were underscored as critical for supervised learning, with techniques for manual and automated annotation, leveraging expert knowledge, and ensuring privacy and ethical compliance.

Privacy and ethical considerations in data collection were addressed, focusing on user privacy, data confidentiality, and adherence to regulations like GDPR and CCPA. Emerging technologies and methodologies such as edge computing, IoT, and blockchain were identified as transformative forces in data collection and preprocessing, promising to enhance efficiency and security.

Automated data cleaning, feature engineering, and enhanced NLP are made possible by preprocessing approaches that evolve with the help of AI and ML. Our ability to recognize and respond to threats proactively is being improved by developments in real-time data processing and predictive analytics.

In conclusion, effective data collection and preprocessing are foundational to robust security measures. By embracing emerging technologies, adhering to ethical standards, and leveraging AI and ML, organizations can develop more effective security systems. Ongoing research and development in this field will continue to enhance the ability to safeguard digital infrastructures against evolving threats.

REFERENCES

1. A.L. Buczak and E. Guven. 2016. A survey of data mining and machine learning methods for cyber security intrusion detection. *IEEE Communications Surveys & Tutorials*, *18*(2), 1153–1176.
2. S. Axelsson. 2000. The base-rate fallacy and its implications for the difficulty of intrusion detection. *ACM Transactions on Information and System Security (TISSEC)*, *3*(3), 186–205.
3. P. Garcia-Teodoro, J. Diaz-Verdejo, G. Macia-Fernández and E. Vazquez. 2009. Anomaly-based network intrusion detection: Techniques, systems and challenges. *Computers & Security*, *28*(1-2), 18–28.
4. M. Jouini, L.B.A. Rabai and A.B. Aissa. 2014. Classification of security threats in information systems. *Procedia Computer Science*, *32*, 489–496.
5. W. Eberle and L. Holder. 2009. Insider threat detection using graph-based approaches. In *Cybersecurity Applications & Technology Conference for Homeland Security* (pp. 237–241). IEEE.
6. D. Shackleford. 2015. *Cyber Threat Intelligence: Lessons from the Front Lines*. SANS Institute.
7. H. Wang, D. Zhang and K.G. Shin. 2004. Change-point monitoring for the detection of DoS attacks. *IEEE Transactions on Dependable and Secure Computing*, *1*(4), 193–208.
8. K. Scarfone and P. Mell. 2007. Guide to intrusion detection and prevention systems (IDPS). *NIST Special Publication*, 800(2007), 94.
9. J. Aycock. 2006. *Computer Viruses and Malware*. Springer.

10. A. Chuvakin, K. Schmidt and C. Phillips. 2010. *Logging and Log Management: The Authoritative Guide to Understanding the Concepts Surrounding Logging and Log Management*. Elsevier.
11. A. Alberto. 2018. *Network Monitoring and Analysis: A Protocol Approach to Capturing and Analyzing Network Traffic*. Springer.
12. J. Stuttgen and I. Cohen. 2013. Anti-forensic resilient memory acquisition. *Digital Investigation, 10*(S1), S105–S115.
13. M. Egele, T. Scholte, E. Kirda and C. Kruegel. 2012. A survey on automated dynamic malware-analysis techniques and tools. *ACM Computing Surveys (CSUR), 44*(2), 1–42.
14. R. Sommer and V. Paxson. 2010. Outside the closed world: On using machine learning for network intrusion detection. In *2010 IEEE Symposium on Security and Privacy* (pp. 305–316). IEEE.
15. W. Stallings. 2017. *Cryptography and Network Security: Principles and Practice*. Pearson.
16. B. Preneel. 2010. Hash functions: Theory, attacks, and applications. In H. M. Heys & K. Nyberg (Eds.), *Selected Areas in Cryptography* (pp. 157–171). Springer. https://doi.org/10.1007/978-3-642-28496-0_13.
17. K. Kent and M. Souppaya. 2006. Guide to computer security log management. *NIST Special Publication*, 800(2006), 92.
18. V. Chandola, A. Banerjee and V. Kumar. 2009. Anomaly detection: A survey. *ACM Computing Surveys (CSUR), 41*(3), 1–58.
19. J. Han, J. Pei and M. Kamber. 2011. *Data Mining: Concepts and Techniques*. Elsevier.
20. G. Chandrashekar and F. Sahin. 2014. A survey on feature selection methods. *Computers & Electrical Engineering, 40*(1), 16–28.
21. I.T. Jolliffe. 2011. *Principal Component Analysis*. Springer.
22. A.K. Elmagarmid, P.G. Ipeirotis and V.S. Verykios. 2007. Duplicate record detection: A survey. *IEEE Transactions on Knowledge and Data Engineering, 19*(1), 1–16.
23. S. Ioffe and C. Szegedy. 2015. Batch normalization: Accelerating deep network training by reducing internal covariate shift. In *Proceedings of the 32nd International Conference on Machine Learning* (pp. 448–456).
24. N. Japkowicz and M. Shah. 2011. *Evaluating Learning Algorithms: A Classification Perspective*. Cambridge University Press.
25. F. Pedregosa, G. Varoquaux, A. Gramfort, V. Michel, B. Thirion, O. Grisel and E. Duchesnay. 2011. Scikit-learn: Machine learning in Python. *Journal of Machine Learning Research, 12*, 2825–2830.
26. R.M. Sakia. 1992. The Box-Cox transformation technique: A review. *The Statistician, 41*(2), 169–178.
27. P. Domingos. 2012. A few useful things to know about machine learning. *Communications of the ACM, 55*(10), 78–87.
28. I. Guyon and A. Elisseeff. 2003. An introduction to variable and feature selection. *Journal of Machine Learning Research, 3*, 1157–1182.
29. N.V. Chawla, K.W. Bowyer, L.O. Hall and W.P. Kegelmeyer. 2002. SMOTE: Synthetic minority over-sampling technique. *Journal of Artificial Intelligence Research, 16*, 321–357.
30. H. He, Y. Bai, E.A. Garcia and S. Li. 2008. ADASYN: Adaptive synthetic sampling approach for imbalanced learning. In *Proceedings of the IEEE International Joint Conference on Neural Networks* (pp. 1322–1328).
31. B. Settles. 2010. Active learning literature survey. *University of Wisconsin, Madison*.
32. P. Kumar, S. Carberry, S. Joshi and H. Forouzesh. 2020. A comprehensive study of data annotation tools for machine learning tasks. *arXiv preprint arXiv:2006.05783*.
33. G. Lample, M. Ballesteros, S. Subramanian, K. Kawakami and C. Dyer. 2016. Neural architectures for named entity recognition. *arXiv preprint arXiv:1603.01360*.

34. L. Sweeney. 2002. k-anonymity: A model for protecting privacy. *International Journal of Uncertainty, Fuzziness and Knowledge-Based Systems, 10*(5), 557–570.
35. General Data Protection Regulation (GDPR), Regulation (EU) 2016/679 (2016).
36. P. Ohm. 2010. Broken promises of privacy: Responding to the surprising failure of anonymization. *UCLA Law Review, 57*(6), 1701–1777.
37. California Consumer Privacy Act (CCPA), Cal. Civ. Code § 1798.100 (2018).
38. Personal Data Protection Act (PDPA), Act 26 of 2012 (2012).
39. D. Doneda and A.F. Almeida. 2020. The Brazil General Data Protection Law (LGPD): A major new data protection regime. *Journal of Data Protection & Privacy*, 3(2), 110–117.

3 Feature Engineering for Threat Detection

Sepideh Bazzaz Abkenar, Mostafa Haghi Kashani, and Mohammad Nikravan

3.1 INTRODUCTION

In recent years, the number of smart devices has significantly increased. As new technologies such as 5G mobile networks and the Internet of Things (IoT) gain popularity, the amount of anomalous traffic in the network increases [1]. Thus, several security issues, including network incidents and intrusions, have been brought on by this growth. Intrusions are defined as attempts or endeavors to jeopardize the privacy, reliability, or accessibility of a computer or network. Consequently, budgets and efforts are allocated to search for new types of attacks or vulnerabilities in computer software or hardware. Security protocols are typically classified as intrusion detection (ID) or prevention systems [2].

Maintenance of cyberspace security in homes, enterprises, and organizations has merged into our daily lives. The term "cybersecurity" relates to a group of procedures and technologies designed to protect systems, data, applications, and networks from threats, illegal access, data loss or destruction, and other concerns [3]. In this respect, machine learning (ML) techniques are commonly employed in malware detection [4]. According to this strategy, malware detection may be performed using traditional pattern recognition or ML approaches, as it is a binary classification problem [5]. ML and data mining approaches are employed to analyze and uncover patterns in traffic data, as well as construct models to categorize each flow to detect anomalies in network traffic more rapidly and accurately [6]. Nonetheless, these unusual flows exhibit high dimensions and high-quantity features. This can lead to difficulties such as overfitting, large computational costs, and extended training times. Hence, the features that are most important should be chosen in order to enhance classifier performance [7].

One of the key strategies for ensuring efficient anomaly-based detection is feature engineering. The main factor determining the effectiveness of ML-based approaches is feature engineering. It is the process of extracting valuable features from dataset to augment the performance of ML models. Feature engineering includes choosing the most beneficial features, transforming current features, developing new features, and managing missing values. Efficient feature engineering may dramatically enhance the effectiveness of ML models by delivering them with more effective, more relevant data from which to train. Application program interfaces (APIs) and permissions [8, 9] are frequently chosen as the features as they provide comprehensive security-related data on which activities may access vital resources. The process of

DOI: 10.1201/9781003521020-3

feature selection removes redundant and inefficient features, which has a noticeable beneficial effect on enhancing performance of IDS, particularly concerning dimension and running time [9].

3.1.1 Threat Detection

Threat detection, or identifying malicious behavior, is one of the most important aspects of cybersecurity [10]. Several strategies have been developed to differentiate between threats and normal traffic. Nonetheless, offering reliable and beneficial threat detection solutions for big data becomes more challenging. The vital component of any network is the IDS which detects various threats. An IDS is a model that applies various techniques to identify these threats. In this regard, a detailed examination of IDS was conducted, and numerous strategies for creating IDS were applied [11]. The development of threat detection models based on ML and deep learning (DL) approaches has become essential because of the recent significant interest in these techniques across various domains [12].

Traditional rules-based IDS cannot always identify complex and evolving threats, so ML and DL are suitable replacements [13]. IDS uses supervised ML algorithms and labeled training data to identify patterns and determine whether network traffic is malicious or benign. In addition, unsupervised learning approaches allow one to recognize threats and incidents even in the absence of prior knowledge about their patterns. As a result, IDS plays an essential role in cybersecurity for defending networks from ever-changing cyber threats [11]. The potential of ML-based IDS to adapt to changing attack strategies is one of its main advantages. As cyber threats change, IDS can be constantly retrained to combat emerging attack trends. Consequently, there is an increasing need for efficient ways to identify and resist evolving threats [12].

This chapter discusses feature engineering for threat detection based on recent research findings. In this chapter, we also illustrate the tools, algorithms, and evaluation parameters, as well as the possible taxonomy of feature engineering for threat detection. This chapter will address the challenges and unresolved issues that researchers must deal with to maximize feature engineering and enhance threat detection. The chapter's remaining sections are arranged as follows: Section 3.2 describes the study's methodology, article selection procedure, and research questions. In Section 3.3, the reviewed papers are summarized, focusing on the main ideas, tools, applied algorithms, advantages, and disadvantages. The findings analysis, open issues, and future directions are explained in Sections 3.4 and 3.5, respectively. Finally, Section 3.6 provides an explanation of the findings.

3.2 RESEARCH METHODOLOGY

Numerous studies have been performed concerning feature engineering for threat detection by researchers. To carry out a thorough analysis, we first define the requirements and issues that inspire this chapter [14, 15]. Answering research questions allows researchers to identify gaps in this subject, which may assist researchers in providing new perspectives and solutions. This chapter's main goal is also to classify feature engineering in terms of potential threat detection. We also developed following research questions:

- *RQ_1*: What evaluation factors are applied in feature engineering for threat detection?
- *RQ_2*: What algorithms and tools are applied in feature engineering for threat detection?
- *RQ_3*: What is the possible classification of feature engineering for threat detection?
- *RQ_4*: What are the challenges and open issues of feature engineering for threat detection?

Next, employing titles and keyword phrases, we searched online in between time range 2019 and May 2024 for articles on this topic from well-known scientific publishers such as IEEE, Springer, ScienceDirect, Wiley, SAGE, Emerald, Inderscience, Taylor & Francis, ACM, and Hindawi. We applied Google Scholar as our primary search engine. The following keywords were used:

("feature engineering" OR feature) AND
("threat detection" OR attack OR risk OR intrusion OR ransomware OR "behavioral analysis" OR vulnerability OR anomaly OR malware OR incident OR endpoint OR hazard OR danger OR "network monitoring")

Furthermore, to extract the most notable publications, we further removed non-peer-reviewed papers, short papers, review papers, theses, non-English articles, and book chapters. We scanned the article abstracts and conclusions. After thoroughly examining the articles' texts, 17 papers were selected for further examination that revealed the methodologies and challenges and adequately addressed our research questions. We propose a feature engineering classification for threat detection regarding the retrieved and reviewed articles. We evaluate the approaches offered, considering their main ideas, advantages, and disadvantages, and we perform analytical and statistical research. This chapter also provides an explanation for the most important unresolved challenges and the main areas where additional research could enhance the approaches employed in the reviewed studies.

3.3 OVERVIEW OF REVIEWED STUDIES

A structured taxonomy of the feature engineering for threat detection is defined in Section 3.4, and a detailed description of each category is provided. Regarding the reviewed papers, feature engineering for threat detection is classified into five main categories: *statistical features, temporal features, content features, structural features, and behavioral features*. Our suggested taxonomy covers a wide range of features engineering in different domains of threat detection. However, each reviewed articles may be included in several feature engineering subcategories, but we considered the main one in this article. In this section, we study the existing articles according to the presented taxonomy in Section 3.4. In addition, Table 3.1 presents an overview of the main ideas, applied tools, algorithms, advantages, disadvantages, and the suggested categories and subcategories.

TABLE 3.1
An Overview of the Reviewed Studies

Category		Ref.	Main Idea	Tools	Applied Algorithms	Advantages	Disadvantages
Statistical features	Descriptive statistics	[16]	Applying statistical features to employ important features and improve the performance of intrusion detection system (IDS)	• Python	• Deep neural network (DNN)	• High accuracy • High F-score • High recall • High precision • Low FPR • Low execution time • Consuming less energy	• Not applying nature-inspired techniques to optimize the neural network design to analyze the resilience of IDS • Not using algorithms inspired by nature for feature selection
	Inferential statistics	[11]	Applying two filter-based feature ranking approaches to extract the pertinent features for IDS	• Python (Scikit-learn)	• ANOVA • Support vector machine (SVM) • K-nearest neighbor (KNN) • Decision tree (DT) • Logistic regression (LR) • Random forest (RF) • SMOTE	• High precision High F-score • High recall • High accuracy • High detection rate	• Not evaluating resource consumption
	Frequency-based statistics	[17]	Introducing a frequency-based method based on system call traces in one-class classification (OCC)	• Not mentioned	• Local outlier factor (LOF) • Isolation forest • OCSVM • KNN	• Low execution time	• Not improving the performance in cross platforms • Not enhancing the performance of feature selection algorithms in cross platforms

(Continued)

TABLE 3.1 ***(Continued)***
An Overview of the Reviewed Studies

Category		Ref.	Main Idea	Tools	Applied Algorithms	Advantages	Disadvantages
Temporal features	Time series analysis	[18]	Proposing a semi-supervised anomaly detection framework for multivariate time series (MTS) data	• Python (Pytorch)	• Long short-term memory (LSTM) • Principal component analysis (PCA) • LightGBM • Heterogeneous feature network (HFN)	• High accuracy	• Not evaluating their proposed method on real heterogeneous datasets • Low scalability • Low applicability
		[19]	Presenting a multivariate time series anomaly detection approach based on probabilistic auto encoder with multi-scale feature extraction	• Python (TensorFlow, Keras)	• Streaming peak over threshold (SPOT) algorithm	• High F-score	• Not evaluating ROC (AUC)
	Inter-arrival times	[20]	Presenting a bit level approximation of time series data, called FCR for time series anomaly detection	• Not mentioned	• Not mentioned	• High accuracy	• Not applying DL to enhance the performance of anomaly

(Continued)

TABLE 3.1 (*Continued*)
An Overview of the Reviewed Studies

Category		Ref.	Main Idea	Tools	Applied Algorithms	Advantages	Disadvantages
Content Features	Signature or pattern detection	[21]	Presenting a malware variant detection system based on opcode and clustering algorithm	• C++ programming language	• FDBC	• High accuracy • High scalability	• Not evaluating ROC (AUC)
	Keyword detection	[22]	Presenting a word embedding feature extraction technique for host-based IDS	• Python	• Extremely randomized trees (ERT)	• High accuracy	• Not addressing the duplicate samples • Not addressing the data imbalance issue • Not evaluating on multi-class scenarios
	Text processing techniques	[23]	Presenting a vulnerability classification framework employing TF-IDF	• Weka	• RF • KNN • DT • Naïve bayes (NB) • SVM • Multilayer perceptron (MLP) • LR	• High precision • High recall • High F-score • Improving vulnerability classification	• Not examining how software vulnerability classification algorithms are affected by wrapper and embedded feature selection techniques

(*Continued*)

TABLE 3.1 (*Continued*)
An Overview of the Reviewed Studies

Category		Ref.	Main Idea	Tools	Applied Algorithms	Advantages	Disadvantages
	Advanced NLP techniques	[24]	Presenting a context-aware feature extraction-based CNN IDS	• Python (Scikit-learn, Keras, TensorFlow)	• CNN • ERT • Select K-best (SKB)	• Reducing the feature space • Reducing computational and classification time • High accuracy • High generalizability	• Not improving the intrusion detection methods in wireless networks • Not optimizing the classification process in both network-based and host-based environments
		[25]	Presenting a framework for detecting attacks in Hindi voice-based systems	• MATLAB • Python (sklearn library)	• eXtreme gradient boosting (Xgboost) • RF • KNN • NB • ResNet27	• Low error • High performance in detecting deep fakes	• Not achieving better results in logical access attacks
Structural features	Network topology	[26]	Proposing an optimal feature selection based on graph convolutional network	• MATLAB	• LSTM • Snake optimizer-based feature selection with optimum graph convolutional network for malware detection (SOFS-OGCNMD) • FPA	• High precision • High recall • High F-score	• Not identifying outlier in the SOFS-OGCNMD approach

(*Continued*)

TABLE 3.1 (*Continued*)
An Overview of the Reviewed Studies

Category	Ref.	Main Idea	Tools	Applied Algorithms	Advantages	Disadvantages
Entity relationships	[27]	Proposing an Android malware detection approach based on *graph-based* feature generation	• Soot (Java bytecode optimization • framework) • Python (Sklearn library)	• RF • KNN • NB • LR • SVM	• High accuracy • High recall	• Cannot distinguish the malware family • Cannot disclose the impact of anomalous payload on the application's behaviors
	[28]	Presenting a feature extraction approach for fraudulent activities based on social network analysis	• Not mentioned	• Hits • PageRank algorithms • BadRank • Gspan	• High accuracy • Low runtime	• Not evaluating ROC (AUC)
	[29]	Suggesting a feature selection approach to detect DDoS attack in SDNs based on incoming flow	• Python (Keras, TensorFlow)	• RF • LSTM • Information gain	• High detection rate • Low latency	• Not detecting other classes of attacks except DDoS attacks • Not evaluating the proposed model on real SDN network • Not detecting attacks in real-time

(*Continued*)

TABLE 3.1 (*Continued*)
An Overview of the Reviewed Studies

Category		Ref.	Main Idea	Tools	Applied Algorithms	Advantages	Disadvantages
Behavioral features	Anomalous user behavior detection	[30]	Presenting an anomaly user behavior detection system by MCF for feature selection	• MATLAB	• Mutation cuckoo fuzzy (MCF) • ENN • Fuzzy C means (FCM) clustering method	• High accuracy • Low execution time	• Cannot be used for multi-class classification problem • Not evaluating on other dataset • Not enhancing the accuracy to detect type of attacks
	Anomalous system behavior detection	[12]	Proposing an optimized feature selection approach for anomalous system behavior	• Python (sklearn, Scikit libraries, Pandas, Keras, RUS Python library)	• Ensemble method • XGBoost • Particle swarm optimization (PSO) • RF • LightGBM • CatBoost classifiers • CNN • SMOTE • RUS	• High F-score • High AUC (ROC)	• Low scalability • Not setting the PSO hyper-parameters

Statistical feature engineering can be divided into descriptive statistics, inferential statistics, and frequency-based statistics. In the category of descriptive statistics, the authors in reference [16] improved the effectiveness of IDS based on deep neural networks (DNNs) by presenting a feature selection method that integrates two statistically significant metrics: the variance between the median and the mean and the standard deviation. Several parameters were used to assess the proposed method, which decreased features based on their rank. The presented strategy outperformed other feature selection techniques, yielding faster execution times and higher performance, according to statistical validation. However, the robustness of IDS was not tested by applying nature-based techniques to improve the neural network architecture. Furthermore, the feature selection methods used by the authors did not use nature-inspired algorithms.

In the category of inferential statistics, to identify and save just the most informative features from datasets, the authors in reference [11] applied filter-based techniques such as one-way ANOVA and Pearson correlation coefficient as part of a feature selection methodology for anomaly-based network intrusion detection systems (NIDS). In addition, optimal features were recovered by applying the theory's union and intersection rules. The assessment showed that the model performed better in detection rates, precision, and recall than traditional ML classifiers. However, neither the resource consumption nor the implementation utilizing additional benchmark datasets was assessed by the authors, nor was the suggested model applied to an IoT gateway for the purpose of identifying and categorizing cyberattacks.

In the category of frequency-based statistics, the authors in reference [17] developed a lightweight feature extraction technique appropriate for cross-platform applications that is meant to function without requiring system call traces. The technique converted system calls into n-gram frequency sequences to extract statistical information, which was used to train a one-class classification model for platform-independent threat detection. Although it performed better than previous approaches, the suggested solution failed to achieve the maximum area under the curve (AUC). In addition, the study could not optimize feature selection techniques for one-class learning or extend the anomaly detection model utilizing sample selection procedures from other platforms.

The temporal feature engineering can be divided into time series analysis and inter-arrival times. In the category of time series analysis, a heterogeneous feature learning for multivariate time series (MTS) was created in reference [18] to improve anomaly detection in real-world datasets. The framework includes three steps: (1) heterogeneous graph structure learning (HGSL) combines sensor embeddings and feature similarities to extract relation subgraphs and model structural information; (2) heterogeneous representation learning embeds variables into vectors, using channel, node, and semantic attention for joint optimization; and (3) abnormal detection and localization calculates deviations between predicted and actual values to detect anomalies. However, the suggested framework was not scalable. It was not tested on complex heterogeneous datasets that included mixed textual and time series data. Furthermore, the methodology failed to investigate the effects of varied sample intervals across datasets, making it inapplicable.

Similarly, a probabilistic autoencoder with multi-scale feature extraction (PAMFE) was presented in reference [19]. It was an unsupervised method for identifying anomalies in multivariate temporal data utilizing a probabilistic autoencoder. The authors created a module leveraging a parallel dilated one-dimensional convolutional neural network (CNN; Conv1D) to efficiently gather comprehensive time series data, as well as a feature fusion module to boost the reconstruction of input data from compressed features. They included multi-level noise during training to advance robustness. PAMFE could assess an observation's abnormality by considering its likelihood of fitting the reconstructed distribution via reconstructing the predicted distribution parameters. Comprehensive experiments revealed that PAMFE outperformed the most advanced techniques.

In the category of inter-arrival times, the authors in reference [20] presented feature-based clipped representation for time series anomaly detection (FCAD), a density-based anomaly detection technique based on feature-based clipped representation (FCR). FCR is a bit-level approach that uses feature-based techniques to identify and apply important turning points (ITP) as crucial features. They also present an FCR similarity metric that keeps the lower boundary constraint of the Euclidean distance so that anomaly detection and time series retrieval procedures do not accidentally discard data. Evaluations showed that compared to benchmark methods, FCR and FCAD detect abnormalities more successfully. However, the authors did not integrate FCR and FCAD with DL techniques to optimize anomaly detection efficacy.

The content feature engineering can be divided into signature or pattern detection, keyword detection, text processing techniques, and advanced natural language processing (NLP) techniques. In the signature or pattern detection category, a method for automatically identifying malware variations by vector representations was presented in reference [21], which were generated by learning and weighing sequences of operation codes (opcodes). The effectiveness of traditional signature-based malware detection techniques was declining due to the explosive growth of dangerous information. To improve the recognition of malware variations, they presented the fast density-based clustering (FDBC) algorithm, which clustered malware instances rapidly and precisely. Studies showed that this method performs better than the most advanced approaches.

In the category of keyword detection, the authors in reference [22] showed that word-embedding techniques like Word2Vec (W2V) and GloVe (GLV) can cause data replicas and reduce diversity in host-based intrusion detectors, leading to overly optimistic results. They experimented with alternative feature sets, adding dimensions and combining W2V and GLV features, which improved model performance by reducing duplication. The findings showed that adding embeddings and counting syscalls improved performance in three datasets. Nevertheless, such issues as managing duplicate samples, generating universally applicable features, validating more datasets, and resolving data imbalance were still unresolved, and these feature sets were not evaluated in multi-class environments.

In the category of text processing techniques, the term frequency-inverse gravity moment (TF-IGM) was utilized in reference [23] to present an automatic vulnerability classification method. The authors evaluated various ML algorithms on ten applications, assessing results with standard metrics. TF-IGM was found to be more effective for classifying vulnerability for feature selection compared to information gain and the classical term-weighting metric (TF-IDF). The evaluation's findings demonstrated that feature selection significantly enhanced classification, even if performance varied throughout datasets. The approach did not, however, investigate the effects of embedding feature selection methodologies and wrappers, nor did it integrate with other vulnerability assessment techniques.

In the category of advanced NLP techniques, the authors in reference [24] suggested a feature extraction method as a preprocessing step for CNN-based multiclass ID. CNN was implemented for picture recognition with colored image inputs or grayscale, and each feature was regarded as a pixel or set of pixels with values ranging from 0 to 255. Their suggested strategy significantly enhanced accuracy but should have focused on developing the ID method for various scenarios, including wireless networks.

In reference [25], regarding the Hindi language, the authors suggested a technique to enhance front-end feature extraction of an audio imitation attack detection framework. The presented approach was executed in three main steps. First, audio samples were turned into spectrograms (Mel, TPAF spectrograms, and Gammatone). This step was comparable to interpreting spectrum patterns over time to find patterns in time series data (audio signals). An NLP text processing method termed feature extraction from spectrograms (audio representations) transformed and examined signal data (audio features). Second, an enhanced residual network (ResNet27) was applied to extract distinctive features from these spectrograms. The implementation of sophisticated models like ResNet27 for feature extraction from audio spectrograms corresponded with modern NLP techniques that aim to extract meaningful features from complex data representations. Twelve systems were developed by applying four binary classifier algorithms to three feature combinations. The Gammatone spectrogram-ResNet27 and XGBoost outperformed previous techniques in attack detection, but not in logical attacks. Moreover, there were inadequate comprehensive datasets available for low-resource languages such as Hindi.

Structural feature engineering can be classified into network topology and entity relationships. In the network topology category, a feature selection approach called SOFS-OGCNMD was developed in reference [26], which combines an optimal graph convolutional network and a snake optimizer as feature selection for malware detection. The proposed model applied the flower pollination algorithm (FPA) to optimize the graph convolutional network (GCN) parameters. The suggested method outperformed the other models regarding precision, accuracy, recall, and F-score. Nevertheless, it could not identify any outliers.

Detecting malicious payloads can be handled as a binary classification problem with traditional ML techniques. Many existing methods ignore program structures, losing important semantic information and reducing accuracy. In reference [27] that aligns with network topology, the authors addressed this by presenting an approach for Android applications via extracting graph-based, semantics-rich features from components and structures, using context-based feature selection from interprocedural control flow graphs (iCFGs). These features are embedded into a feature vector space to train a highly accurate malware detector. However, this method only provided binary classification and could not distinguish between malware families or assess the impact of malicious payloads.

Regarding a reviewed paper in the category of entity relationships, the quick development of e-business and Internet technologies has led to a rise in fraudulent activity. Fraud detection is vital in combating fraudulent activities, with a focus on speed and accuracy. The authors in reference [28] proposed a feature extraction mechanism called FEMBSNA that employs preprocessing at user and network level features. In this approach, various features were obtained by setting up and evaluated weighted directed financial interaction networks. The findings in the evaluations showed that FEMBSNA significantly improves accuracy of fraud identification with acceptable runtime durations.

Similarly, software-defined networking (SDN) centralizes network management, simplifying the management of complex infrastructures. While it improves security and threat detection via open APIs, it also presents new challenges like distributed denial-of-service (DDoS) attacks. Detecting DDoS attacks in SDNs is difficult due to numerous network features and the overhead of ML. In this regard, the authors in reference [29] proposed a DL technique with long short-term memory (LSTM) and autoencoder, and they employed information gain (IG) and random forest (RF) to understand the relationships between network entities and their interactions. This approach was not tested on a real SDN network for real-time intrusion handling, but it successfully detected DDoS attacks with high accuracy and few false alarms.

Behavioral feature engineering can be categorized into anomalous user behavior detection and anomalous system behavior detection. Due to the enormous amount of data generated from several networks throughout the digital revolution, data security has become crucial. IDSs are capable of discriminating between internal and external threats. In this respect, in the category of anomalous user behavior detection, the authors in reference [30] focused on enhancing IDS efficiency by selecting significant features from large datasets to reduce detection execution time. Using the modified cuckoo search algorithm (CSA) for feature selection to detect unusual patterns or attacks and an evolutionary neural network (ENN), the presented model improved accuracy and reduced execution time. Validated with the NSL-KDD dataset, results showed enhanced IDS performance, though it lacked focus on multi-class attack detection.

In the category of anomalous system behavior detection, Chameleon, a combination of swarm intelligence and ensemble learning was suggested in reference [12] to improve the feature selection parameters. The network logs were classified into benign and anomalous through ensemble models combining classifiers based on DL and ML. Every particle in the swarm used ensemble classifiers to iteratively converge toward optimal solutions. Features selected by the ensemble model were used to build an anomaly detection auto-encoder, refined iteratively to surpass existing models. However, the presented model evaluated limited hyper-parameters for optimization algorithms like RF and XGBoost. In addition, it has low scalability and the need to explore adaptive PSO variants to optimize hyper-parameters is another limitation.

3.4 DISCUSSION

We investigate the papers in this section after considering the principles given in Section 3.2. An overview of the reviewed articles has also been provided in Section 3.3. In addition, a comparison between the studies is discussed in this section by answering the research questions that were previously mentioned.

RQ_1: What evaluation factors are applied in feature engineering for threat detection?

Researchers have employed different evaluation factors based on RQ_1. The highest percentage of evaluation factors (17%) is accounted for accuracy, as seen in Figure 3.1. Recall, precision, and F-score come next with 15% each. Time was applied to evaluate the proposed approaches at 10%. Figure 3.1 indicates that most approaches attempted to improve accuracy, precision, recall, and F-score as well as reduce time.

RQ_2: What algorithms and tools are applied in feature engineering for threat detection?

As shown in Figure 3.2, the majority of classifiers and approaches used in the reviewed articles are ensemble algorithms and DL. Concerning RQ_2, the statistical illustration of the proportion of applied tools in the studies is presented in Figure 3.3. Notably, Python emerges as the dominant tool, accounting for 63% of all usage, while MATLAB follows with 19%.

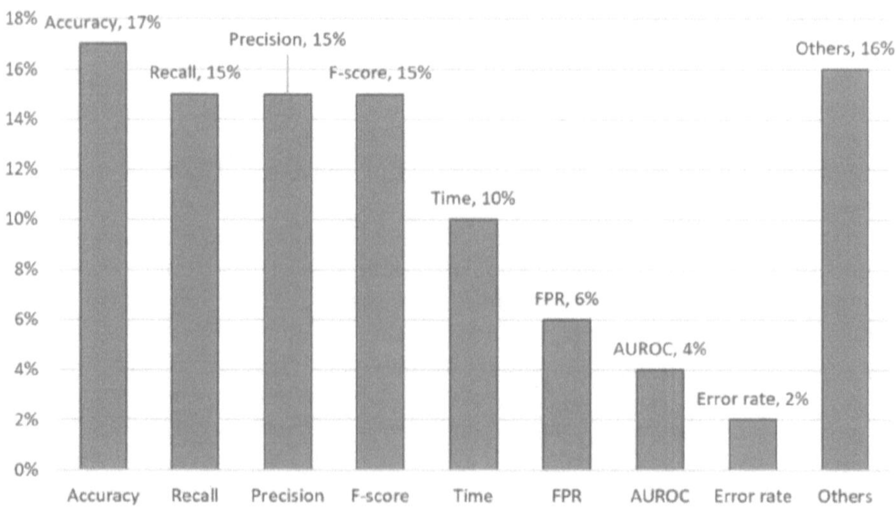

FIGURE 3.1 The percentage of evaluation factors in feature engineering for threat detection.

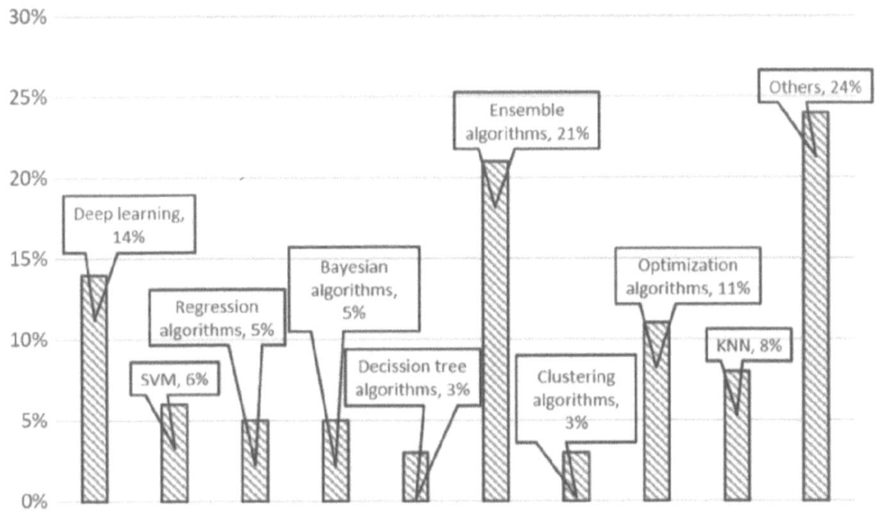

FIGURE 3.2 The percentage of applied algorithms in feature engineering for threat detection.

RQ_3: What is the possible classification of feature engineering for threat detection?

Figure 3.4 displays the suggested categorization in which the reviewed papers are classified into five main categories: *statistical features, temporal features, content features, structural features*, and *behavioral features.* Different taxonomies may be available and possible, although offering a comprehensive study on feature

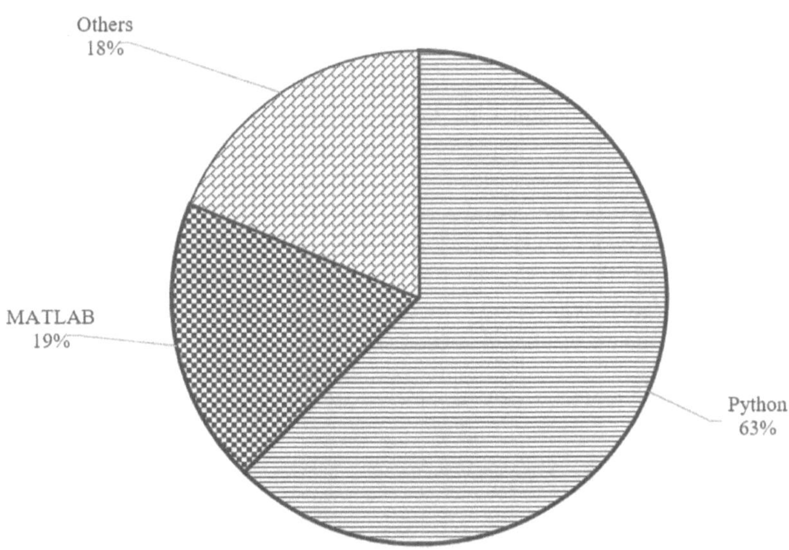

FIGURE 3.3 The percentage of evaluation tools in feature engineering for threat detection.

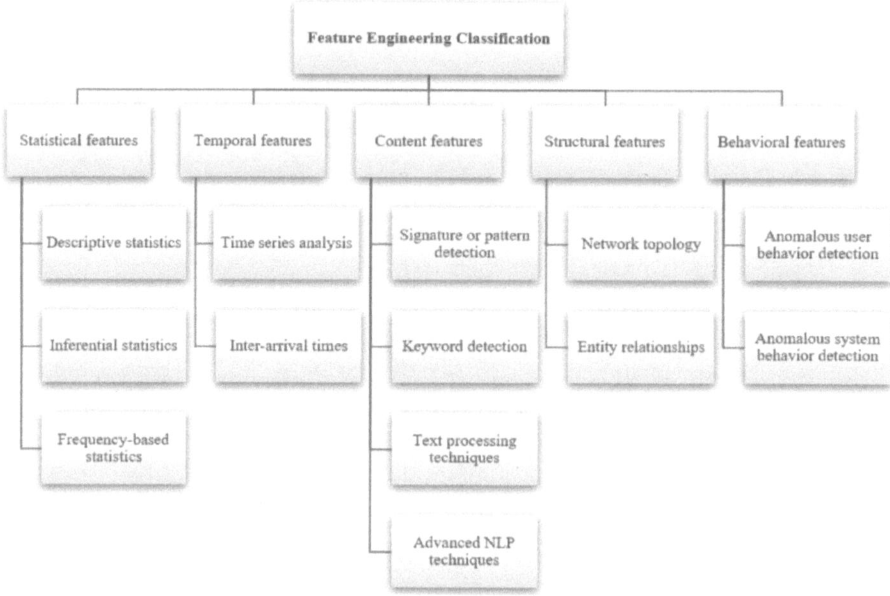

FIGURE 3.4 Feature engineering taxonomy for threat detection.

engineering for threat detection is challenging. The category of statistical characteristics comprises three subcategories: frequency-based statistics, which track the frequency of specific occurrences or events; inferential statistics, such as correlation and ANOVA; and descriptive statistics, which calculate variance, mean, median, mode, and standard deviation. Two subclasses of temporal characteristics include time series analysis, which focuses on patterns over time, and inter-arrival times, which examines the time periods between events.

Content and textual features include NLP for keyword detection in logs and messaging; text processing techniques like tokenization, stemming, and lemmatization; and signature or pattern discovery in payloads. Sentiment analysis and named entity recognition (NER), two more sophisticated NLP approaches, are also available in this category. Two kinds of structural features have been identified: entity relationships, which emphasize social network analysis and user-device interactions, and network topology, which leverages graph-based metrics like centrality and clustering coefficient. Finally, two categories of behavioral features are identified: anomalous user behavior detection, which detects unusual or suspicious variations in the user behavior (unusual access times or access patterns), and anomalous system behavior detection, which focuses on system or networks' unusual behavior such as unexpected resource usage or unusual network.

3.5 OPEN CHALLENGES

Regarding the guidelines outlined in Section 3.2, we investigated the reviewed papers in Section 3.3. In addition, a discussion of them is provided in Section 3.4 by considering the research questions. The detection of abnormal behavior can provide valuable information at critical times, allowing researchers to react to incidents in a targeted manner to prevent or eliminate abnormal events. Professionals are interested in anomaly detection in various fields, including robotics, multi-agent systems, finance, healthcare, insurance, biological systems, and so on. Due to the complexity and constant change of threats, feature engineering for threat detection encounters several issues and unresolved problems.

RQ_4: What are the challenges and open issues of feature engineering for threat detection?

Although the reviewed approaches in Section 3.3 have achieved good performance on some datasets, they still face major challenges. Considering RQ_4, the challenges and open issues of feature engineering for threat detection are discussed in this section as follows:

- *High dimensionality of data*: Finding the most relevant features in security data can be challenging because it is frequently high-dimensional. A model that includes too many irrelevant features may overfit the training set. Feature selection and dimensionality reduction are the strategies that can be applied to address such issues. Although this issue has been addressed in such studies as reference [26], additional research is required to determine feature selection and dimensionality reduction techniques that maintain valuable data.
- *Incomplete data and labeling*: It can be challenging to guarantee data quality and collect precisely labeled datasets for model training, particularly in cybersecurity, where attacks can be misleading and complex. An incomplete log or missing values are common scenarios that may harm feature extraction and impact the features' reliability. In addition, noise and incomplete information in security data may lead the model to be misled. Although some researchers [21, 30] mitigate these challenges by employing semi-supervised and unsupervised learning techniques for better labeling, it remains a significant open issue.
- *Real-time analysis*: Real-time processing is required for threat detection to identify and combat threats. Real-time threat detection requires optimizing feature extraction procedures and minimizing latency with efficient algorithms. Developing real-time feature extraction frameworks that can process streaming data for instantaneous behavioral analysis or anomaly detection is still a major challenge.
- *Dynamic feature extraction and drift concept*: Zero-day attacks offer a challenge as existing features might not capture the features of these threats. The nature of threats and even data can change over time, making previous features less relevant or obsolete. The features must be adaptable to new types of threats. Implementing adaptive learning models and continuous monitoring of feature relevance can help tackle the drift concept. Thus, developing methods for dynamically adjusting feature extraction based on evolving data patterns or changes in the environment is another challenge.

- *Imbalanced data*: Datasets are greatly imbalanced because security incidents are uncommon compared to regular activities. This might lead algorithms used in ML to neglect the minority class. To address this issue, researchers employed such approaches as oversampling and undersampling. The synthetic minority over-sampling technique (SMOTE) was employed by the authors in references [11, 12] to produce more minority sample instances through replication, provide a balanced dataset, and decrease training time. Consequently, creating an appropriate testbed is often highly challenging.
- *Ethical and privacy-preserving*: As data-driven techniques become more pervasive, ethical considerations regarding data privacy and the ethical implications of detecting anomalies or behaviors need careful attention. It is essential to ensure that feature engineering processes respect user privacy and comply with data protection regulations. So, developing techniques for secure computation of features to protect sensitive data during the feature extraction process is an ongoing concern.
- *Scalability*: Scalability in feature engineering for threat detection handles huge amounts of data effectively. As data volume increases, extracting relevant features in real time becomes more challenging. Continuous feature upgrades are required to guarantee precise models. Therefore, to maintain detection efficacy, it is essential to have strong infrastructure and optimization techniques. So, scalability is still an issue that must be fully solved.
- *Automated feature engineering*: Research on automated feature engineering methods and methodologies that are adaptable to react to evolving threats and new features remains unsolved in many areas. Adding more automated processes, possibly employing methods like reinforcement learning for feature extraction or selection, could improve the efficiency and flexibility of the taxonomy.

3.6 CONCLUSION

Cybersecurity is a daily practice that safeguards computers, networks, and data from attacks and intrusions. ML is, therefore, widely used in two domains: threat detection and network traffic analysis. Selecting the most relevant features is essential to improve detection accuracy and efficiency while preventing overfitting and additional processing costs. As a result, developing and choosing the most pertinent features is necessary to maximize the efficiency of threat detection models. This chapter aimed to analyze and present a classification of feature engineering for threat detection. We offered a taxonomy based on papers reviewed in response to RQ_3. The offered taxonomy is categorized into five main categories: *statistical features, temporal features, content features, structural features*, and *behavioral features.* According to RQ_1, accuracy accounts for the most significant percentage of evaluation factors at 17%. Recall, precision, and F-score are ranked second with 15% each. With regard to RQ_2, ensemble methods and DL are the most typically utilized classifiers in the reviewed studies. Based on the statistical percentage of applied tools, Python has a 63% utilization rate compared to 19% for MATLAB. Although we provide an extensive taxonomy of feature engineering for threat detection, its future development and practical application will depend on how well we handle issues with real-time analysis, scalability, ethical and privacy preservation, and automated feature engineering. The taxonomy will be refined and expanded over time due to continuing technological and methodological advancements.

REFERENCES

1. Qiao, M., A.H. Sung, and Q. Liu. 2016. Merging permission and API features for android malware detection. 2016 5th IIAI International Congress on Advanced Applied Informatics (IIAI-AAI). 566–571.
2. Arp, D., M. Spreitzenbarth, M. Hubner, H. Gascon, K. Rieck, and C. Siemens. 2014. Drebin: Effective and explainable detection of android malware in your pocket. NDSS. 23–26.
3. Buczak, A.L. and E. Guven. 2015. A survey of data mining and machine learning methods for cyber security intrusion detection. IEEE Communications Surveys and Tutorials. 18(2): 1153–1176.
4. Skovoroda, A. and D. Gamayunov. 2017. Automated static analysis and classification of android malware using permission and API calls models. 2017 15th Annual Conference on Privacy, Security and Trust (PST). 243-24309.
5. Aswathy, A., T. Amal, P. Swathy, M. SHOJAFAR, and P. Vinod. 2020. SysDroid: a dynamic ML-based android malware analyzer using system call traces. Cluster Computing. 23(4): 2789–2808.
6. Bazzaz Abkenar, S., M. Haghi Kashani, M. Akbari, and E. Mahdipour. 2023. Learning textual features for Twitter spam detection: A systematic literature review. Expert Systems with Applications. 228: 120366.
7. Mahindru, A. and P. Singh. 2017. Dynamic permissions based android malware detection using machine learning techniques. Proceedings of the 10th innovations in software engineering conference. 202–210.
8. Chen, L., M. Zhang, C.-Y. Yang, and R. Sahita. 2017. POSTER: Semi-supervised classification for dynamic android malware detection. Proceedings of the 2017 ACM SIGSAC Conference on Computer and Communications Security. 2479–2481.
9. Kang, B., S.Y. Yerima, S. Sezer, and K. McLaughlin. 2016. N-gram opcode analysis for android malware detection. arXiv preprint arXiv:1612.01445.
10. Reshma, R. and A.J. Anand. 2023. Predictive and comparative analysis of LENET, ALEXNET and VGG-16 network architecture in smart behavior monitoring. 2023 Seventh International Conference on Image Information Processing (ICIIP). 450–453.
11. Walling, S. and S. Lodh. 2024. Network intrusion detection system for IoT security using machine learning and statistical based hybrid feature selection. Security and Privacy: e429. https://doi.org/10.1002/spy2.429.
12. Chohra, A., P. Shirani, E.B. Karbab, and M. Debbabi. 2022. Chameleon: Optimized feature selection using particle swarm optimization and ensemble methods for network anomaly detection. Computers and Security. 117: 102684.
13. Hemalatha, S., M. Mahalakshmi, V. Vignesh, M. Geethalakshmi, D. Balasubramanian, and A.J. Anand. 2023. Deep learning approaches for intrusion detection with emerging cybersecurity challenges. 2023 International Conference on Sustainable Communication Networks and Application (ICSCNA). 1522–1529.
14. Haghi Kashani, M., M. Madanipour, M. Nikravan, P. Asghari, and E. Mahdipour. 2021. A systematic review of IoT in healthcare: Applications, techniques, and trends. Journal of Network and Computer Applications. 192: 103164.
15. Ahmadi, Z., M. Haghi Kashani, M. Nikravan, and E. Mahdipour. 2021. Fog-based healthcare systems: A systematic review. Multimedia Tools and Applications. 80(30): 36361–36400.

16. Thakkar, A. and R. Lohiya. 2023. Fusion of statistical importance for feature selection in deep neural network-based intrusion detection system. Information Fusion. 90: 353–363.
17. Liu, Z., N. Japkowicz, R. Wang, Y. Cai, D. Tang, and X. Cai. 2020. A statistical pattern based feature extraction method on system call traces for anomaly detection. Information and Software Technology. 126: 106348.
18. Zhan, J., C. Wu, C. Yang, Q. Miao, and X. Ma. 2024. HFN: Heterogeneous feature network for multivariate time series anomaly detection. Information Sciences. 670: 120626.
19. Zhang, G., X. Gao, L. Wang, B. Xue, S. Fu, J. Yu, Z. Huang, and X. Huang. 2023. Probabilistic autoencoder with multi-scale feature extraction for multivariate time series anomaly detection. Applied Intelligence. 53(12): 15855–15872.
20. Zhan, P., H. Xu, and L. Chen. 2020. FCAD: Feature-based clipped representation for time series anomaly detection. 2020 IEEE 3rd International Conference on Information Systems and Computer Aided Education (ICISCAE). 206–210.
21. Yin, H., J. Zhang, and Z. Qin. 2020. A malware variants detection methodology with an opcode-based feature learning method and a fast density-based clustering algorithm. International Journal of Computational Science and Engineering. 21(1): 19–29.
22. Mvula, P.K., P. Branco, G.-V. Jourdan, and H.L. Viktor. 2023. Evaluating word embedding feature extraction techniques for host-based intrusion detection systems. Discover Data. 1(1): 2.
23. Chen, J., P.K. Kudjo, S. Mensah, S.A. Brown, and G. Akorfu. 2020. An automatic software vulnerability classification framework using term frequency-inverse gravity moment and feature selection. Journal of Systems and Software. 167: 110616.
24. Shams, E.A., A. Rizaner, and A.H. Ulusoy. 2021. A novel context-aware feature extraction method for convolutional neural network-based intrusion detection systems. Neural Computing and Applications. 33(20): 13647–13665.
25. Chakravarty, N. and M. Dua. 2024. An improved feature extraction for Hindi language audio impersonation attack detection. Multimedia Tools and Applications: 1–26.
26. Daniel, A., R. Deebalakshmi, R. Thilagavathy, T. Kohilakanagalakshmi, S. Janakiraman, and B. Balusamy. 2023. Optimal feature selection for malware detection in cyber physical systems using graph convolutional network. Computers and Electrical Engineering. 108: 108689.
27. Liu, X., Q. Lei, and K. Liu. 2020. A graph-based feature generation approach in android malware detection with machine learning techniques. Mathematical Problems in Engineering. 2020(1): 3842094.
28. Zandian, Z.K. and M.R. Keyvanpour. 2019. Feature extraction method based on social network analysis. Applied Artificial Intelligence. 33(8): 669–688.
29. El Sayed, M.S., N.-A. Le-Khac, M.A. Azer, and A.D. Jurcut. 2022. A flow-based anomaly detection approach with feature selection method against DDoS attacks in SDNs. IEEE Transactions on Cognitive Communications and Networking. 8(4): 1862–1880.
30. Sarvari, S., N.F.M. Sani, Z.M. Hanapi, and M.T. Abdullah. 2020. An efficient anomaly intrusion detection method with feature selection and evolutionary neural network. IEEE Access. 8: 70651–70663.

4 Anomaly Detection with Artificial Intelligence

Mohammad Nikravan, Mostafa Haghi Kashani, and Sepideh Bazzaz Abkenar

4.1 INTRODUCTION

Monitoring network traffic to detect anomalies has been widely addressed in research since 1965. The anomaly detection method is designed to detect unusual patterns and irregularities in network traffic or stored datasets that deviate from normal conditions. These deviations could be signs of a problem, such as unexpected errors, system performance decreases, or security threats and intrusions. The growing use of anomaly detection in different areas, such as security, healthcare systems, financial applications, and smart city applications, has made it an important issue. It is especially outstanding in network security, data security, data mining, statistical applications, and computer vision. Identifying phishing fraud by detecting unusual transactions [1], identifying body injuries by detecting abnormal areas in radiographic images [2], and detecting device mis-operation through monitoring abnormal network traffic [3] are some examples of anomaly detection applications.

The rapid growth of the internet has led to a considerable increase in big data and network traffic. Network traffic usually carries complex, private, important, and sensitive data vulnerable to various security threats. As a consequence, system security becomes a critical issue, and network traffic anomaly detection emerges as an important mechanism for providing security, which merits more in-depth research and investigation. Anomaly detection in data and network traffic is a necessary solution that assists organizations and enterprises in dealing with network failures and network security vulnerabilities and applying appropriate responses effectively. Anomaly detection techniques could rapidly detect data leakage and data robbery, enhance network performance, and protect user data against security threats.

Traditional anomaly detection approaches are rule-based mechanisms that generate alerts when certain conditions are met [4]. In traditional systems, the experts should manually set thresholds and periodically fine-tune them to adapt the system to the changing data patterns. Therefore, ever-changing anomaly patterns, new unseen anomaly patterns, velocity, variety, and volume of generated data, high-dimensional and complex data structures, and rare data types render traditional anomaly detection methods ineffective and inappropriate for the dynamic and complex nature of modern networks [5]. These systems also have less accuracy in conditions where various parameters affect the anomaly.

The anomaly detection with artificial intelligence (AI) uses machine learning (ML) and AI algorithms to detect unusual patterns. This anomaly detection approach

DOI: 10.1201/9781003521020-4

does not leverage only predefined thresholds, fixed rules, and simple models but also uses complex models that continuously learn from network traffic and stored data. Therefore, AI-based anomaly detection approaches can dynamically adapt to new and constantly changing patterns, be better compatible with dynamic environments in detecting complex and subtle irregularity patterns, and be suitable for the dynamic and complex nature of modern networks [6]. In addition, they continuously improve themselves over time through reinforcement learning (RL). In brief, the advantages of AI-based anomaly detection include proactivity, effective recognition, real-time detection, and capability of processing large datasets. However, the fast development of AI technologies in recent years has led to an increase in industrial and academic investigations relevant to dealing with complex data structures, such as time series data definition and high-dimensional data representation [7]. Many AI-based approaches have been proposed in anomaly detection that prove AI is a promising way of solving many real-world issues [8, 9].

This chapter presents a holistic study of recent AI-based network traffic anomaly detection methods and discusses related challenges, considering the latest research results. It continues by classifying the proposed approaches related to AI-based network traffic anomaly detection methods and raising challenges. It also highlights the applied evaluation factors, algorithms, and tools. Finally, it highlights the vital research roadmaps, limitations, challenges, and open issues to enhance the efficiency and practicality of AI-based network traffic anomaly detection.

The remainder of the chapter is structured as follows: Section 4.2 discusses a few points on anomaly detection. Section 4.3 details the methodology and questions of the research and the paper selection process. Section 4.4 categorizes and analyzes the selected articles in detail, pointing out their pros and cons. Sections 4.5 and 4.6 discuss the results analysis, future trends, and open issues. Finally, Section 4.7 concludes the chapter.

4.2 ANOMALY DETECTION

Anomaly detection tries to recognize patterns in network traffic or stored data that do not match normal activities and expected patterns. In various application domains, these unexpected patterns are usually interpreted as exceptions, outliers, anomalies, or deviations that could be signs of a problem, such as errors, system performance decreases, or security threats and intrusions. In the literature on anomaly detection, the terms outliers and anomalies have been used more than the others. Anomaly detection has been widely used in various applications, such as forgery detection in insurance systems, electronic healthcare systems, fraud detection in financial transactions, cyber-security solutions, and intrusion detection systems (IDSs). The anomaly detection is important because the detected anomalies could give us critical information [10]. For instance, an abnormal network traffic pattern could be a sign that a compromised node is receiving data from an unauthorized source or sending data to an unauthorized destination. Two main network anomaly classes are anomalies related to security threats and anomalies related to performance. Attackers could cause security anomalies through malicious activities such as injecting flood traffic into the network and blocking the network services for legal users. In addition, server failure, broadcast floods, and temporary congestion could generate performance anomalies. In the abstract, anomalies are patterns that do not match normal activities. A straight way to recognize anomalies is to determine a set of normal behaviors, and any observed activity outside of this set is considered an anomaly. However, this straightforward idea becomes challenging due to the following factors:

- Determining a set that includes all possible normal behaviors is hard, and the border between abnormal and normal activities may be imprecise. Therefore, behaviors that lie near the normal–malicious border may be misinterpreted.
- Attackers and intruders use masquerading and change themselves to make abnormal activities seem normal. This makes the process of determining normal behaviors more complicated.
- In many fields, the domain of normal behaviors is still evolving, and the current definitions of normal behaviors may be insufficient for the future.
- The concept of anomaly and normal deviation range are strongly application-dependent. For example, a data deviation that indicates an abnormality in a medical application may be normal in a financial application. This necessitates the development of application-specific deviation definition strategies.
- Labeled data availability to train/validate AI-based anomaly detection methods is still a challenging problem.
- Sometimes, the noise in the data is similar to the real anomalies, which makes it difficult to detect and delete the noises.

Due to these issues, the anomaly detection process is complex, especially through traditional approaches. Therefore, the researchers have combined concepts from different fields, such as data mining, ML, AI, spectral theory, information theory, and statistics, to formulate and solve application-specific anomaly detection problems.

4.3 RESEARCH METHODOLOGY

Researchers have performed many investigations on the application of AI in anomaly detection in networks and its issues and challenges. First, we elucidate the reasons and needs that motivated us to conduct this research. Responding to the research questions identifies the research gaps and provides a roadmap for researchers to develop innovative solutions. This chapter, in particular, studies the application of AI in anomaly detection in networks and the challenges met. To this end, it compares and classifies the proposed approaches. To satisfy the research goals, the following research questions are defined:

- RQ_1: What is the probable classification of the proposed approaches of AI in anomaly detection in networks?
- RQ_2: What are the evaluation techniques, evaluation factors, methods, and tools used in the proposed approaches of AI in anomaly detection in networks?
- RQ_3: What are the current research gaps and challenges related to approaches of AI in anomaly detection in networks?

Then, we did an online search from 2023 to 2024 using Google Scholar as the search engine on well-known scientific databases, including IEEE, ACM, ScienceDirect, Springer, SAGE, Emerald, Inderscience, Wiley, Hindawi, and Taylor & Francis. The search has considered the title, keywords, and abstract, and the below search string was used:

(Anomaly OR Outlier) AND (Detection OR Identification OR Recognition) AND (AI OR "Deep Learning" OR "Machine Learning" OR "Artificial Intelligence")

In addition, we dropped review or survey papers, book chapters, short or editorial papers, non-peer-reviewed papers, non-English articles, and theses to obtain the highest-quality papers related to the subject. After that, we studied the full texts of the papers, evaluated their qualities, and selected 26 more relevant papers (JCR-indexed) that explicitly explained their method and evaluation details and covered the research scope. The next step involves categorizing the 26 selected into four classes: unsupervised learning, supervised learning, hybrid learning, and RL. Finally, we studied selected papers, extracted and discussed the main ideas, and described the advantages and drawbacks. The data were extracted, analyzed, and compared, and utilizing this data, the results were discussed, and research questions were answered. The study performed on the selected papers discloses the research gaps related to the subject and reveals open issues and remaining challenges that merit more in-depth research and investigation in applying AI in network anomaly detection. It will provide a roadmap for researchers to develop new and innovative ideas.

4.4 A CLASSIFICATION OF APPLICATIONS OF AI IN ANOMALY DETECTION IN NETWORKS

This section details the 26 selected articles and provides their features, advantages, weak points, and distinctions. Since the literature on the applications of AI in anomaly detection in networks is widely diverse, structuring systematic research is hard. Since the authors have applied four ML models for anomaly detection, including unsupervised learning, supervised learning, hybrid learning, and RL, we have classified the selected articles into these four classes, as shown in Figure 4.1. The subsections 4.4.1, 4.4.2, 4.4.3, and 4.4.4 explains in detail these categories.

4.4.1 Supervised Learning Class

Supervised ML models have been widely used in different applications, especially for anomaly detection, due to their high accuracy, high speed, diverse algorithms, and ability to learn from past data. In supervised learning, the machine is given labeled data to learn a function to predict the expected output for new inputs. They are trained on labeled data, enabling them to differentiate between abnormal and normal patterns based on past experiences. For example, labeled data for anomaly recognition may include normal and abnormal network traffic patterns. However, they are relatively expensive, require a supervisor, and cannot counter unknown patterns. This subsection explains, analyzes, and compares the supervised learning-based approaches in anomaly detection and summarizes the results in Table 4.1.

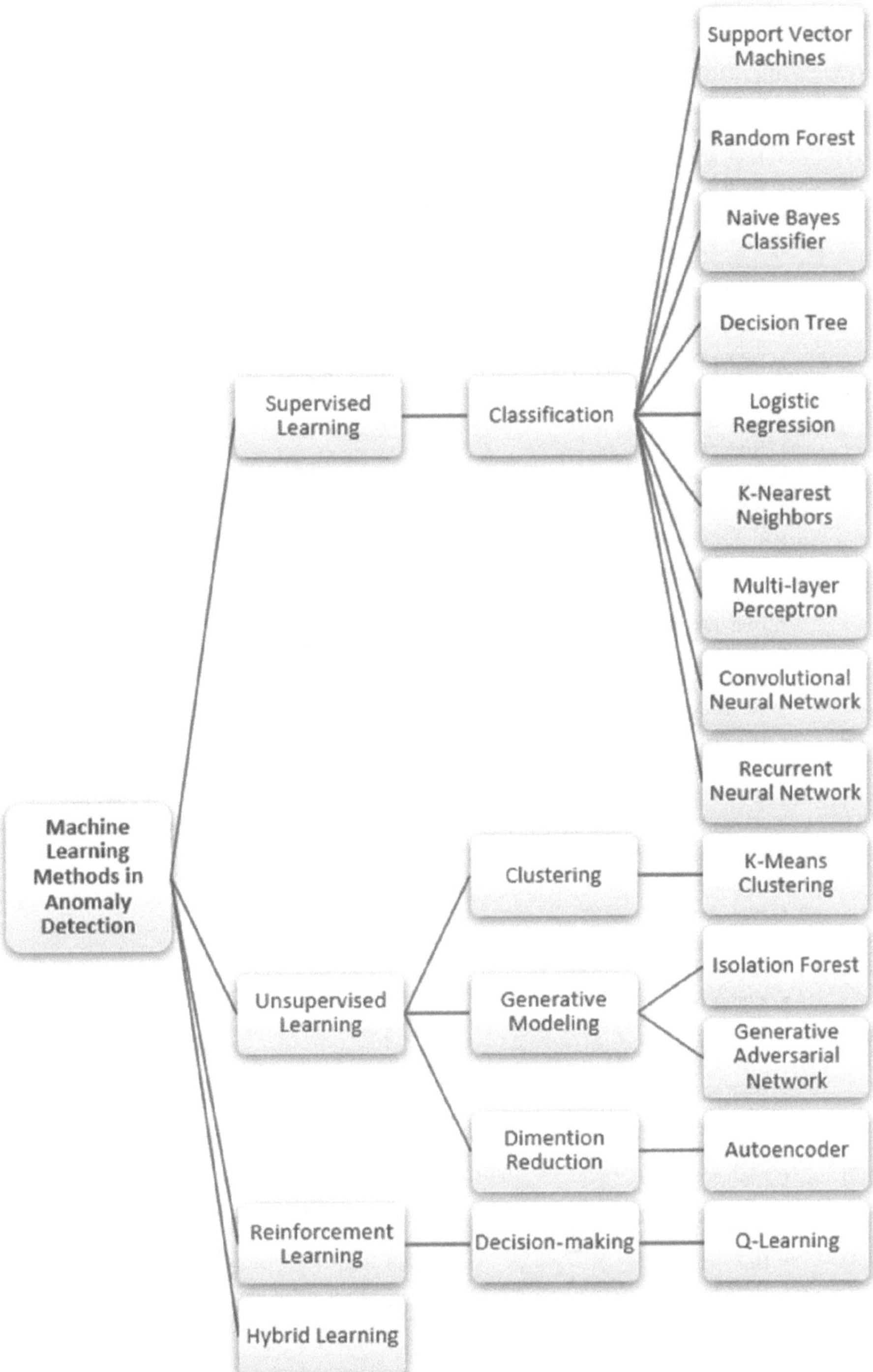

FIGURE 4.1 Classification of approaches related to the AI in anomaly detection in networks.

TABLE 4.1
Reviewing and Comparing the Supervised Learning-Based Approaches

Ref.	Advantage(s)	Disadvantage(s)	Applied models
[11]	• High accuracy • High precision • High recall • High F1-score • High specificity • High efficiency • High security • High robustness • High detection rate	• The article does not discuss the scalability of the proposed approach • The article does not discuss the potential impact of false positives and false negative	• CNN • LSTM • GBM
[12]	• High accuracy • High precision • High recall • High F1-score • High efficiency • High security	Low scalability	• KNN • Quantum autoencoder
[13]	• High accuracy • High precision • High recall • High F1-score • High adaptability	Low detection rate	• DL • Attention machine
[14]	• High accuracy • High precision • High recall • High F1-score • High adaptability	Low scalability	• RF • Ensemble technique
[15]	• High accuracy • High sensitivity • High specificity • High accuracy • High F1-score	Low scalability	ELM
[16]	• High accuracy • High precision • High recall • High F1-score • High detection rate	• Detects a few numbers of attacks • Low scalability	• DNN • LSTM • Recurrent DL
[17]	• High accuracy	• Detects a few numbers of attacks • Low scalability	• Adaptive boosting • RF, LR DNN, Perception
[18]	• High security • High accuracy • High efficiency • High trust	• The used binary class anomaly detection can limit the generalization of the findings	• DNN • SVM • KNN • RF, DT • Adaptive boosting

(Continued)

TABLE 4.1 (*Continued*)
Reviewing and Comparing the Supervised Learning-Based Approaches

Ref.	Advantage(s)	Disadvantage(s)	Applied models
[19]	• High accuracy • High precision • High recall • High F1-score	• Offline anomaly detection • Low scalability	Convolutional LSTM
[20]	• High accuracy • High precision • High recall • High F1-score • High efficiency • Low training time • Low false positive rate • Low computational cost	• Lack of analysis scalability • Potential overfitting	• DT • RF
[21]	• High accuracy • High precision • High recall • High F1-score • High true positive rate	• Offline anomaly detection • Low scalability	• LR • NB • DT • RF • ANN
[22]	• High accuracy • High specificity • High sensitivity • High precision • High detection rate • Low mean square error • Low root mean square error • Low false negative rate	• Sometimes fails to analyze the compressed data	• ELM • Heuristic optimizer
[23]	• High accuracy • High F1-score • High KAPPA coefficient • High NMI coefficient • Low training time	• Low performance during test phase Requires to be examined with new types of attacks	DT
[24]	• High accuracy • High precision • High sensitivity • High F1-score	• Detects a few numbers of attacks • Low scalability	• LSTM • Residual network
[25]	• High accuracy • High precision • High recall • High F1-score	• Slow convergence rate • Poor learning efficiency	Recurrent neural network
[26]	• High accuracy • High precision • High recall • High F1-score	• Biased or one-sided perspective • Insufficient analysis of counterarguments	• LSTM • Convolutional LSTM

In an era when energy networks are moving toward digitalization, protecting cyber-physical microgrids against security vulnerabilities, anomalies, and breaches has become crucial. The authors in reference [11] combined the capabilities of long short-term memory (LSTM) and convolutional neural networks (CNN) and suggested a method for microgrid systems that efficiently detects anomalies and breaches and identifies intrusions with high accuracy in real time. This method was integrated with gradient-boosting machines (GBM) to improve the total detection capability. Experimental results show the proposed ML-based method's efficiency, precision, and accuracy while enhancing microgrid flexibility, although using GBM improves the efficiency. Hybrid models were created by reference [12] using the abilities of quantum deep learning (DL) and quantum ML in conjunction with quantum autoencoders. The three anomaly detection schemes were constructed by combining the quantum autoencoder with a quantum one-class support vector machine (SVM), a quantum k-nearest neighbor (KNN), and a quantum random forest (RF), respectively. Evaluations showed that all schemes accurately and efficiently detect network traffic anomalies, but the highest accuracy is achieved by combining the quantum KNN and the quantum autoencoder. This indicates that the development of quantum schemes offers a promising attack and anomaly detection capability and provides network security.

A feature subset selection method and an anomaly detection mechanism were proposed by integrating DL methods with attention mechanisms [13] to secure the cloud environment. The feature selection method includes a grasshopper optimization algorithm for reducing features' dimensions and selecting a subset of features. The approach also applies attention convolutional bidirectional LSTM for classification and anomaly detection and uses a deer hunting optimizer system to fine-tune hyperparameter selection. It can also recognize complex data dependencies and patterns and aims to accurately identify and classify the anomalies of the cloud platform with improved adaptability and performance.

With the rapid growth of connected diverse Internet of Things (IoT) devices, considerable security concerns have been raised. Therefore, reference [14] suggested a scheme to profile the behaviors dynamically and detect anomalies in software-defined IoT networks (SD-IoT). The proposed scheme creates a dynamic profile of IoT device behavior through a precise and gradual process to grab evolving features over time, representing real-time device communication and interaction patterns. Next, ML-based algorithms analyze the profiles to detect anomalies and deviations from correct patterns. Once the anomaly is detected, the proper adaptive policies are triggered. Eventually, the SDN controller dynamically applies adaptive policies to prevent anomaly diffusion and provide network integrity. The scheme could effectively detect anomalies and security vulnerabilities and mitigate their effect, and thanks to SDN advantages, increases the resilience and security of the IoT environment.

IDSs could tackle privacy and security concerns in IoT networks. Thereby, reference [15] applied the kernel principal component analysis algorithm to choose the main features from the decreased features' vector and presented an IDS based on anomaly detection to protect the IoT ecosystem against different cyber-attacks. The proposed IDS employs the kernel extreme learning machine (ELM) classifier to detect malicious and safe traffic flows for binary categorization and categorize attacks in specific classes for multiclass categorization to specify the attack type. Evaluations showed the proposed method's efficiency, enhanced performance, and accuracy.

Since traditional IDS often have constraints that decrease noise sensitivity, anomaly detection efficiency, and detection rates, reference [16] presented an anomaly detection and network protection scheme for the IoT edge computing environment. The scheme uses instance-level horizontal reduction and nested moving sliding windows to reduce data complexity and dimensions and applies recurrent DL methods for anomaly detection and protection against network attacks. The sliding windows proceed with a specified step in the data and, based on anomaly type in the data, create various numbers of histograms.

Leveraging the IoT in healthcare systems has improved patients' care, but serious security concerns are raised. To tackle the concerns, reference [17] balanced the Canadian Institute for Cybersecurity (CIC) IoT dataset and used it to train different supervised ML techniques, including adaptive boosting, RF, DNN, perceptron, and logistic regression (LR). Next, the results were compared to find the most efficient network traffic anomaly detection technique in IoT-based healthcare systems. In addition, the ML algorithms were evaluated across multiclass and two-class dataset representations, the computational response time of the ML algorithms was measured, and the essential features for the extension of ML schemes were determined. The RF was found optimal for binary and multiclass classification with an approximate accuracy of 99.55%.

A major concern in autonomous driving is cyber-attacks in which autonomous vehicles (AVs) are vulnerable to various types of anomalies. Thus, reference [18] proposed an end-to-end explainable AI scheme to interpret the anomaly detection decisions made by AI methods in AV networks. In addition, the scheme includes two new explainable AI-based methods for feature selection to identify the rank and contribution of important features influencing an AV's anomaly categorization and for taking necessary prudence. The scheme offers local and global interpretations for anomaly detection AI methods in AVs. It generates justifications and explanations that are understandable to humans to clarify the decision-making process of AI methods when an abnormal AV is detected.

To counter cyber-security attacks threatening intelligent cyber-physical transportation systems (ICTS), reference [19] presented a DL-based IDS to secure ICTSs and designed an LSTM method based on DL to detect malicious activities in AV networks. In addition, a hybrid convolutional LSTM method is proposed that combines the advantages of LSTM and CNN in simultaneously investigating the temporal and spatial aspects of data packets. Simulations showed the proposed IDS's accuracy. To solve the problem of feature selection and extraction difficulty within IoT networks, reference [20] used locality-sensitive hashing techniques to demonstrate raw network data packets as vectorized data appropriate for machine-learning modeling and remove the burden of feature extraction and choosing. Furthermore, the RF and decision tree (DT) models were used for ML modeling to identify anomalies within IoT networks. The proposed mechanism doesn't need feature extraction and selection steps.

To detect anomalies and vulnerabilities in IoT smart devices, in two scenarios, five supervised ML algorithms, including Naive Bays (NB), LR, artificial neural network (ANN), RF, and DT, were applied to the same dataset [21]. In the first scenario, the whole dataset is fed to all algorithms, while in the second, the data records having 0 and 1 values are excluded from the dataset, then the dataset is fed to all classifiers. Simulations showed that the DT, RF, LR, and ANN are similar and more efficient than the NB in scenario 1, while the RF and DT overcome the other applied algorithms in scenario 2. A comparative study of the proposed method and similar works shows its superiority in terms of accuracy and detection rate.

A three-step sensor data anomaly detection approach for wireless sensor networks (WSNs), including data compression, prediction, and anomaly detection, is presented in [22]. The first step involves data pre-processing, eliminating duplicate values from the dataset, and applying the piecewise aggregate approximation method, which accurately extracts low-dimensional features, for data compression. Reducing data dimension enhances detection efficiency. The second step uses the ELM for prediction. The enhanced transient search arithmetic optimization was utilized to optimize the ELM parameters. Finally, in the third step, the data anomalies are identified utilizing the dynamic thresholding method, which defines a set of threshold values to distinguish the abnormal and normal data.

For the IoT network, reference [23] presented a data-driven anomaly and intrusion detection method. The proposed method balances the dataset using random undersampling and synthetic minority oversampling technique algorithms (SMOTE). This prevents bias in ML models and enhances their detection performance. The feature selection process is based on the mutual information index, in which the less relevant features to the output class are discarded, and more relevant features remain. This decreases the dataset size, reducing training time and computational cost. Next, the auto-ML algorithm is used to find the most efficient model producing optimal results and fine-tune the classification hyper-parameters, which in this work is a set of DTs. Finally, a set of DTs is used for anomaly detection.

Another work [24] designed a DL-based anomaly detection method to prevent the data anomalies of automated and connected vehicles caused by data failures or network attacks. The proposed method adds a wavelet convolutional layer as the network's initial input layer for extracting the most frequent data features from the input signal. Moreover, an Omni-scale block extracts impressive information adaptively. Therefore, the more relevant data features remain from the huge initial data. Next, it abstracts the extracted features using the LSTM and residual network block. Finally, it realizes particular categorization. The experimental results show the proposed method's detection performance, flexibility, and accuracy in mixed anomaly scenarios. Also, reference [25] designed a DL model based on the gated recurrent unit (GRU) neural network called SEMI-GRU to detect anomalies in vehicular adhoc networks (VANET) traffic. The model deploys semi-supervised learning and leverages data oversampling. First, the SEMI-GRU converts the data into binary features. Next, it oversamples the minority class by applying the STMOE algorithm. Then, the symmetrical reduction feed-forward neural network is applied to extract features. Finally, it uses the simplified version of the MixMatch model. The simulations showed that the proposed model overcomes existing approaches in accuracy and low false positive rate. Finally, reference [26] proposed an IDS based on DL using convolutional LSTM to protect autonomous connected vehicles, demonstrating more detection accuracy than the existing methods.

4.4.2 Unsupervised Learning Class

The unsupervised learning models in anomaly detection automatically examine the data and identify normal and abnormal patterns, structures, or clusters in data without labels just by accessing the input data without any labels or external information since no explicit data labels exist. Unsupervised learning aims to discover meaningful anomaly patterns and allow us to extract useful information. This approach can help to discover hidden and unknown anomaly patterns in data and improve the quality of anomaly detection decisions. However, they have low learning accuracy. This subsection explains, analyzes, and compares the unsupervised learning-based approaches in anomaly detection and summarizes the results in Table 4.2.

Since the mobile ad-hoc networks (MANET) used with IoT sensors are vulnerable to security threats, reference [27] combined the firefly algorithm and genetic style and proposed an efficient hybrid optimization method to select efficient, trustworthy, and safe routes, detect anomalies, and prevent Blackhole and Grayhole attacks in the MANETIoT sensor network. The optimization method uses the unsupervised K-means ML algorithm. The recommender filter of K-means calculates the trustworthiness through the security monitor; additionally, the security monitor calculates the nodes' trust values used to plan the route.

4.4.3 Reinforcement Learning Class

RL is an ML method that tries to model the behavior of the environment and, through communication and interaction with that environment, learn more about the environment's behaviors to detect anomalies. This method is based on the idea that the model independently learns from its experiences without needing labeled data. The model makes detection decisions and then uses the rewards or punishments it receives due to these decisions to improve its performance. Passing time and repeating this process, the model learns which detection decisions increase the reward or decrease the punishment and gradually identifies the normal and abnormal

TABLE 4.2
Reviewing and Comparing the Unsupervised Learning-Based Approaches

Ref.	Advantage(s)	Disadvantage(s)	Applied models
[27]	• High packet delivery ratio • High throughput • Low delay • High detection rate • Low energy consumption • High security	• The efficiency should be improved	• K-means

TABLE 4.3
Reviewing and Comparing the Reinforcement Learning-Based Approaches

Ref.	Advantage(s)	Disadvantage(s)	Applied model
[28]	• High data confidentiality • High data integrity • Low network access time • High accuracy • High security • Low false-positive rate	• High computational cost • Low scalability	• Reinforcement learning
[29]	• High accuracy • High precision • High recall • High F1-score • High efficiency • Low training time	• Support limited data structures • Low scalability	• Deep Q-Network • Autoencoder

behavioral patterns. This subsection explains, analyzes, and compares the RL-based approaches in anomaly detection and summarizes the results in Table 4.3.

Zero-trust security has become important in the industrial IoT(IIoT) and current methods are time-consuming and inefficient because they require continuous device verification every time a node joins. Therefore, reference [28] proposed an anomaly detection solution in zero-trust security networks, which provides data security and a vigorous authentication mechanism. The solution includes three phases: compression function to ensure data integrity and confidentiality, device profiling based on device features using deep RL to decrease device verification and authentication, and anomaly detection through RL. Detecting anomalies through device profiling will amplify the accuracy and performance of the IIoT networks, while DL improves system management in anomaly detection.

A deep RL-based anomaly detection mechanism to mitigate cyber-attacks in cyber-physical systems is proposed in reference [29].It applies a deep Q-network to model the thresholds in detecting anomalies as a Markov decision process and aims to make a balance between computational cost and detection efficiency. The proposed mechanism enables dynamically defining thresholds and adaptive anomaly recognition. The mechanism hybrid architecture contains an autoencoder module to learn the features and score the anomalies and a deep Q-network module to make sequential detection decisions. The simulation showed the mechanism's high efficiency, performance, and robustness.

4.4.4 Hybrid Learning Class

Hybrid learning classes combine different ML models to leverage their benefits and cover the weak points. This subsection explains, analyzes, and compares the hybrid learning-based approaches in anomaly detection and summarizes the results in Table 4.4.

TABLE 4.4
Reviewing and Comparing the Hybrid Learning-Based Approaches

Ref.	Advantage(s)	Disadvantage(s)	Applied model
[30]	• High accuracy • High precision • High recall • High F1-score • Low false positive rate • Low false negative rate	• High computational cost • High detection time • Inappropriate for real-time detection • Low scalability	• Deep belief network • LSTM • GRU • RF
[31]	• High accuracy • High precision • High recall • High F1-score • Low false positive rate	High training time	• K-means • SMOTE • Autoencoder • GBM
[32]	• High accuracy • High precision • High recall • High F1-score • Low communication cost	• Oversimplification of complex issues • Biased or one-sided perspective	• SVM • NB • GBM • Isolation forest
[33]	• High accuracy • High precision • High recall • High F1-score • Low false positive rate • High precision-recall curve	• Insufficient details on the parameter tuning and sensitivity analysis • Lack of comprehensive evaluation with diverse datasets	• SVM • Isolation forest
[34]	• High accuracy • High precision • High recall • High F1-score	• Require to consider the processing power of edge devices • Implementation complexity	• DL • Convolutional autoencoder
[35]	• High accuracy • High precision • High recall • High F1-score	• Lack of hyperparameter tuning exploration • Lack of discussion on scalability	• SMOTE, Autoencoder • Adaptive boosting • DT, RF • LSTM • ANN
[36]	• High accuracy • High precision • High recall • High F1-score	Offline anomaly detection	DNN

A two-step anomaly detection approach and a V2V attack detection system based on DL models are presented in reference [30] that use two ML classifiers from two changed prepared datasets, capable of simultaneously detecting all kinds of attacks. Simulations showed that the RF and GRU models have higher accuracy in detecting attacks, while the LSTM model has higher sensitivity in detecting types of attacks. The deep belief network (DBN) model has the lowest accuracy. The RF and DBN models are the fastest, while the GRU and LSTM models are the slowest. However, the two-step anomaly detection method's superiority over the one-step method is obvious. Since data imbalance problems affect network anomaly detection solutions, reference [31] proposed a hybrid anomaly detection scheme to tackle the anomaly detection problem in imbalanced network traffic, combining the K-means clustering algorithm and SMOTE. The K-means performs undersampling, while the SMOTE conducts over-sampling of the minority class. The denoising autoencoder also selects the most important features and decreases the data dimension. An improved version of the GBM model is applied to detect anomalies, and the Shapley additive explanation method offers explanations. The scheme balances the data with minimum information loss, doesn't increase data size, and detects anomalies accurately.

The authors in reference [32] focused on using ML methods to detect anomalies caused by compromised sensors in the network of IoT devices. To this end, they applied unsupervised (one-class SVM, local outlier factor, isolation forest) and supervised (Gaussian Naive Bayes, XGboost) methods. The unsupervised methods demonstrated admirable accuracy, but accuracy alone isn't always the final metric for effectively detecting outliers. When the main objective is detecting all outliers (maximizing recall rather than maximizing precision), the F1-score and accuracy should be considered. Simulations showed that one-class SVM is more efficient than isolation forest and local outlier factor in outlier detection, and supervised methods represent higher performance, accuracy, efficiency, and F1-score than unsupervised methods.

Since a wide range of IoT applications depend on the accuracy and reliability of the data gathered by wireless sensors, reference [33] proposed a hybrid model combining isolation forest and one-class SVM to flag abnormal sensor data and the generation source in WSNs. The model has two steps. First, the raw unlabeled data collected from the real world is labeled using one-class SVM. Then, using isolation forest, it detects anomalies, identifies abnormal data, and flags the anomalous sensors producing this abnormal data. Similarly, reference [34] presented a dynamic DL-based scheme for anomaly recognition in the Fog-assisted Internet of Vehicles (IoVs). The proposed method uses an autoencoder and convolutional layers for effective anomaly detection and feature extraction. In the comparative study, the proposed method demonstrates a higher F1-score and lower false alarms than existing schemes, which leads to secure communication. Moreover, reference [35] used different classifiers and presented an ML-based anomaly detection method for smart homes, which improves accuracy, F1-score, recall, and precision. Finally, reference [36] utilized DL models and presented an IDS based on anomaly detection for IoT networks. A feature selection based on a deep neural network model is specially designed to select more relevant data features effectively.

4.5 ANALYSIS OF RESULTS

According to the guidelines introduced in Section 4.2, this section investigates the studied papers. Moreover, we analyze and compare the studied articles to answer the research questions.

RQ_1: What is the probable classification of the proposed approaches of AI in anomaly detection in networks?

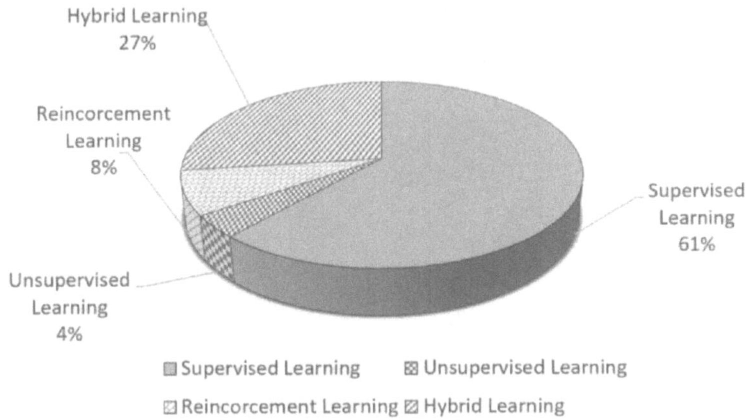

FIGURE 4.2 The percentage of machine-learning models used for anomaly detection in networks.

Based on the RQ_1 and investigated papers, the proposed approaches to AI applications in network anomaly detection are classified into four basic categories: unsupervised learning, supervised learning, hybrid learning, and RL. Figure 4.2 shows that the majority (61%) of investigated papers used supervised learning-based models, while 27% applied hybrid learning-based models, 8% preferred to utilize RL-based models, and 4% leveraged unsupervised learning-based models.

RQ_2: What are the evaluation techniques, evaluation factors, methods, and tools used in the proposed approaches of AI in anomaly detection in networks?

According to the RQ_2, researchers have used various evaluation factors. Figure 4.3 shows that accuracy has been considered more than the other evaluation factors (21%). The next considered factors are the F1-score, precision, and recall, with 16%,

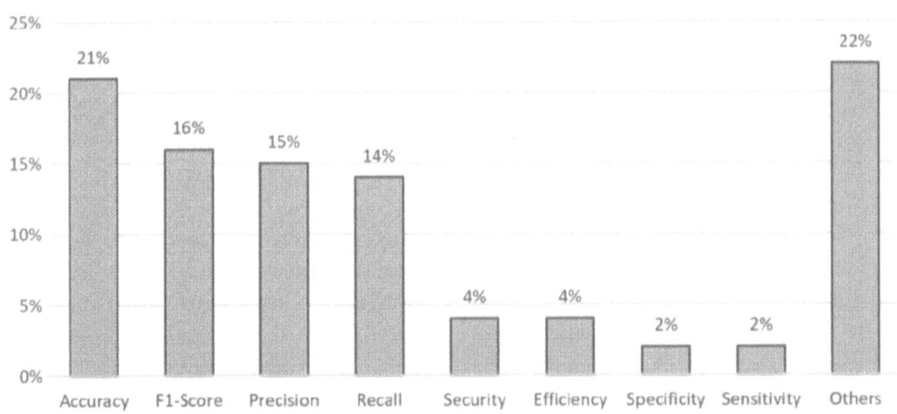

FIGURE 4.3 The percentage of evaluation factors used to evaluate the proposed approaches.

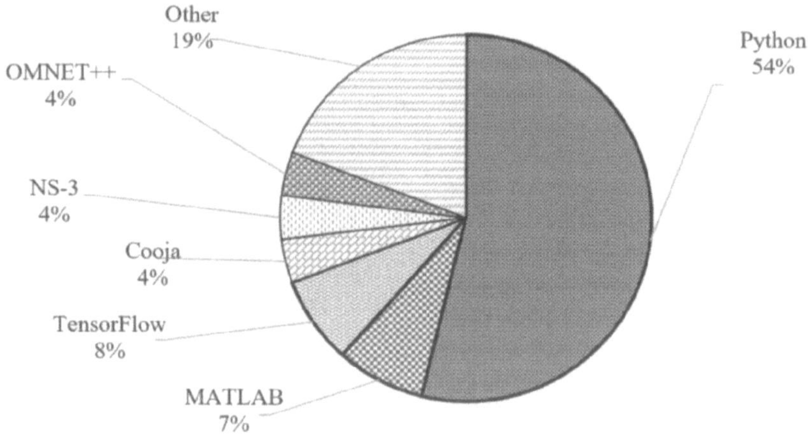

FIGURE 4.4 The percentage of evaluation tools used to evaluate the proposed approaches.

15%, and 14%, respectively. Security, efficiency, specificity, and sensitivity are the next attractive evaluation factors, with 4%, 4%, 2%, and 2%. As Figure 4.3 explains, most solutions attempt to enhance accuracy, F1-score, precision, and recall while improving efficiency and security.

Considering the RQ_2, Figure 4.4 depicts that Python, TensorFlow, MATLAB, Cooja, NS-3, and OMNET++ have been used to evaluate the proposed approaches by 54%, 8%, 7%, 4%, 4%, and 4%, respectively. Finally, all the authors evaluated their proposed solution through simulation.

4.6 OPEN ISSUES AND CHALLENGES

Investigating the selected articles highlights some research challenges that deserve more in-depth study in the future. Therefore, this section explains the challenges, considering RQ_3.

RQ_3: What are the current research gaps and challenges related to approaches of AI in anomaly detection in networks?

- *Imbalanced data problem*: When the dataset contains imbalanced classes, the classifier is more attracted to the majority classes, and the minority classes are disregarded or assumed as noisy data [37]. The anomalies are often infrequent data instances, whereas normal instances form the majority classes. Therefore, with imbalanced data, typical evaluation factors such as detection accuracy or rate may be unsuitable. However, it is crucial to solve the imbalanced data problem. Therefore, in the presence of imbalanced data, typical evaluation factors such as detection accuracy or rate may be unsuitable. However, solving the imbalanced data problem is crucial.
- *Computation efficiency*: Offline solutions could process large data volumes and be optimized for higher detection accuracy, but they do not easily adapt

to ever-changing network traffic and anomaly patterns. Online solutions require online training. On the other hand, online training is prone to noise, and training time, algorithm speed, and required storage become big challenges. Therefore, effective methods must be developed to balance computation efficiency and detection accuracy.

- *Labeling and data quality issues*: The accuracy of ML and AI schemes seriously depends on input data quality. Collecting high-quality data representing normal activities is critical for anomaly detection. This process requires recognizing the "normal" activities that could be conceptual and differ over various networks. In addition, data labeling needs time and skill, which complicates the data preparation process. This issue emphasizes the significance of robust data collection techniques in developing efficient anomaly detection systems. However, data collecting and pre-processing are challenging.
- *Resource requirements and computational complexity*: Executing the ML and AI models for real-time anomaly recognition requires considerable computational and storage resources. The computational complexity of models, such as DL, requires efficient software platforms and powerful processing resources. In addition, adapting to new anomaly patterns requires continual management and updating models, which increases resource demands [38]. Therefore, designing scalable schemes is necessary to ensure that the anomaly detection system continues normal operation even under constrained or unstable resources [39]. Thus, resource requirements and computational complexity are challenging issues.
- *Model explainability and interpretability*: While ML and AI schemes provide advanced anomaly detection capabilities, they operate like black boxes, making it hard to perceive how they make decisions [40]. This non-transparency prevents administrators and users from trusting the system since they require explanations for identified anomalies [41]. Therefore, providing model explainability and interpretability is crucial since it enables the users to understand the reasons behind detected anomalies and adapt their activities based on the rules [42]. Attempts to develop more transparent schemes, such as integrating explainable AI methods, could help solve this issue.
- *Adaptability to new vulnerabilities*: The networks have a dynamic environment with changing anomaly patterns and new threats. Ensuring all ML and AI-based anomaly detection strategies could effectively detect these new patterns and threats is a challenging issue [43]. Training the model with historical data may cause it to fail to detect new anomaly patterns, which shows the necessity of continuous adaptation and training. In addition, the rapid growth and complexity of cyber threats quickly make traditional static anomaly detection strategies obsolete. Therefore, adaptive machine-learning models are needed to address this challenge. These models must be able to dynamically update the concept of normal activities and learn from new data.
- *Real testbed environment*: Most studied papers have been evaluated through simulations and simulation does not reflect all real-world conditions. The proposed solutions should be implemented in the real world to obtain actual results. Constructing a suitable real testbed is important since realizing the

proposed solutions in a real testbed reveals to what extent the solutions can effectively detect anomalies and provide security. It also discloses the challenges and shortcomings that researchers should attempt to solve.

4.7 CONCLUSION

This chapter aimed to study, analyze, and classify the proposed applications of AI in anomaly detection in networks. The presented classification includes four main classes: unsupervised learning, supervised learning, hybrid learning, and RL. In addition, this chapter tried to study the evaluation parameters, advantages, weak points, and tools applied by the selected papers. Considering RQ2, accuracy has been considered more than the other evaluation factors (21%). The next considered factors are the F1-score, precision, and recall, with 16%, 15%, and 14%, respectively. As depicted in Figure 4.2, the majority (61%) of investigated papers used supervised learning-based models, while 27% applied hybrid learning-based models, 8% preferred to utilize RL-based models, and 4% leveraged unsupervised learning-based models. Based on the statistics, Python, TensorFlow, MATLAB, Cooja, NS-3, and OMNET++ have been used to evaluate the proposed approaches by 54%, 8%, 7%, 4%, 4%, and 4%, respectively. Moreover, all the authors evaluated their proposed solution through simulation. Finally, to answer RQ3, we presented a detailed explanation of challenges and future trends and highlighted related research gaps.

REFERENCES

1. G. M. Rao and K. Srinivas. 2022. RNN-BD: An approach for fraud visualisation and detection using deep learning. International Journal of Computational Science and Engineering, 25(2): 166–173.
2. S. M. Hussain, D. Buongiorno, N. Altini, F. Berloco, B. Prencipe, M. Moschetta, V. Bevilacqua and A. Brunetti. 2022. Shape-based breast lesion classification using digital tomosynthesis images: The role of explainable artificial intelligence. Applied Sciences, 12(12): 6230.
3. I. Stellios, K. Mokos and P. Kotzanikolaou. 2022. Assessing smart light enabled cyber-physical attack paths on urban infrastructures and services. Connection Science, 34(1): 1401–1429.
4. M. Nikravan and M. Haghi Kashani. 2022. A review on trust management in fog/edge computing: Techniques, trends, and challenges. Journal of Network and Computer Applications, 204: 103402.
5. S. Thudumu, P. Branch, J. Jin and J. Singh. 2020. A comprehensive survey of anomaly detection techniques for high dimensional big data. Journal of Big Data, 7(1): 42.
6. S. Hemalatha, M. Mahalakshmi, V. Vignesh, M. Geethalakshmi, D. Balasubramanian and J. A. A. 2023. Deep learning approaches for intrusion detection with emerging cybersecurity challenges. In 2023 International Conference on Sustainable Communication Networks and Application (ICSCNA), 1522–1529.
7. R. Wang, K. Nie, T. Wang, Y. Yang and B. Long. 2020. Deep learning for anomaly detection. In Proceedings of the 13th international conference on web search and data mining, 894–896.
8. D. Kwon, H. Kim, J. Kim, S. C. Suh, I. Kim and K. J. Kim. 2019. A survey of deep learning-based network anomaly detection. Cluster Computing, 22(1): 949–961.

9. R. Reshma and A. J. Anand. 2023. Predictive and Comparative Analysis of LENET, ALEXNET and VGG-16 Network Architecture in Smart Behavior Monitoring. in 2023 Seventh International Conference on Image Information Processing (ICIIP), 450–453.
10. V. Kumar. 2005. Parallel and distributed computing for cybersecurity. IEEE Distributed Systems Online, 6(10): 1.
11. K. Gokulraj and C. B. Venkatramanan. 2024. Advanced machine learning-driven security and anomaly identification in inverter-based cyber-physical microgrids. Electric Power Components and Systems, 1–18.
12. M. Hdaib, S. Rajasegarar and L. Pan. 2024. Quantum deep learning-based anomaly detection for enhanced network security. Quantum Machine Intelligence, 6(1): 26.
13. V. S. Bai and M. Punithavalli. 2024. Leveraging feature subset selection with deer hunting optimizer based deep learning for anomaly detection in secure cloud environment. Multimedia Tools and Applications, 83, 65949–65966.
14. S. P. K. Palaniappan, B. Duraipandi and U. M. Balasubramanian. 2024. Dynamic behavioral profiling for anomaly detection in software-defined IoT networks: A machine learning approach. Peer-to-Peer Networking and Applications, 17(4): 2450–2469.
15. S. Bacha, A. Aljuhani, K. B. Abdellafou, O. Taouali, N. Liouane and M. Alazab. 2024. Anomaly-based intrusion detection system in IoT using kernel extreme learning machine. Journal of Ambient Intelligence and Humanized Computing, 15(1): 231–242.
16. N. Abbasi, M. Soltanaghaei and F. Zamani Boroujeni. 2024. Anomaly detection in IOT edge computing using deep learning and instance-level horizontal reduction. The Journal of Supercomputing, 80(7): 8988–9018.
17. M. M. Khan and M. Alkhathami. 2024. Anomaly detection in IoT-based healthcare: Machine learning for enhanced security. Scientific Reports, 14(1): 5872.
18. S. Nazat, L. Li and M. Abdallah. 2024. XAI-ADS: An explainable artificial intelligence framework for enhancing anomaly detection in autonomous driving systems. IEEE Access, 12: 48583–48607.
19. H. N. AlEisa, F. Alrowais, R. Allafi, N. S. Almalki, R. Faqih, R. Marzouk, M. M. Alnfiai, A. Motwakel and S. S. Ibrahim. 2024. Transforming transportation: Safe and secure vehicular communication and anomaly detection with intelligent Cyber–Physical system and deep learning. IEEE Transactions on Consumer Electronics, 70(1): 1736–1746.
20. M. L. Hernandez-Jaimes, A. Martinez-Cruz and K. A. Ramírez-Gutiérrez. 2024. A machine learning approach for anomaly detection on the Internet of Things based on locality-sensitive hashing. Integration, 96: 102159.
21. I. Mukherjee, N. K. Sahu and S. K. Sahana. 2023. Simulation and modeling for anomaly detection in IoT network using machine learning. International Journal of Wireless Information Networks, 30(2): 173–189.
22. C. Ravindra, M. R. Kounte, G. S. Lakshmaiah and V. N. Prasad. 2023. ETELMAD: Anomaly detection using enhanced transient extreme machine learning system in wireless sensor networks. Wireless Personal Communications, 130(1): 21–41.
23. H. Xu, Z. Sun, Y. Cao and H. Bilal. 2023. A data-driven approach for intrusion and anomaly detection using automated machine learning for the Internet of Things. Soft Computing, 27(19): 14469–14481.
24. Z. He, Y. Chen, H. Zhang and D. Zhang. 2023. WKN-OC: A new deep learning method for anomaly detection in intelligent vehicles. IEEE Transactions on Intelligent Vehicles, 8(3): 2162–2172.
25. G. ALMahadin, Y. Aoudni, M. Shabaz, A. V. Agrawal, G. Yasmin, E. S. Alomari, H. M. R. Al-Khafaji, D. Dansana and R. R. Maaliw. 2024. VANET network traffic anomaly detection using GRU-based deep learning model. IEEE Transactions on Consumer Electronics, 70(1): 4548–4555.

26. P. Mansourian, N. Zhang, A. Jaekel and M. Kneppers. 2023. Deep learning-based anomaly detection for connected autonomous vehicles using spatiotemporal information. IEEE Transactions on Intelligent Transportation Systems, 24(12): 16006–16017.
27. S. Alangari. 2024. An Unsupervised Machine Learning Algorithm for Attack and Anomaly Detection in IoT Sensors. Wireless Personal Communications.
28. R. K. Dhanaraj, A. Singh and A. Nayyar. 2024. Matyas–Meyer oseas based device profiling for anomaly detection via deep reinforcement learning (MMODPAD-DRL) in zero trust security network. Computing, 106(6): 1933–1962.
29. X. Yang, E. Howley and M. Schukat. 2024. ADT: Time series anomaly detection for cyber-physical systems via deep reinforcement learning. Computers & Security, 141: 103825.
30. N. C. Kushardianto, S. Ribouh, Y. El Hillali and C. Tatkeu. 2024. Vehicular network anomaly detection based on 2-step deep learning framework. Vehicular Communications, 49: 100802.
31. M. K. Hooshmand, M. D. Huchaiah, A. R. Alzighaibi, H. Hashim, E.-S. Atlam and I. Gad. 2024. Robust network anomaly detection using ensemble learning approach and explainable artificial intelligence (XAI). Alexandria Engineering Journal, 94: 120–130.
32. H. Bilakanti, S. Pasam, V. Palakollu and S. Utukuru. 2024. Anomaly detection in IoT environment using machine learning. SECURITY AND PRIVACY, 7(3): e366.
33. A. Srivastava and M. R. Bharti. 2023. Hybrid machine learning model for anomaly detection in unlabelled data of wireless sensor networks. Wireless Personal Communications, 129(4): 2693–2710.
34. S. Yaqoob, A. Hussain, F. Subhan, G. Pappalardo and M. Awais. 2023. Deep learning based anomaly detection for fog-assisted IoVs network. IEEE Access, 11: 19024–19038.
35. N. Sarwar, I. S. Bajwa, M. Z. Hussain, M. Ibrahim and K. Saleem. 2023. IoT network anomaly detection in smart homes using machine learning. IEEE Access, 11: 119462–119480.
36. B. Sharma, L. Sharma, C. Lal and S. Roy. 2023. Anomaly based network intrusion detection for IoT attacks using deep learning technique. Computers and Electrical Engineering, 107: 108626.
37. D. Devi, S. K. Biswas and B. Purkayastha. 2019. Learning in presence of class imbalance and class overlapping by using one-class SVM and undersampling technique. Connection Science, 31(2): 105–142.
38. G. Muruti, F. A. Rahim and Z. A. Ibrahim. 2018. A Survey on Anomalies Detection Techniques and Measurement Methods. in 2018 IEEE Conference on Application, Information and Network Security (AINS), 81–86.
39. M. H. Bhuyan, D. K. Bhattacharyya and J. K. Kalita. 2014. Network anomaly detection: Methods, systems and tools. IEEE Communications Surveys & Tutorials, 16(1): 303–336.
40. G. Pang, C. Shen, L. Cao and A. V. D. Hengel. 2021. Deep learning for anomaly detection: A review. ACM Computing Surveys (CSUR, 54(2): 1–38.
41. H. Zenati, M. Romain, C. S. Foo, B. Lecouat and V. Chandrasekhar. 2018. Adversarially Learned Anomaly Detection. in 2018 IEEE International Conference on Data Mining (ICDM), 727–736.
42. G. Tripathi, M. Abdul Ahad and S. Paiva. 2020. Sms: A secure healthcare model for smart cities. Electronics, 9(7): 1135.
43. W. Ullah, A. Ullah, I. U. Haq, K. Muhammad, M. Sajjad and S. W. Baik. 2021. CNN features with bi-directional LSTM for real-time anomaly detection in surveillance networks. Multimedia Tools and Applications, 80(11): 16979–16995.

5 Signature-Based Security in Wireless Communication

J. S. Prasath, S. Benjamin Arul, A. Vijaya Lakshmi, and A. Jose Anand

5.1 INTRODUCTION

Internet of Things (IoT) is a rapidly evolving technique that allows a huge number of parameters to communicate information without the need for manual intervention. IoT reflects to actual physical collections of objects having sensing elements, interfaces, power supplies, programmes, and other concepts to share data with other electronic gadgets over any transmission technology. The term IoT has been criticized because gadgets must only be individually readable and linked to a network, not the general internet [1]. Depending on sensed data and the controlling information applied to develop the process, the control action must be carried out. For constrained and diverse network environments, security concerns must be discussed. The design goals, characteristics and options for industry-based wireless sensor networks (WSN) are mentioned in reference [2]. The present methodology and industry procedures are reconsidered. The problems in WSN reduce the parameters in industry-based system. The industrial IoT (IIoT) has the probability to boost manufacturing productivity significantly. Through predictive maintenance and remote management, the IIoT improves operational efficiency [3]. IoT security services address a variety of energy-efficient mechanisms [4]. The deployment environment and the target protocol are both subjected to energy-saving mechanisms. For risk assessment of cybersecurity, IoT security issues and simulating platforms are used. The issue of cyber threats in IoT settings is addressed in reference [5]. The smart home security case study is completed and evaluated by the Small World platform. By controlling the process information, attackers can disrupt the network. A novel hardware device is proposed to identify the denial-of-service attacks (Dos) [6] by completely representing signals in the circuits [7]. Intrusion detection system for IoT trends, issues, and future research are discussed [8]. The focus of IoT research is to look into different detection procedures and placement strategies, improve the attack identification values, provide better IoT concepts in medical applications, improve verification and alerting traffic, and improve security provisions [9]. To ensure the security and reliability of transmitted messages, a directional security gateway concept is proposed [10]. To prevent sensitive industry-based plant information from unauthorized access, security issues must be addressed. In the IIoT, a position privacy safety concept is decided that satisfies the differential personal constraint

DOI: 10.1201/9781003521020-5

and enhances the utilization of data and algorithm while protecting location data privacy [11].

The challenge-based security on task of structures protects from unauthorized entry and ensures the safety of transmission channels. As a novel public-key encryption procedure, the encryption process named Cramer-Shoup with minimal ciphertexts are proposed [12]. The Diffie–Hellman (DDH) assumption, which is a simple decisional assumption, underpins security [13]. Strong key management and security algorithms must be proposed. Reference [14] discusses the various security threats and vulnerabilities associated with IoT. To ensure security in IoT applications, the universal IoT security architecture can be implemented [15]. In a wireless industry-based automated system, an energy-efficient security system is suggested in reference [16]. For battery-operated vehicles, packet protection on encryption consumes energy [17]. Structure stage attacks provide those that use channels, components, programmes, logics, clocks, and supply requirements. The detection of the programme's improper behaviour is suggested using a self-organizing approach [18]. Unnecessary codes are inserted at unknown location within the network using code-based injection attack [19]. A low-cost procedure to preserve the side channel issues using embedded programmes are proposed [20]. Software attacks on the protocols can conclude with malicious behaviour such as packet latency, deadlock, or unknown destination. IoT attacks target hardware, software, and networks. The lightweight hash function is proposed, which reduces hardware implementation complexity while maintaining standard security [21]. Wireless sensors and embedded systems are examples of limited devices for which the lightweight hash function is crucial. Various IoT access mechanism solutions are emphasized [22]. The most common internet protocols are incompatible with constrained environments. The random seed circulation is combined with fleeting master key apparatuses in a key management procedure [23]. It is suitable for static networks because nodes are incapable to inaugurate novel keys afterward the specified passé. To guarantee confidentiality and integrity, the key management machineries used to protect IoT data would be robust. These algorithms [24] are recommended to deliver end-to-end confidentiality for data sharing. To protect the data from brute force attacks, the key size has been increased. The challenges of energy competence, real-time enactment, cohabitation, interoperable needs, security, and confidentiality are discussed [25]. For IIoT devices, the symmetric algorithm can provide a lightweight solution. Routing algorithms [26] and procedures are required to conclude secure message communication [27]. The procedures and tools for protected routing in the IoT are examined [28]. For IoT devices, the standard secure routing algorithm is required. The hardware assured safety schemes are considered with a hybrid cryptographic algorithm to provide process information authentication and data confidentiality [29]. It is the most cost-effective method for monitoring sensitive plant data over the internet while also providing a high level of security. IoT networks can self-execute and attend without the need for manual intervention. The advantages and disadvantages of the distributed IoT tactic are discussed [30]. Security mechanisms become more complex as a result of the distributed approach. The IoT relies on wireless transportations that are susceptible to a variety of outbreaks such as DoS, man-in-the-middle, snooping, camouflage, and fullness [31].

The key security challenges in IIoT include physical device assaults, attackers listening in on process information, unauthorized monitoring of process data, and restricting access to procedure info to sanctioned users. Many available mechanisms for securing connectivity and sanctuary in IoT are discussed [32]. The low-rate wireless PAN (LoWPAN) adaptation level, which allows IPv6 pack communication above IEEE standard 802.15.4, the IPv6 for route identification and CoAP; Constrained Application Protocol which facilitates transportations at the application level, are examples of current IoT protocols [33]. Scalability, privacy, security, and bandwidth are some of the networking-related difficulties in the IoT [34] examine the safety concerns for distributed industrial control systems. Sanctuary at the structural design stage, safety at the terminal stage for sanctuary evaluation, and advanced safety measures in IoT schemes are all factors to consider. The difficulties in securing the connection of sensor devices to the internet are discussed, in coverage in industry-oriented circumstances [35]. The incorporation of the internet in computerization and controller devices has increased the amount of sanctuary breaches linked to pressures and susceptibilities. The challenges associated with the IoT's distributed approach are examined [36]. It improves security mechanisms like access privileges, authentication, identification, and security procedures, among other things. The common security approach is essential with consideration of the majority of attacks and to ensure secure process data transmission and provides safety to plant equipment.

5.2 PROPOSED HYBRID CRYPTOGRAPHY ALGORITHM

5.2.1 Advanced Encryption Standard Algorithm

Advanced Encryption Standard (AES) uses a symmetric block cypher that can be used in hardware and software to encrypt sensitive data, and is used to safeguard classified information. It is crucial for government computer security, cybersecurity, and the protection of electronic data. Each cypher uses cryptographic keys of 128, 192, or 256 bits to encrypt and decrypt data in blocks of 128 bits are used. Symmetric cyphers employ the same key for both transmitter and receiver encryption and decryption process. The key size considered in this projected effort for I/O chunks is 128-bit AES. It is created on the state process and delivers the transitional assessment of AES encoding and decoding. The AES encryption/decryption procedure is shown in Figure 5.1.

AES S-box matrix has 256 elements, 16 rows and 16 columns, and its value ranges from 0 to 15 or 0 to F in hexadecimal. They are usually employed in block cyphers to hide the connection between the key and the ciphertex. Sub Bytes () or S-Box with 256 data elements, shown in Figure 5.2, performs nonlinear byte replacement on every byte of the state. To recite this table, the byte input is divided into two 4-bit splits.

Next is the row shift operation, in this the first row is left unaltered. The bytes in the state's latter three rows are shifted by a dissimilar quantity of bytes. The second row's bytes are shifted one to the left. The third and fourth rows are also displaced by two and three offsets, respectively as shown in Figure 5.3. In each row, the cipher's 128-bit internal state is shifted.

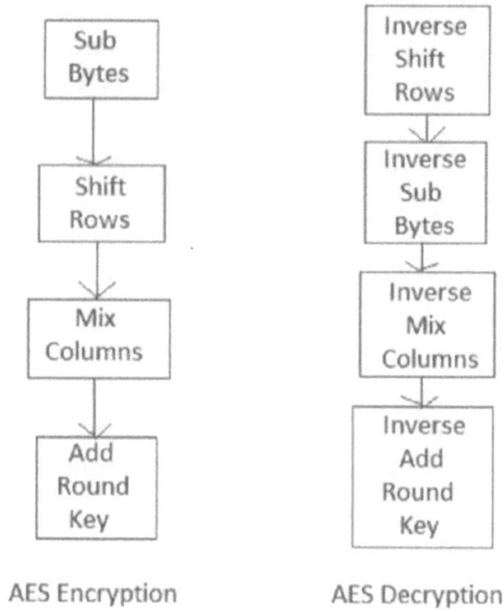

FIGURE 5.1 AES algorithm architecture.

	0	1	2	3	4	5	6	7	8	9	A	B	C	D	E	F
0	63	7C	77	7B	F2	6B	6F	C5	30	01	67	2B	FE	D7	AB	76
1	CA	82	C9	7D	FA	59	47	F0	AD	D4	A2	AF	9C	A4	72	C0
2	B7	FD	93	26	36	3F	F7	CC	34	A5	E5	F1	71	D8	31	15
3	04	C7	23	C3	18	96	05	9A	07	12	80	E2	EB	27	B2	75
4	09	83	2C	1A	1B	6E	5A	A0	52	3B	D6	B3	29	E3	2F	84
5	53	D1	00	ED	20	FC	B1	5B	6A	CB	BE	39	4A	4C	58	CF
6	D0	EF	AA	FB	43	4D	33	85	45	F9	02	7F	50	3C	9F	A8
7	51	A3	40	8F	92	9D	38	F5	BC	B6	DA	21	10	FF	F3	D2
8	CD	0C	13	EC	5F	97	44	17	C4	A7	7E	3D	64	5D	19	73
9	60	81	4F	DC	22	2A	90	88	46	EE	B8	14	DE	5E	0B	DB
A	E0	32	3A	0A	49	06	24	5C	C2	D3	AC	62	91	95	E4	79
B	E7	C8	37	6D	8D	D5	4E	A9	6C	56	F4	EA	65	7A	AE	08
C	BA	78	25	2E	1C	A6	B4	C6	E8	DD	74	1F	4B	BD	8B	8A
D	70	3E	B5	66	48	03	F6	0E	61	35	57	B9	86	C1	1D	9E
E	E1	F8	98	11	69	D9	8E	94	9B	1E	87	E9	CE	55	28	DF
F	8C	A1	89	0D	BF	E6	42	68	41	99	2D	0F	B0	54	BB	16

FIGURE 5.2 AES S-box with 256 elements.

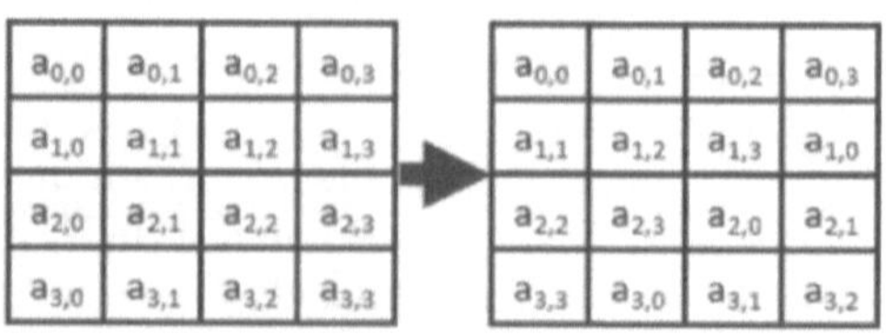

FIGURE 5.3 Row shift operations in AES.

The complex operation in AES process involves the multiplication of input matrix with the maximum distance separable (MDS) matrix. The MDS matrix is the most expensive part of the cypher process and it serves as a perfect diffusion primitive as well. In this transformation, each column is preserved as a polynomial with four terms that execute on column-by-column states as shown in Figure 5.4. Regarding the wide trail approach of the cypher, this modification is crucial and is an essential aspect of the cipher's diffuser.

The last stage in the AES encryption is the round key operation. Round key is affixed to the state that performs the bitwise XOR process as shown in Figure 5.5. For 128-bit AES encryption, 10 rounds are performed. At the finish of the 10th stout, the cypher text is obtained.

In order for encryption to function, plain text must be transformed into cypher text, which is composed of seemingly random characters. It can only be unlocked by those who possess the magical key. AES uses symmetric key encryption, which encrypts and decrypts data using just one secret key. AES decryption is the inverse process of encryption. In the middle of the cypher and modified key expansion, the reverse add round key is executed. All operations, with the exception of inverse mix columns (IMC), inverse shift rows (ISR), inverse sub-bytes (ISB), and inverse add round key (IARK), are carried out in order to produce the original plain text at the last iteration. The shift rows transformation is the reverse of the ISR operation. As seen in Figure 5.6, the fluctuating procedure takes place when the latter three rows of bytes in the stage are cycled with a dissimilar quantity of bytes.

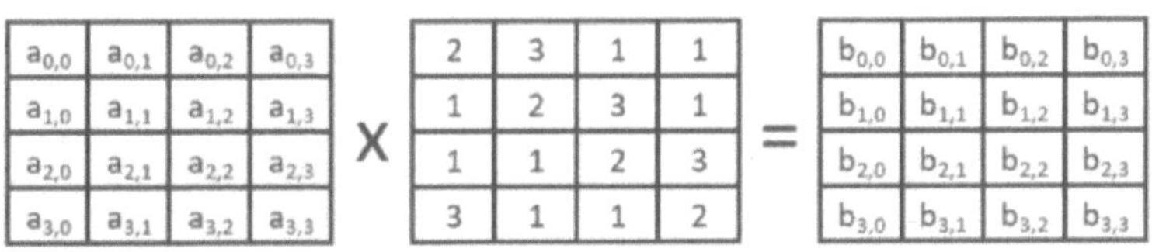

FIGURE 5.4 Mix column operations in AES.

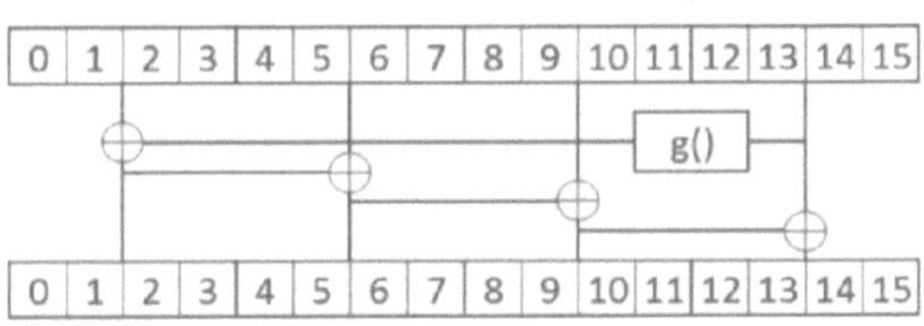

FIGURE 5.5 Round key operations in AES.

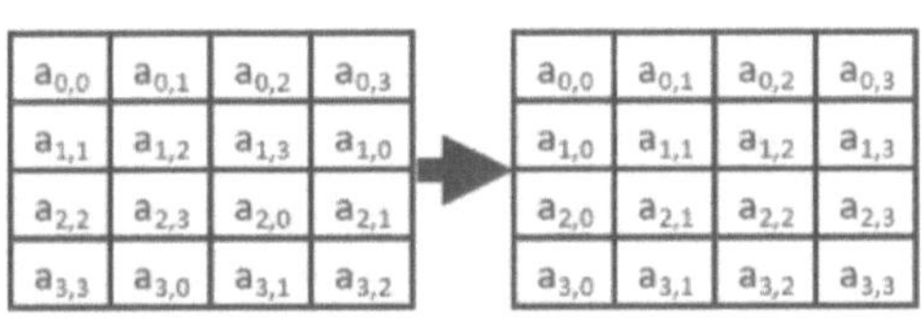

FIGURE 5.6 Inverse shift row operations in AES.

	1	2	3	4	5	6	7	8	9	0a	0b	0c	0d	0e	0f	
0	52	9	6a	d5	30	36	a5	38	bf	40	a3	9e	81	f3	d7	fb
10	7c	e3	39	82	9b	2f	ff	87	34	8e	43	44	c4	de	e9	cb
20	54	7b	94	32	a6	c2	23	3d	ee	4c	95	0b	42	fa	c3	4e
30	8	2e	a1	66	28	d9	24	b2	76	5b	a2	49	6d	8b	d1	25
40	72	f8	f6	64	86	68	98	16	d4	a4	5c	cc	5d	65	b6	92
50	6c	70	48	50	fd	ed	b9	da	5e	15	46	57	a7	8d	9d	84
60	90	d8	ab	0	8c	bc	d3	0a	f7	e4	58	5	b8	b3	45	6
70	d0	2c	1e	8f	ca	3f	0f	2	c1	af	bd	3	1	13	8a	6b
80	3a	91	11	41	4f	67	dc	ea	97	f2	cf	ce	f0	b4	e6	73
90	96	ac	74	22	e7	ad	35	85	e2	f9	37	e8	1c	75	df	6e
a0	47	f1	1a	71	1d	29	c5	89	6f	b7	62	0e	aa	18	be	1b
b0	fc	56	3e	4b	c6	d2	79	20	9a	db	c0	fe	78	cd	5a	f4
c0	1f	dd	a8	33	88	7	c7	31	b1	12	10	59	27	80	ec	5f
d0	60	51	7f	a9	19	b5	4a	0d	2d	e5	7a	9f	93	c9	9c	ef
e0	a0	e0	3b	4d	ae	2a	f5	b0	c8	eb	bb	3c	83	53	99	61
f0	17	2b	4	7e	ba	77	d6	26	e1	69	14	63	55	21	0c	7d

FIGURE 5.7 Inverse sub-bytes operations in AES.

For each of the final three rows, the ISR transformation applies circular shifts in the opposite direction. After the ISR operation comes the ISB operation. It uses the byte substitution to perform an inverse operation as shown in Figure 5.7. It is calculated by first determining the input value's inverse affine translation, then the multiplicative inverse. The inverse S-box is applied to each byte of the state. Next is the IMC operation and it uses the Mix Columns function to perform the inverse operation as shown in Figure 5.8. Every column is saved as a polynomial that operates on the formal column by column. The last step in the decryption is the IARK operation, where the round key would be chosen in the opposite manner. The XOR operation is carried out by its inverse function.

Cipher block chaining (CBC) mode is proposed as an advanced form of block cypher encryption that adds complexity to the encrypted data. A countermeasure

$$\begin{bmatrix} a_{0,0} & a_{0,1} & a_{0,2} & a_{0,3} \\ a_{1,0} & a_{1,1} & a_{1,2} & a_{1,3} \\ a_{2,0} & a_{2,1} & a_{2,2} & a_{2,3} \\ a_{3,0} & a_{3,1} & a_{3,2} & a_{3,3} \end{bmatrix} \times \begin{bmatrix} 14 & 11 & 13 & 9 \\ 9 & 14 & 11 & 13 \\ 13 & 9 & 14 & 11 \\ 11 & 13 & 9 & 14 \end{bmatrix} = \begin{bmatrix} b_{0,0} & b_{0,1} & b_{0,2} & b_{0,3} \\ b_{1,0} & b_{1,1} & b_{1,2} & b_{1,3} \\ b_{2,0} & b_{2,1} & b_{2,2} & b_{2,3} \\ b_{3,0} & b_{3,1} & b_{3,2} & b_{3,3} \end{bmatrix}$$

FIGURE 5.8 Inverse mix column operations in AES.

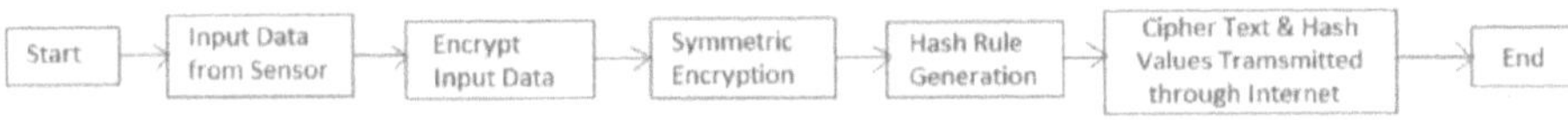

FIGURE 5.9 Hybrid encryption flowchart.

technique based on liability space alteration is proposed to defend AES-128 bits from prejudiced liability attacks [37]. It is forbidden to use collision-based outbreaks. Before encryption, each plain text block is XORed with the preceding cypher text block, and the consequence is translated with the key. By changing the initialization vector, CBC mode can generate different cypher texts for identical input messages. The combination of an asymmetric and secure hash algorithm is proposed for monitoring the process data in the wastewater treatment plants using IoT [38]. The liquefied oxygen and the pH rate are encrypted and monitored through IoT.

5.2.2 Proposed Hybrid Cryptographic Algorithm

Figure 5.9 depicts the flowchart of the proposed encryption algorithm. The data from the temperature and gas sensors are used as input. When the plant data in the input changes, the hash value changes as well. In parameters such as power values, unit energy values, terms of speed, a multi-model examination structure for cryptographic procedures is utilized [39–47]. According to the results of the experiment and analysis, the plain text size is not proportional to the energy consumption and time expenditures of cryptographic procedures. Figure 5.10 depicts the flowchart of the proposed hybrid decryption algorithm.

5.3 RESEARCH METHOD

The projected hybrid safety procedure is developed in entrenched hardware, with progression restrictions sent over a wireless network. Both the transmitter and the receiver can monitor the process data via the internet. The block illustration of protected nursing process data using embedded systems and IoT is shown in Figure 5.11. Wi-Fi is used to send the encrypted data to the receiver nodule. The IP location is required to take care of the sensed data from sensors via the internet on both the transmitter and receiver sides. This architecture is developed with three nodes, which are used for protected transmission and reception of development data. The transceiver nodel in Figure 5.12 performs encryption to defend the course information from unlawful admittance.

The data from the gas-sensing element is read by transceiver node 1 that will create the necessary message in cipher form, and the entire encryption algorithm codes

FIGURE 5.10 Hybrid decryption flowchart.

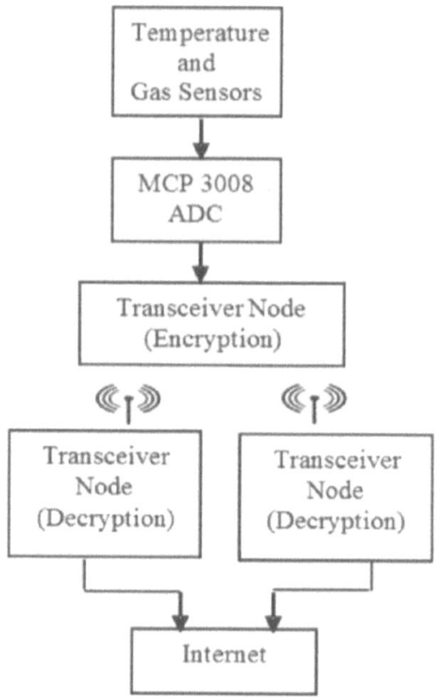

FIGURE 5.11 IoT-based process monitoring.

are written using Python language. The node 2 transceiver that acts as a receiving node to collect the sensed gas parameters in secure manner is depicted in Figure 5.13. The transceivers are connected using a Wi-Fi module to obtain uninterrupted connectivity to the internet.

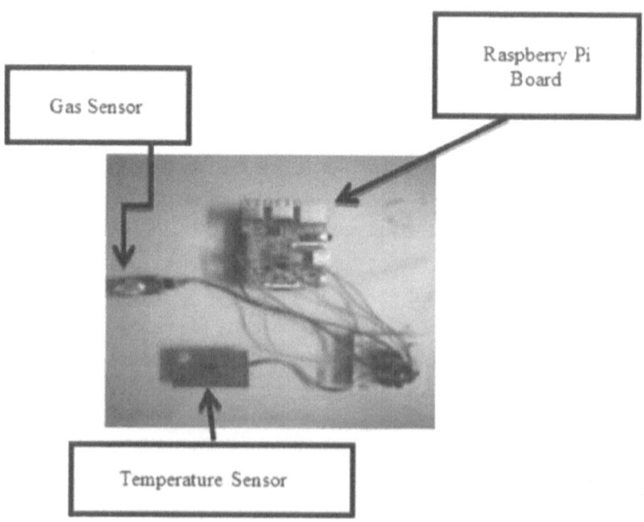

FIGURE 5.12 Node 1 transceiver.

FIGURE 5.13 Raspberry Pi-based transceiver node.

At the receiver, the transceiver node 3 is used to empower safe checking of development data via the internet in isolated areas. This planned mix cryptography procedure fortifies sensible process information during wireless network transmission.

5.4 RESULTS AND DISCUSSION

Most of the existing cryptography algorithms were developed and tested only using simulation software. In addition, the existing algorithms are verified for IT (Information Technology) security applications. Due to the usage of the internet widely in-process monitoring and control applications, security threats increase. Industrial types of equipment were not designed with security as a major concern. The security mechanisms are essential for industrial operations due to the extensive use of IoT. The security algorithms are necessary to shield the highly valuable process instruments from unauthorized access and modifications of progression data. The symmetric AES 128-bit architecture system suggested is detailed below. All the values mentioned are in hexadecimal form.

128-bits Plain Text

64	77	4F	60	8F	6D	75	40	8E	39	2E	85	90	44	47	5E

AES Algorithm – First Round Key
128-bits Key

84	28	71	24	93	70	3D	49	80	6B	35	2E	97	50	36	95

5.4.1 (37, 7A, 4D, 55) Substitution as S-box

5.4.1.1 Initial Round Key

B2	72	EC	F6	94	17	41	83	B7	79	E6	5B	D4	89	6C	7B

5.4.1.2 Byte Substituting

Initial matrix is

$$\begin{pmatrix} 00 & 3C & 6E & 47 \\ 1F & 4E & 22 & 74 \\ 0E & 08 & 1B & 31 \\ 54 & 59 & 0B & 1A \end{pmatrix}$$

State matrix substitution is made for every byte by the consistent admission in AES S-box. These clues to the novel state matrix as

$$\begin{pmatrix} 63 & EB & 9F & A0 \\ C0 & 2F & 93 & 92 \\ AB & 30 & AF & C7 \\ 20 & CB & 2B & A2 \end{pmatrix}$$

5.4.1.3 Row Shifting

Here the state matrix is shifted for the last three rows in the matrix and the initial row of the matrix is not shifted. The new state matrix obtained is:

$$\begin{pmatrix} 63 & EB & 9F & A0 \\ 2F & 93 & 92 & C0 \\ AF & C7 & AB & 30 \\ A2 & 20 & CB & 2B \end{pmatrix}$$

5.4.1.4 Mix Columns

The fixed matrix is multiplied against the current state matrix as,

$$\begin{pmatrix} 02 & 03 & 01 & 01 \\ 01 & 02 & 03 & 01 \\ 01 & 01 & 02 & 03 \\ 03 & 01 & 01 & 02 \end{pmatrix} X \begin{pmatrix} 63 & EB & 9F & A0 \\ 2F & 93 & 92 & C0 \\ AF & C7 & AB & 30 \\ A2 & 20 & CB & 2B \end{pmatrix} X \begin{pmatrix} BA & 84 & E8 & 1B \\ 75 & A4 & 8D & 40 \\ F4 & 8D & 06 & 7D \\ 7A & 32 & 0E & 5D \end{pmatrix}$$

5.4.1.5 Add Round Key

The round key is added to the state in which the XOR operation is performed. The encrypted AES output after the initial round is given by:

D8	87	C8	3B	F5	B2	9C	EA	39	A4	E7	5D	8D	79	2F	FB

Similarly, all iterations are executed in the same way, and at the finish of 10th iteration the cipher text will be as follows:

69	53	5A	8F	C7	74	30	F3	70	23	79	D3	5A	6B	D4	7E

Thus the MD5 hash procedure is utilized to guarantee data truthfulness and is articulated as a 32-digit hexadecimal number. The span of the memorandum previously stuffing is attached as a 64-bit unit and produces the hash number for a specified message input.

5.4.1.6 128-bits Input Data

The following be the 128-bit input data:

54	68	61	78	33	28	4D	83	60	6D	B5	9E	52	50	84	47

5.4.1.7 MD5 Hash Value

The following will be the MD5 hash value:

b7	29	66	be	9e	f8	9d	4c	fa	13	f1	ae	d3	c3	8c	ca

Figure 5.14 shows the encrypted data value structure of a gas sensor that was seen online at the transmitter. By compiling the well-known encryption technique created in the Python programming language, this value is obtained. To monitor and operate the cypher text monitoring of the gas sensor data, an IP address is required. The decrypted gas sensor data that was viewed online at the receiver is shown in Figure 5.15. The IP address is essential to observe and control the gas parameters in an arithmetical manner.

The sensitive progression data is monitored using the internet in an unreadable form. Figure 5.16 shows the sample data input to be encrypted with the 16-bit key and the encrypted encryption text. In accumulation, the hash procedure is proposed in this work which ensures data integrity. The unauthorized parties cannot read and alter the process data transmitted across the internet. This proposed cryptography structure is applicable in a wide range of industries, including power plants,

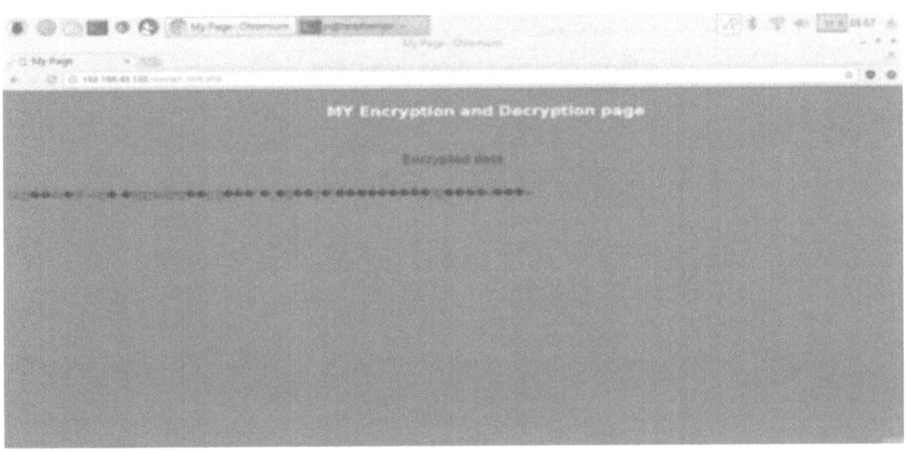

FIGURE 5.14 Monitoring of encrypted data through the internet.

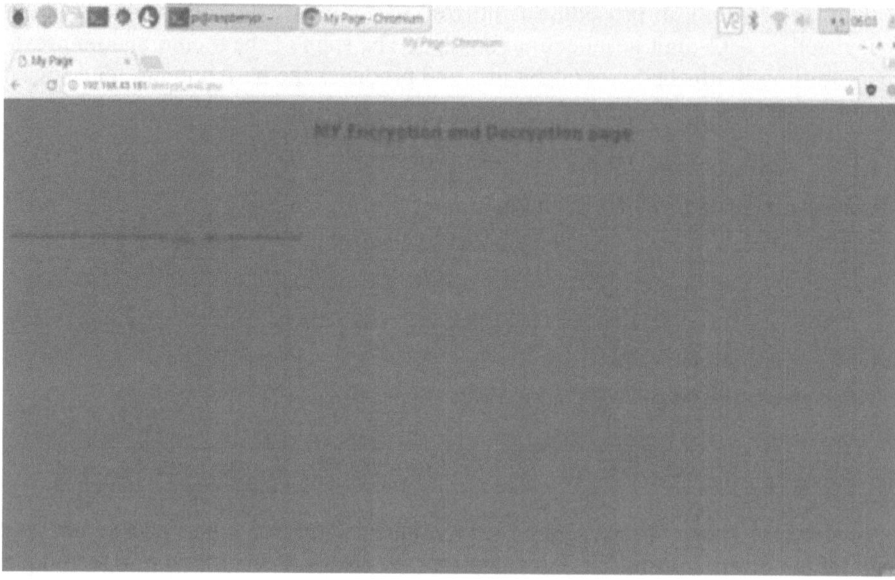

FIGURE 5.15 Monitoring of decrypted data through the internet.

petrochemical, oil and gas, sugar, chemical plants, etc., for monitoring plant information over the internet. The improvements from the suggested architecture are implemented using various combinations of cryptography algorithms in real-time industrial applications. The key size and the number of rounds in security algorithms can be increased to enhance the security level of the procedure to monitor

Text to be Encrypted

This is a test data to check the encryption using Symmetric and Hash Algorithms for Security in Wireless Communication

Key Size - 16-bits

Secret Key - 123456789ABCDEF0

Encrypted Data

+3E7EdhlDXzhCN5dIVtQUEhGzOuJXFKjdl
+eF1u0d/dPl+b1g4zdcxjSc6mh62XWmO
14ri6BSCwSdZ44tuo23wvjXjUAMXDDow
loNaGEe7byAS6fX1vzBjNffc5tVTOwh49lj
ayXU3kSAw9wFJ4LMDa9/qklRGqvtfbBLz
L7iel=

FIGURE 5.16 Input data and encrypted data using the 16-bit key.

and control the application. Security and safety can be assured for a variety of plant operations through hybrid cryptography algorithms.

5.5 CONCLUSION

To protect the privacy and accuracy of procedure data during wireless communication, this intended compound cryptography technique combines symmetric and hash procedures. Data that has been processed is sent through wireless networks while the scheduled safety procedure is implemented in an embedded system. The symmetric block cypher is employed in CBC mode, which makes the encrypted data more complex. The hashing process guarantees data authenticity and integrity. In addition, it provides industrial managers and engineers with discretion when checking on the position of private plant data. The material for encryption and decryption to be transmitted is managed by the internet and a recipient. The implementation of a hybrid security algorithm guarantees efficient plant operations, and offers plant operators great security and safety. Highly expensive industrial gadgets are shielded from intruders by it. When used in conjunction with wireless networks, embedded systems become more beneficial. This suggests compound sanctuary procedure offers trustworthy safety for all manufacturing engineering processes to save private information about industrial plants. This is crucial to examine security threats and put sanctuary measures connected with contemporary manufacturing engineering mechanization schemes into place. The adoption of a hybrid safety procedure verifies that plant operations proceed without hiccups and it offers plant personnel strong security and safety. Highly expensive industrial gadgets are shielded from intruders by it. When combined with wireless networks, the utilization of embedded structures becomes more cost-effective. The suggested compound safety algorithm offers trustworthy safety for all industrial processes to protect private information about industrial plants. During industrial revolutions, using recent technologies safety mechanisms are to be designed for utilizing the defenses promptly.

REFERENCES

1. J. Granjal, E. Monteiro, and J. S. Silva, "Security for the Internet of Things: A Survey of Existing Protocols and Open Research Issues," IEEE Communications Surveys and Tutorials, Vol. 17, No. 3, pp. 1294–1312, 2015.
2. M. Raza, N. Aslam, H. Le-Minh, S. Hussain, Y. Cao, and N. M. Khan, "A Critical Analysis of Research Potential, Challenges, and Future Directives in Industrial Wireless Sensor Networks," IEEE Communications Surveys & Tutorials, Vol. 20, No. 1, pp. 39–95, 2018.
3. H. Hellaoui, M. Koudil, and A. Bouabdallah, "Energy Efficient Mechanisms in Security of the Internet of Things", Computer Networks: The International Journal of Computer and Telecommunications Networking, Vol. 127, No. C, pp. 173–189, 2017.
4. R. Vaishnavi, J. Anand, and R. Janarthanan, "Efficient Security for Desktop Data Grid using Cryptographic Protocol", Proceedings of the IEEE International Conference on Control, Automation, Communication, and Energy Conservation, Vol. 1, pp. 305–311, 4-6 June 2009.
5. A. Furfaro, L. Argento, A. Parise, and A. Piccolo, "Using Virtual Environments for the Assessment of Cyber Security Issues in IoT Scenarios", Simulation Modeling Practice and Theory, Vol. 73, pp. 43–54, 2017.

6. D. Srinath, J. Janet, and J. Anand, "A Survey of Routing Instability with IP Spoofing on the Internet,' Asian Journal of Information Technology, Medwell Journals Scientific Research Publishing Company, Vol. 9, No. 3, pp. 154–158, 2010.
7. R. Jinnai, A. Inomata, I. Arai, and K. Fujikawa, "Proposal of Hardware Device Model for IoT Endpoint Security and its Implementation," IEEE International Conference on Pervasive Computing and Communications Workshops, pp. 91–93, 2017.
8. B. Z. Bruno, S. M. Rodrigo, T. K. Claudio, and A. Sean C de, "A Survey of Intrusion Detection in the Internet of Things," Journal of Network and Computer Applications, Vol. 84, pp. 25–37, 15 April 2017.
9. R. R. Aditya, H. Ajay, M. Balavanan, R. Lalit, and A. Jose, "A Novel Cardiac Arrest Alerting System Using IoT," International Journal of Science Technology & Engineering, Vol. 3, No. 10, pp. 78–83, 2017.
10. Y. Heo, B. Kim, D. Kang, and J. Na, "A Design of Unidirectional Security Gateway for Enforcement Reliability and Security of Transmission Data in Industrial Control Systems," 18th International Conference on Advanced Communication Technology, pp. 310–313, 2016.
11. C. Yin, J. Xi, R. Sun, and J. Wang, "Location Privacy Protection Based on Differential Privacy Strategy for Big Data in Industrial Internet of Things," IEEE Transactions on Industrial Informatics, Vol. 14, No. 8, pp. 3628–3636, 2018.
12. A. Michel, B. Fabrice, and P. David, "Public-Key Encryption Indistinguishable Under Plaintext-Checkable Attacks," IET Information Security, Vol. 10, No. 6, pp. 288–303, 2016.
13. N. Li, "Research on Diffie-Hellman Key Exchange Protocol," 2nd International Conference on Computer Engineering and Technology, pp. V4-634–V4-637, 2010.
14. A. A. Fadele, O. Mazliza, A. T. H. Ibrahim, and A. Faiz, "Internet of Things Security: A Survey," Journal of Network and Computer Applications, Vol. 88, pp. 10–28, 2017.
15. J. Anand, D. Srinath, R. Janarthanan, and C. Uthayakumar, "Efficient Security for Desktop Data Grid Using Fault Resilient Content Distribution," International Journal of Engineering Research and Industrial Applications, Vol. 2, No. VII, pp. 301–313, 2009.
16. R. Muradore, and D. Quaglia, "Energy-Efficient Intrusion Detection and Mitigation for Networked Control Systems Security," IEEE Transactions on Industrial Informatics, Vol. 11, No. 3, pp. 830–840, 2015.
17. K. M. Manoj, D. Akash, H. P. J. Allen, and A. Jose, "Electric Motorcycle with Clutchless Multi-Speed Gear Reduction System," International Research Journal of Engineering and Technology, Vol. 7, No. 4, pp. 1431–1434, 2020.
18. X. Zhai, K. Appiah, S. Ehsan, G. Howells, H. Hu, and D. Gu, et al., "A Method for Detecting Abnormal Program Behavior on Embedded Devices," in IEEE Transactions on Information Forensics and Security, Vol. 10, No. 8, pp. 1692–1704, Aug. 2015.
19. Z. C. S. S. Hlaing, and M. Khaing, "A Detection and Prevention Technique on SQL Injection Attacks," IEEE Conference on Computer Applications, pp. 1–6, 2020.
20. G. Agosta, A. Barenghi, G. Pelosi, and M. Scandale, "The MEET Approach: Securing Cryptographic Embedded Software Against Side-Channel Attacks," IEEE Transactions on Computer-Aided Design of Integrated Circuits and Systems, Vol. 34, No. 8, pp. 1320–1333, 2015.
21. M. M. Puliparambil, M. Sindhu, S. Chungath, and S. Madathil, "Hash-One: A Lightweight Cryptographic Hash Function," IET Information Security, Vol. 10, No. 5, pp. 225–231, 2016.
22. O. Aafaf, M. Hajar, A. E. Anas, and A. O. Abdellah, "Access Control in the Internet of Things: Big Challenges and New Opportunities," Computer Networks, Vol. 112, pp. 237–262, 2017.
23. F. Gandino, M. Bartolomeo, and R. Maurisio, "Key Management for Static Wireless Sensor Networks with Node Adding," IEEE Transactions on Industrial Informatics, Vol. 10, No. 2, pp. 1133–1143, 2014.

24. J. Anand, H. Gowtham, R. Lingeshwaran, J. Ajin, and J. Karthikeyan, "IoT Based Smart Electrolytic Bottle Monitoring," Advances in Parallel Computing Technologies and Applications. IOS Press, Vol. 40, pp. 391–399, 2021.
25. E. Sisinni, A. Saifullah, S. Han, U. Jennehag, and M. Gidlund, "Industrial Internet of Things: Challenges, Opportunities, and Directions," IEEE Transactions on Industrial Info, Vol. 14, No. 11, pp. 4724–4734, 2018.
26. J. Anand, J. R. P. Perinbam, and D. Meganathan, "Performance of Optimized Routing in Biomedical Wireless Sensor Networks Using Evolutionary Algorithms," Comptes Rendus De l'Academie Bulgare Des Sciences, Tome, Vol. 68, No. 8, pp. 1049–1054, 2015.
27. J. Anand, J. R. P. Perinbam, and D. Meganathan, "Q-Learning-Based Optimized Routing in Biomedical Wireless Sensor Networks," IETE Journal of Research, Vol. 63, No. 1, pp. 89–97, 2017.
28. A. David, G. Jairo, and K. R. Sayan, "Secure Routing for Internet of Things: A Survey," Journal of Network and Computer Applications, Vol. 66, pp. 198–213, 2016.
29. J. S. Prasath, U. Ramachandraiah, and G. Muthukumaran, "Modified Hardware Security Algorithms for Process Industries Using Internet of Things," Journal of Applied Security Research, Vol. 16, No. 1, pp. 1–14, 2021.
30. R. Rodrigo, Z. Jianying, and L. Javier, "On the Features and Challenges of Security and Privacy in Distributed Internet of Things," Computer Networks, Vol. 57, pp. 2266–2279, 2013.
31. Y. Zhou, and K. Zhang, "DoS Vulnerability Verification of IPSec VPN," 2020 IEEE International Conference on Artificial Intelligence and Computer Applications, pp. 698–702, 2020.
32. D. Aakash, and P. Shanthi, "Lightweight Security Algorithm for Wireless Node Connected with IoT," Indian Journal of Science and Technology, Vol. 9, pp. 1–8, 2016.
33. N. Makarem, W. Bou Diab, I. Mougharbel, and N. Malouch, "Performance Study of the Constrained Application Protocol in Lossy Networks," 2019 IFIP Networking Conference (IFIP Networking), pp. 1–2, 2019.
34. M. Cheminod, L. Durante, and A. Valenzano, "Review of Security Issues in Industrial Networks," IEEE Transactions on Industrial Informatics, Vol. 9, No. 1, pp. 277–293, 2013.
35. C. Alcaraz, R. Roman, P. Naiera, and J. Lopez, "Security of Industrial Sensor Network-Based Remote Substations in the Context of the Internet of Things," Ad Hoc Networks, Elsevier, Vol. 11, No. 3, pp. 1091–1104, 2013.
36. R. Roman, J. Zhou, and J. Lopez, "On the Features and Challenges of Security and Privacy in Distributed Internet of Things," Computer Networks, Vol. 57, No. 10, pp. 2266–2279, 2013.
37. S. Patranabis, A. Chakraborty, D. Mukhopadhyay, and P. P. Chakrabarti, "Fault Space Transformation: A Generic Approach to Counter Differential Fault Analysis and Differential Fault Intensity Analysis on AES-Like Block Ciphers," IEEE Transactions on Information Forensics and Security, Vol. 12, No. 5, pp. 1092–1102, 2017.
38. J. S. Prasath, S. Jayakumar, and K. Karthikeyan, "Real-Time Implementation for Secure Monitoring of Wastewater Treatment Plants Using Internet of Things," International Journal of Innovative Technology and Exploring Engineering, Vol. 9, No. 1, pp. 2997–3002, 2019.
39. W. Jiang, Z. Guo, Y. Ma, and N. Sang, "Measurement-Based Research on Cryptographic Algorithms for Embedded Real-Time Systems," Journal of Systems Architecture, Vol. 59, pp. 1394–1404, 2013.
40. M. Yaseen, P. Durai, P. Gokul, S. Justin, and A. J. Anand, "Artificial Intelligence Based Automated Appliances in Smart Home," 2023 Seventh International Conference on Image Information Processing (ICIIP), Solan, India, 2023, pp. 442–445.

41. Anuradha T., Jasphin Jeni Sharmila P., Kanimozhiraman, K. Kalaiselvi, C. K. Shruthi and Jose A. A., "Automatic Categorization of Emails into Folders Based on the Content of the Messages," 2024 Third International Conference on Intelligent Techniques in Control, Optimization and Signal Processing (INCOS), Krishnankoil, Virudhunagar district, Tamil Nadu, India, 2024, pp. 1–6.
42. S. Hemalatha, M. Mahalakshmi, V. Vignesh, M. Geethalakshmi, D. Balasubramanian, and A. A. Jose., "Deep Learning Approaches for Intrusion Detection with Emerging Cybersecurity Challenges," 2023 International Conference on Sustainable Communication Networks and Application (ICSCNA), Theni, India, 2023, pp. 1522–1529.
43. G. Ashok, and A. A. Jose., "Modified Image Encryption Algorithm Based on Chaotic Cryptography," 2023 7th International Conference on Electronics, Communication and Aerospace Technology (ICECA), Coimbatore, India, 2023, pp. 1506–1512.
44. P. Bhambri, R. Kaur, and S. Kaur, Hosiery Management System: An Automation Software (Vol. 1). Lap Lambert Academic Publishing, 2015. ISBN: 9783659675782.
45. P. Bhambri, and F. Goyal, Development of Phylogenetic Tree Based on Kimura's Method: Based on Un-Weighted Pair Group Method with Arithmetic Mean (UPGMA) and Neighbor Joining (NJ) Scoring Techniques (Vol. 1). Lap Lambert Academic Publishing, 2013. ISBN: 9783659336539.
46. P. Bhambri, and V. Paika, Image Recognition Using Neuro-Fuzzy Techniques: Developing a Mamdani's Fuzzy Inference System in MATLAB Using Fuzzy Logic Toolbox (Vol. 1). Lap Lambert Academic Publishing, 2013. ISBN: 9783659460838.
47. P. Bhambri, and D. Garg, Enhanced Model for Fusion of Multi-Modality Images: Discrete Wavelet Transformation Using Region Based Fusion Rules (Vol. 1). Lap Lambert Academic Publishing, 2012. ISBN: 9783659208089.

6 Behavioral Analysis for Threat Detection

Satya Subrahmanyam

6.1 INTRODUCTION

In an increasingly digital and security-conscious world, behavioral analysis for threat detection has emerged as a critical approach for identifying potential risks before they escalate. This method leverages patterns in human behavior, network activity, and system interactions to detect anomalies that may indicate malicious intent, fraud, or cyber threats. Unlike traditional rule-based security models, behavioral analysis employs advanced technologies such as machine learning (ML), artificial intelligence (AI), and predictive analytics to establish baselines of normal behavior and flag deviations in real time. This proactive strategy is widely applied in cybersecurity, law enforcement, fraud detection, and national security, helping organizations mitigate threats effectively. By continuously adapting to evolving attack patterns, behavioral analysis enhances both physical and digital security, making it an indispensable tool in modern threat intelligence frameworks.

6.1.1 Overview of Behavioral Analysis in Cybersecurity

When it comes to cybersecurity, behavioral analysis is all about keeping an eye on how devices, apps, and users interact with a network in order to spot any suspicious activity. Behavioral analysis aims to spot out-of-the-ordinary occurrences that could be signs of malicious activity, as opposed to the signature-based identification of known threats that is the main emphasis of conventional security procedures. Advanced persistent threats (APTs), insider threats, and zero-day vulnerabilities are examples of complex and ever-changing attacks that may elude conventional security mechanisms [1].

In order to construct an exhaustive profile of typical behavior, behavioral analysis uses of a number of data sources, such as user actions, system logs, and network traffic. Security systems are able to identify potential dangers by constantly monitoring and analyzing this data for even the most minute changes or trends. Organizations may improve their security posture by using a proactive strategy to detect and mitigate risks before they cause substantial impact [2].

6.1.2 Importance of Behavioral Analysis in Modern Threat Detection

The increasing complexity and sophistication of cyber threats necessitate advanced detection methods that go beyond traditional security solutions. Behavioral analysis is pivotal in modern threat detection for several reasons:

DOI: 10.1201/9781003521020-6

1. *Detection of unknown threats*: Conventional security protocols are vulnerable to emerging attacks since they are based on previously identified patterns of attack. As an alternative, behavioral analysis may spot risks that haven't been seen before by identifying odd behaviors that don't follow the norm.
2. *Identification of insider threats*: Insider threats pose significant risks as they originate from within the organization, often bypassing traditional security controls. Behavioral analysis can detect abnormal activities by legitimate users, such as accessing sensitive data without authorization or engaging in unusual communication patterns, which may indicate malicious intent.
3. *Adaptive security posture*: Cyber threats are constantly evolving, requiring security systems to adapt quickly. Behavioral analysis enables continuous learning and adaptation by analyzing new data and updating behavioral baselines, ensuring that security measures remain effective against emerging threats.
4. *Comprehensive threat detection*: By monitoring a wide range of activities across different layers of the network, behavioral analysis provides a holistic view of the security landscape. This comprehensive approach enhances the ability to detect multi-stage attacks that involve a series of coordinated activities across different systems.
5. *Enhanced incident response*: By using behavioral analysis to uncover irregularities early on, investigations and responses may be initiated without delay. Organizations can lessen the likelihood of hazards and the severity of security events if they can spot such dangers early on.

6.1.3 How Artificial Intelligence and Machine Learning Can Improve Behavioral Analysis

With their enhanced data processing, pattern recognition, and anomaly detection capabilities, AI and ML technologies have completely transformed behavioral analysis in cybersecurity. Behavioral analysis is improved with the combination of AI and ML in several important ways that are discussed as follows:

1. *Automation and scalability*: ML and AI algorithms can handle massive volumes of data in real time. For big companies dealing with intricate networks and massive amounts of data, this scalability is vital [3].
2. *Advanced pattern recognition*: Algorithms trained by ML can spot connections and patterns in data that humans would miss. These algorithms may improve threat detection accuracy by learning from past data to differentiate between harmless and harmful actions.
3. *Adaptive learning*: AI and ML models can continuously learn and adapt to new data, enhancing their ability to detect evolving threats. By updating behavioral baselines and refining detection criteria, these models ensure that security measures remain effective over time.
4. *Contextual analysis*: To improve the precision of threat detection, AI and ML may take into account contextual information including user roles,

network setups, and past behavior. This contextual analysis helps to reduce false positives and ensures that alerts are relevant and actionable.

5. *Predictive analytics*: ML models may use previous data and established trends to identify prospective dangers, which brings us to this point, predictive analytics. Proactive threat mitigation is made possible by this predictive capacity, which may discover vulnerabilities and potential attack vectors before they are exploited.
6. *Integration with security systems*: AI and ML may be synced with other security systems to improve threat detection and response capabilities. This includes solutions for security information and event management (SIEM) and intrusion detection systems (IDS). By combining the best features of several systems, this integration allows for a more cohesive strategy for security [2].

In today's cybersecurity landscape, behavioral analysis is important for detecting threats in a proactive and adaptable manner. Organizations can now identify and react to complex risks in real time, thanks to behavioral analysis that is enhanced by AI and ML integration. A strong cybersecurity plan must include behavioral analysis and the use of AI and ML to improve it in order to keep up with the ever-changing nature of cyber threats.

6.2 THEORETICAL FOUNDATIONS OF BEHAVIORAL ANALYSIS FOR THREAT DETECTION

6.2.1 Definition and Key Concepts of Behavioral Analysis

When discussing cybersecurity, the term "behavioral analysis" is used to describe the process of methodically checking a network for patterns and actions that might reveal a security risk. The usual course of action is defined by this method, which centers on learning and simulating the habits of users, devices, and apps. Malicious actions, such as insider assaults or malware infections, may be detected when this baseline deviates from the norm [2].

Key concepts in behavioral analysis include:

- Anomaly detection, or the finding of data patterns that do not match predicted behavior, is a central idea in behavioral analysis. When things don't add up, it can mean there's a security risk.
- Establishing a baseline allows one to compare future actions to a predetermined level of normalcy. In order to spot changes that might indicate danger, this is vital.
- Using algorithms to learn from data, ML may enhance threat detection accuracy over time. To enable systems to adapt to new dangers, ML has become an essential part of contemporary behavioral analysis [4].
- User and entity behavior analytics (UEBA) is a subfield of behavioral analysis that aims to identify compromised accounts and insider threats by studying how devices and people interact with one another [5].

6.2.2 Historical Development and Evolution of Behavioral Analysis in Security

The concept of behavioral analysis has its roots in early IDSs that were developed in the 1980s. Dorothy Denning's model of an IDS, proposed in 1987, is one of the foundational works in this area. Denning's model emphasized the importance of detecting anomalies in system behavior to identify potential security breaches [6]. In the 1990s, the focus shifted toward more sophisticated techniques, including statistical methods and rule-based systems for detecting anomalies. Advanced technologies that use ML and AI to evaluate massive amounts of data in real time were built upon these early systems [2].

The advent of big data analytics in the 2000s significantly enhanced the capabilities of behavioral analysis. With the ability to process and analyze massive datasets, security systems could now build more accurate and comprehensive models of normal behavior. This period also saw the rise of ML algorithms that could automatically detect and adapt to new threats [7]. Recent years have seen a dramatic shift in threat identification thanks to behavioral analysis that incorporates AI and ML. Thanks to these innovations in technology, we can detect ever-more-complex dangers since they allow for constant learning and development. To keep security systems successful in spite of ever-changing cyber threats, improved pattern recognition and predictive analytics enable proactive threat identification and mitigation [3].

6.2.3 Core Principles and Methodologies

Behavioral analysis for threat detection is grounded in several core principles and methodologies that guide its implementation and effectiveness. These are as follows:

1. *Data collection and preprocessing*: The first stage of behavioral analysis involves collecting data from a variety of sources, including user behaviors, system logs, and network traffic. In order to guarantee that the analysis is conducted on top-notch data, this data is then preprocessed to eliminate any unnecessary information and noise.
2. *Feature extraction and selection*: It involves selecting the most important characteristics or qualities from the gathered data. To make detection algorithms more accurate and efficient, feature selection is key.
3. *Modeling normal behavior*: This step involves using either ML or statistical approaches to develop a model of typical behavior. When evaluating subsequent actions, this model is used as a reference point. This is the stage when techniques like grouping, classification, and regression analysis come into play.
4. *Anomaly detection and classification*: This process involves comparing the observed behavior with the set baseline in order to identify any unusual occurrences. Next, the severity and possible effect of the anomalies are evaluated to determine the best course of action. It is common practice to use methods like support vector machines, decision trees, and neural networks.

5. *Continuous learning and adaptation*: Implementing systems that learn from fresh data and update their models to increase detection accuracy is called continuous learning and adaptation. Given the ever-changing nature of cybersecurity threats, this is of utmost importance. The detection system's efficacy is maintained throughout time by means of adaptive learning.
6. *Integration with security systems*: It is recommended that behavioral analysis be used with other security measures like SIEM solutions and IDSs. A holistic security strategy that makes use of the capabilities of several technologies is possible, thanks to this integration [3].

Many different ideas, advancements in history, and approaches make up the theoretical underpinnings of behavioral analysis for threat detection. Organizations may strengthen their security posture and identify and react to advanced cyber-attacks by learning and implementing these concepts.

6.3 APPROACHES AND PROCEDURES IN BEHAVIORAL ANALYSIS FOR DETECTING THREATS

6.3.1 Gathering and Preparing Data for Behavioral Analysis

Behavioral analysis for threat detection is fundamentally dependent on the meticulous gathering and preparation of data. The core objective is to construct a comprehensive dataset that can be scrutinized to pinpoint potential threats. This involves aggregating relevant data from a multitude of sources, such as user activities, system logs, and network traffic. The collected data is instrumental in building models that can effectively differentiate between normal and abnormal network behaviors.

6.3.1.1 Collecting Data

A thorough understanding of network activities necessitates the compilation of data from various sources. One crucial source is network traffic monitoring, which entails examining network traffic to detect unusual patterns or spikes that may signify security issues [2]. For example, an unexpected surge in data flow could indicate a potential breach. System logs also play a vital role, providing detailed information on system events and user actions from numerous network devices, including servers and firewalls. These logs can reveal access patterns and potential security breaches [3]. In addition, tracking user activity by monitoring how users interact with the system, such as the files they access and their login attempts, can highlight anomalies in behavior that may suggest insider threats or compromised accounts [5].

6.3.1.2 Data Preprocessing

Once data collection is complete, preprocessing activities are necessary to clean and prepare the data for analysis. Data cleaning involves removing extraneous or noisy information from the dataset, addressing missing values, correcting errors, and eliminating irrelevant data. Data transformation is another critical step, which includes converting unstructured data into a structured format suitable for analysis. This process may involve normalization, standardization, and aggregation to create consistent and comparable datasets. Feature extraction is also essential, focusing on identifying and selecting the most relevant features for analysis, thereby enhancing the accuracy and efficiency of detection algorithms.

6.3.2 Behavioral Analysis Machine Learning Algorithms

ML algorithms are crucial in behavioral analysis for detecting threats, as they enable the identification of complex patterns and anomalies within large datasets. The choice of ML algorithm depends on the specific requirements of threat detection and the nature of the data being analyzed.

6.3.2.1 Supervised Learning

Supervised learning utilizes labeled datasets to train models with known outcomes. These algorithms learn to map inputs to outputs based on historical data, allowing them to predict future behavior. Common supervised learning algorithms include decision trees, which are hierarchical models that answer yes/no questions and are effective with both numerical and categorical data [8]. Support vector machines (SVMs) are another example, excelling in high-dimensional datasets by finding the optimal boundary between data classes, whether linear or nonlinear [7]. Neural networks, which mimic the structure and function of the human brain, are capable of detecting complex patterns and relationships within data, making them suitable for identifying sophisticated threats [3].

6.3.2.2 Unsupervised Learning

Unsupervised learning algorithms do not require labeled data, instead these identify structures and patterns based on similarities and differences within the data. These algorithms are particularly valuable for detecting new or unexpected threats. Clustering techniques, such as K-means and density-based spatial clustering of applications with noise (DBSCAN), group data points with similar features, making them effective for handling large datasets [4]. Principal component analysis (PCA) reduces data dimensionality, facilitating the identification of significant features and anomalies [9]. Autoencoders, which use neural networks to compress and reconstruct data, detect anomalies by monitoring reconstruction errors [5].

6.3.2.3 Reinforcement Learning

Reinforcement learning trains models to maximize a reward signal through iterative decision-making, making it suitable for adaptive security systems that learn and respond to new threats in real time. Common reinforcement learning methods include Q-learning, which learns the value of actions based on observed rewards and is effective in dynamic environments [3]. Deep Q-networks (DQNs) combine Q-learning with deep neural networks to handle high-dimensional data and complex decision-making processes, enabling the development of sophisticated threat detection and response strategies [8].

6.3.3 Behavioral Pattern Recognition

Behavioral pattern recognition aims to identify and replicate the typical actions of network nodes, clients, and applications, providing a baseline for detecting anomalies. Statistical analysis uses techniques such as mean, variance, and correlation analysis to model normal behavior and establish thresholds for anomaly detection [2]. Time series analysis examines temporal patterns in data, using methods like moving averages and autoregressive models to identify trends and seasonal patterns [10]. Graph analysis represents network activities as graphs, using techniques like community detection and centrality measures to uncover hidden structures and detect anomalies [7].

6.3.4 Ways to Spot Abnormalities

The primary goal of behavioral analysis is to detect anomalies, which are deviations from normal behavior that may indicate a security threat. Various anomaly detection methods, each with its strengths and weaknesses, can be employed. Statistical techniques, such as Z-score, Chi-square, and Bayesian networks, use predefined criteria to identify outliers [9]. ML techniques apply algorithms that learn patterns from data to detect anomalies, utilizing supervised, unsupervised, or reinforcement learning methods depending on the type of threats and availability of labeled data [4]. Hybrid approaches combine multiple methods to reduce false positives and enhance detection accuracy, leveraging the strengths of both statistical and ML techniques [5].

6.3.4.1 Anomaly Detection Under Supervision

Supervised anomaly detection uses labeled datasets to train algorithms to distinguish between normal and abnormal behavior. Common methods include classification algorithms like neural networks, decision trees, and SVMs, which learn patterns in the data and apply them to new data. Regression techniques model relationships between variables to predict future behavior, with anomalies identified by deviations between predicted and actual data.

6.3.4.2 Unsupervised Detection of Abnormalities

Unsupervised anomaly detection is effective for identifying unknown threats as it does not require labeled data. Clustering methods group data points based on similarities, identifying outliers that do not fit into any group [4]. Dimensionality reduction techniques, such as PCA and t-distributed Stochastic neighbor embedding (t-SNE), reduce the number of features, making it easier to detect anomalies based on deviations from principal components [9].

6.3.4.3 Reinforcement Learning for Anomaly Detection

Reinforcement learning is increasingly used for adaptive anomaly detection, where models learn to identify and respond to anomalies based on rewards and penalties. Q-Learning learns the value of actions from environmental rewards, making it suitable for dynamic and uncertain environments [3]. Deep reinforcement learning combines deep neural networks with reinforcement learning to handle complex decision-making and high-dimensional data, optimizing strategies for detecting and responding to anomalies [8].

Behavioral analysis for threat detection encompasses data collection, preprocessing, ML, pattern recognition, and anomaly detection. These comprehensive approaches enable organizations to build robust systems capable of identifying and mitigating complex cyber threats.

6.4 APPLICATIONS OF BEHAVIORAL ANALYSIS IN THREAT DETECTION

Behavioral analysis plays a critical role in modern cybersecurity strategies by leveraging patterns and anomalies in user and network behavior to identify potential threats. Its applications are diverse and span various aspects of threat detection, including insider threats, phishing attacks, malware and ransomware behavior, user activity monitoring, and network security.

6.4.1 Identifying Insider Threats

Insider threats are particularly challenging to address due to the legitimate access insiders have to sensitive data and systems. Differentiating between legitimate and malicious activity becomes difficult when dealing with insiders. Behavioral analysis excels in identifying insider threats by establishing a baseline of typical user behavior and then detecting deviations from this norm.

6.4.1.1 Behavioral Indicators

Observing access patterns can help identify suspicious trends, such as an employee logging in at unusual hours or accessing sensitive files they typically would not. Such behavior might indicate malicious intent. In addition, detecting anomalies in the use of applications and systems is crucial. Sudden changes in the frequency or type of activities performed by a user can signal potential insider threats.

6.4.1.2 Techniques

UEBA technologies employ ML to see trends in user actions across several platforms, which might indicate insider threats [5]. Anomaly detection, using statistical methods and ML models, identifies deviations from established behavior patterns, alerting security teams to potential insider threats [4].

6.4.2 Detecting Phishing Attacks

Phishing attacks are a prevalent and effective tactic used by attackers to trick individuals into divulging sensitive information. Behavioral analysis enhances the detection of phishing attempts by examining network traffic, user activity, and email content.

6.4.2.1 Behavioral Indicators

Email analysis can identify characteristics of phishing emails, such as unusual sender addresses, suspicious attachments, and abnormal language patterns. Analyzing user response patterns, such as clicking on links or opening attachments, can also detect unusual behaviors indicative of phishing attacks.

6.4.2.2 Techniques

Natural language processing (NLP) algorithms can scan email content to detect phishing attempts by identifying language patterns and terms commonly associated with phishing [11]. ML classifiers, trained on labeled datasets of both phishing and genuine emails, can detect and alert possible phishing attempts [12].

6.4.3 Recognizing Malware and Ransomware Behavior

Ransomware and other forms of malware pose significant threats to organizations' data and infrastructure. Behavioral analysis aids in detecting such malicious software by observing system and network activity.

6.4.3.1 Behavioral Indicators

Anomalies in network traffic, such as unexpected outbound connections or data transfers, may indicate the presence of malware or ransomware. Sudden changes in system configurations, file encryptions, or the appearance of ransom notes are strong indicators of ransomware attacks.

6.4.3.2 Techniques

Signature-based detection involves identifying known malware signatures through behavioral analysis of system and network activities [13]. Behavior-based detection uses ML algorithms to detect behaviors commonly associated with malware and ransomware, such as file encryption and unauthorized data exfiltration [3].

6.4.4 Monitoring User Activities and Identifying Unusual Patterns

Continuous monitoring of user activities is essential for detecting suspicious behavior that may indicate security threats. Behavioral analysis provides a robust framework for analyzing user interactions with systems and applications to identify unusual patterns.

6.4.4.1 Behavioral Indicators

Unusual login patterns, such as several unsuccessful attempts, logins from unknown locations, or access at odd hours, can indicate an account breach. Anomalies in the use of applications, such as accessing restricted areas or performing unauthorized actions, can also signal potential threats.

6.4.4.2 Techniques

Time series analysis examines temporal patterns in user activities to detect deviations from normal behavior over time. Graph analysis visualizes user interactions as graphs, helping to identify and mitigate security risks more effectively.

6.4.5 Behavioral Analysis in Network Security

Behavioral analysis is an essential component of network security, helping to identify and counteract various types of network-based threats. By examining network traffic and activities, behavioral analysis aids in detecting suspicious activity that may indicate security breaches.

6.4.5.1 Behavioral Indicators

Monitoring network traffic for unusual patterns, such as sudden spikes, unexpected connections, or data exfiltration, can reveal potential threats. Detecting deviations in the use of network protocols, such as unexpected protocol combinations or unusual port usage, can also indicate malicious activities.

6.4.5.2 Techniques

Flow analysis, using techniques such as NetFlow and IPFIX, detects anomalies in traffic patterns and volumes [14]. Deep packet inspection (DPI) examines the content of data packets to identify malicious payloads and unauthorized communications, making it effective for detecting sophisticated network-based threats [10].

Behavioral analysis is an essential tool in the cybersecurity arsenal, offering robust methods for detecting a wide range of threats. From identifying insider threats and phishing attacks to recognizing malware behavior and monitoring user activities, behavioral analysis leverages advanced techniques to enhance security. By continuously analyzing patterns and anomalies in user and network behavior, organizations can proactively detect and mitigate security threats, ensuring a stronger defense against cyber-attacks.

6.5 REAL-WORLD CASE STUDIES IN BEHAVIORAL ANALYSIS FOR THREAT DETECTION

Behavioral analysis has become an essential part of contemporary cybersecurity, offering advanced ways to detect and lessen the impact of different kinds of attacks. This section delves into three practical examples where behavioral analysis played a crucial role in thwarting data breaches, countering insider threats, and identifying APTs.

6.5.1 Case Study 1: Behavioral Analysis in Preventing Data Breaches

6.5.1.1 Background

Sophisticated cybercriminals were trying to hack the sensitive client data of a big financial institution, and the danger was growing. The ever-changing nature of these threats rendered ineffective the use of conventional security measures like firewalls and signature-based detection systems. In order to strengthen its defenses, the organization chose to install a behavioral analysis system.

6.5.1.2 Implementation

The organization set up UEBA technology, which track and analyze network activity using ML techniques, to keep tabs on people and devices. This technology was able to identify any security breaches by creating a baseline of typical operations and then detecting any deviations from that.

6.5.1.3 Results

The behavioral analysis system found many instances of suspicious network activity in the first few months of implementation. As an example, it found that one user account had an unusually high amount of data access outside of typical business hours. An external attacker had infiltrated the account and was trying to steal data, according to the inquiry. Before any data was taken, the security team was able to limit the incident thanks to the quick discovery [5].

6.5.1.4 Lessons Learned

The usefulness of behavioral analysis in discovering and averting data breaches is shown in this case study. Organizations can react to possible dangers before they do major harm by constantly monitoring user and network activity.

6.5.2 Case Study 2: Using Behavioral Analysis to Combat Insider Threats

6.5.2.1 Background

Multiple instances of insider theft of intellectual property occurred at a global technology corporation. The company's image and finances took a serious hit as a consequence of these episodes. In order to detect and counteract insider threats, the organization required a strong solution.

6.5.2.2 Implementation

The company integrated behavioral analysis tools into its existing security infrastructure. These tools employed ML models to analyze employee activities and identify patterns indicative of insider threats. Key indicators included unusual access to sensitive information, changes in work patterns, and deviations from established behavioral baselines.

6.5.2.3 Results

The behavioral analysis system successfully identified multiple cases of potential insider threats. At one instance, an employee who had recently resigned began accessing and downloading large volumes of sensitive data. The system flagged this activity as anomalous, and the security team intervened, preventing the exfiltration of valuable intellectual property [15].

6.5.2.4 Lessons Learned

The significance of behavioral analysis in identifying and reducing insider risks is shown in this case study. Organizations may prevent themselves from falling victim to insider threats by keeping tabs on user activity and evaluating it.

6.5.3 Case Study 3: Successful Application of Behavioral Analysis in Detecting Advanced Persistent Threats

6.5.3.1 Background

A government agency responsible for national security was targeted by an APT group. The attackers used sophisticated techniques to infiltrate the agency's network, remaining undetected for an extended period. The agency needed a solution capable of identifying and neutralizing these stealthy threats.

6.5.3.2 Implementation

The agency implemented a comprehensive behavioral analysis system that combined ML algorithms with threat intelligence feeds. The system continuously monitored network traffic, user activities, and system processes to detect patterns associated with APTs. It also incorporated anomaly detection techniques to identify deviations from normal behavior.

6.5.3.3 Results

The behavioral analysis system detected several indicators of compromise that traditional security tools had missed. For example, it identified unusual lateral movement within the network and unauthorized access to critical systems. These anomalies were linked to the APT group, allowing the security team to take swift action to contain and eradicate the threat [10].

6.5.3.4 Lessons Learned

In order to identify and counteract APTs, behavioral analysis is crucial, as this case study demonstrates. Behavioral analysis is an effective method for detecting and preventing advanced threats, which are able to elude conventional security procedures.

When protecting against cyberattacks, behavioral analysis is becoming a must-have tool. This document presents real-world case studies that show how successful it is in identifying APTs, preventing data breaches, and fighting insider attacks. Organizations may strengthen their defenses against various cyber threats by using ML algorithms and constantly monitoring user and network activity.

6.6 INTEGRATING BEHAVIORAL ANALYSIS WITH OTHER SECURITY MEASURES

In cybersecurity, behavioral analysis is essential for understanding user and network activities, identifying suspicious patterns, and detecting potential threats. However, its effectiveness is greatly enhanced when combined with other security measures, creating a comprehensive and multi-layered defense strategy. This section explores the benefits of integrating behavioral analysis with traditional security measures, enhancing threat intelligence, and its role in proactive threat hunting.

6.6.1 When Threat Intelligence and Behavioral Analysis Are Used Together

Threat intelligence involves gathering and analyzing data about potential or existing threats to improve security measures. By combining behavioral analysis with threat intelligence, organizations can significantly enhance their ability to detect and respond to threats.

6.6.1.1 Benefits

The integration of behavioral data with threat intelligence provides a more comprehensive understanding of the threat landscape. This approach enables organizations to monitor emerging threats that behavioral analysis alone might miss [3]. In addition, threat intelligence can provide indicators of compromise (IoCs) that, when combined with behavioral anomalies, improve the accuracy and speed of threat detection [5].

6.6.1.2 Implementation

Integrating threat intelligence feeds with behavioral analysis systems allows for the correlation of IoCs with observed behavioral patterns. This approach helps in identifying sophisticated threats that use advanced evasion techniques. Moreover, organizations can develop automated response mechanisms that are triggered when behavioral anomalies match known threat signatures from threat intelligence databases.

6.6.2 Enhancing Traditional Security Measures with Behavioral Insights

Traditional security measures, such as firewalls, IDSs, and antivirus software, are the backbone of an organization's security framework. Integrating behavioral analysis into these conventional measures can significantly enhance their effectiveness.

6.6.2.1 Benefits

Behavioral analysis adds an extra layer of context, reducing the false positives and negatives that are common with traditional security tools. It also provides real-time insights, enabling quicker identification and mitigation of threats.

6.6.2.2 Implementation

Traditional security measures can be enhanced by incorporating behavioral-based rules and policies. For instance, an IDS can be programmed to flag activities that deviate from established behavioral baselines [16]. In addition, insights from behavioral analysis can be used to dynamically adjust security configurations, such as modifying firewall rules in real time to block suspicious activities identified through behavioral monitoring [17].

6.6.3 Leveraging Behavioral Analysis for Proactive Threat Hunting

Proactive threat hunting involves security professionals actively searching for vulnerabilities and threats within an organization's network before they can cause damage. Behavioral analysis is a critical component of this preventive approach.

6.6.3.1 Benefits

Behavioral analysis helps in identifying subtle indicators of potential threats, allowing security teams to act before an attack fully manifests. By analyzing behavior, threat hunters can uncover patterns and techniques used by attackers, improving the organization's ability to anticipate and defend against future threats.

6.6.3.2 Implementation

Organizations can use behavioral analysis to identify anomalies and investigate potential threats continuously. This method involves monitoring and analyzing user and network behavior to spot deviations [4]. Establishing dedicated threat hunting teams that utilize behavioral analysis tools can focus on identifying and investigating suspicious behaviors that automated systems might miss [13].

Integrating behavioral analysis with other security measures significantly strengthens an organization's cybersecurity posture, creating a multi-layered defense strategy. By combining threat intelligence, proactive threat hunting, and traditional security measures with behavioral analysis, organizations can detect and mitigate threats more effectively. This integration enhances threat detection accuracy, speeds up response times, and provides a deeper understanding of the threat landscape, keeping organizations one step ahead of cyber attackers.

6.7 CHALLENGES AND ETHICAL CONSIDERATIONS IN BEHAVIORAL ANALYSIS

Behavioral analysis has become a cornerstone of modern cybersecurity strategies, offering the ability to detect anomalies and potential threats based on user and network behavior. Despite its benefits, the implementation of behavioral analysis comes with significant challenges and ethical considerations. These include data privacy and ethical concerns, addressing biases in AI and ML models, and overcoming technical challenges in implementation.

6.7.1 Data Privacy and Ethical Concerns in Behavioral Analysis

6.7.1.1 Data Privacy Issues

Data collection and analysis are the backbone of behavioral analysis, which often involves handling private and sensitive information. This raises several data privacy issues.

Collecting behavioral data often requires monitoring users' activities, which can be perceived as invasive. Organizations must ensure they have obtained explicit consent from users before collecting their data, as failure to do so can lead to legal repercussions and damage the organization's reputation [18]. In addition, organizations that store large volumes of behavioral data become attractive targets for cybercriminals. Strong data security measures must be put in place to prevent breaches and unauthorized access to sensitive information.

6.7.1.2 Ethical Considerations

Ethical issues also arise in the context of behavioral analysis. Continuous monitoring of user behavior can be seen as a form of surveillance, leading to concerns about the right to privacy. Organizations need to balance security needs with respecting individuals' privacy rights. Moreover, the ethical use of behavioral analysis requires transparency in how data is collected, analyzed, and used. Organizations must be accountable for their actions and ensure that users are informed about the purposes of data collection.

6.7.2 Addressing Biases in AI and Machine Learning Models

6.7.2.1 Sources of Bias

AI and ML models used in behavioral analysis can inadvertently perpetuate or even exacerbate existing biases:

If ML models are trained with biased data, they are likely to produce biased results. For example, the models could miss certain users' anomalies if their behaviors aren't well captured in the training data. Also, the algorithms themselves can introduce biases, particularly if they are designed or implemented without considering potential bias sources [18].

6.7.2.2 Mitigation Strategies

Several approaches can help mitigate bias in AI and machine learning models. These are discussed in subsequent text.

One approach is to use data that is representative of all user groups and varied in nature. This involves considering a wide range of demographic variables, including gender, age, and ethnicity. Methods can also be employed to identify and eliminate biases in models. Maintaining the models' objectivity may require frequent reviews and upgrades. Additionally, creating and following AI frameworks with an emphasis on ethics can help better incorporate ethical considerations into the development process of ML models.

6.7.3 Overcoming Technical Challenges in Implementing Behavioral Analysis

6.7.3.1 Scalability

One of the primary technical challenges in implementing behavioral analysis is scalability. Behavioral analysis can require significant resources due to the massive volumes of data that must be processed in real time. It is critical for organizations to have the proper infrastructure in place to manage massive amounts of data. In addition, the computational power required for real-time analysis can be substantial. Leveraging cloud-based solutions can provide the necessary scalability and flexibility to meet these demands [19].

6.7.3.2 Accuracy and Precision

Ensuring the accuracy and precision of behavioral analysis models is another significant challenge. Both alert fatigue and missed threats may be caused by high rates of false positives and false negatives, respectively. Models must be fine-tuned to increase their precision and accuracy. Moreover, cyber threats are constantly evolving, requiring behavioral analysis models to adapt and update regularly.

Implementing adaptive learning mechanisms can help models stay effective against new and emerging threats.

6.7.3.3 Integration with Existing Systems

Integrating behavioral analysis with existing security systems presents several challenges. Ensuring compatibility between behavioral analysis tools and legacy systems can be difficult. Organizations may need to invest in new technologies or upgrade existing systems to achieve seamless integration. In addition, achieving data interoperability between different security tools is essential for effective behavioral analysis. Implementing standard data formats and protocols can facilitate better data sharing and analysis.

Despite its usefulness in identifying and mitigating cyber risks, behavioral analysis comes with significant challenges and ethical considerations. To successfully use behavioral analysis, it is crucial to address data privacy and ethical problems, mitigate biases in AI and ML models, and overcome technological hurdles. Organizations can improve their security posture without sacrificing ethics or personal privacy if they handle these concerns with care.

6.8 FUTURE TRENDS AND INNOVATIONS IN BEHAVIORAL ANALYSIS FOR THREAT DETECTION

Behavioral analysis has become an essential component of modern cybersecurity strategies. As cyber threats continue to evolve, there is a growing demand for more advanced and efficient behavioral analysis methods. This section explores the latest developments in behavioral analysis for threat detection, the role of behavioral analysis in next-generation cybersecurity, and potential future research directions and advancements.

6.8.1 Emerging Technologies in Behavioral Analysis for Threat Detection

6.8.1.1 Machine Learning and Artificial Intelligence

ML and AI are driving significant advancements in behavioral analysis. These technologies enable systems to improve their threat detection capabilities over time by learning from large volumes of data.

Deep learning (DL), a subset of ML, utilizes multi-layer neural networks to understand complex patterns in data. This approach is particularly effective at identifying sophisticated threats that may evade traditional security measures [20]. Reinforcement learning (RL) involves algorithms learning optimal actions through trial and error in an environment. This approach is well-suited for dynamic threat environments where threats are constantly evolving [21].

6.8.1.2 Behavioral Biometrics

Behavioral biometrics focuses on analyzing patterns in human behaviors, such as typing rhythms, mouse movements, and touchscreen interactions. These patterns are unique to individuals and can be used for authentication and anomaly detection.

Keystroke dynamics involves monitoring the way users type to identify deviations from typical typing patterns, which may indicate compromised accounts or malicious insiders. Similarly, analyzing mouse movements can reveal unusual behaviors that signal potential threats, such as remote access attacks [22].

6.8.1.3 Quantum Computing

Quantum computing has the potential to revolutionize behavioral analysis by providing unprecedented computational power. Although still in its early stages, quantum computing could enable real-time analysis of vast amounts of behavioral data.

Quantum algorithms could analyze complex patterns and relationships in behavioral data more efficiently than classical algorithms, leading to faster and more accurate threat detection [23]. Quantum machine learning (QML) combines quantum computing with ML techniques to further enhance behavioral analysis capabilities [24].

6.8.2 Role of Behavioral Analysis in Next-Generation Cybersecurity

Behavioral analysis is poised to play a crucial role in next-generation cybersecurity by enhancing threat detection and response capabilities. This section highlights the impact of behavioral analysis on proactive defense strategies, adaptive security measures, and automated response systems.

6.8.2.1 Proactive Defense Strategies

Behavioral analysis enables organizations to adopt proactive defense strategies by identifying and mitigating threats before they cause harm.

Behavioral analytics tools can continuously monitor user and network activities, flagging suspicious behaviors that may indicate impending attacks. This approach allows security teams to intervene early and prevent potential breaches [4]. In addition, by analyzing historical behavioral data, organizations can identify trends and patterns that help predict future threats, enabling more effective threat prevention [25].

6.8.2.2 Adaptive Security Measures

Next-generation cybersecurity relies on adaptive security measures that can dynamically respond to changing threat landscapes. Behavioral analysis plays a key role in this approach.

Behavioral analytics can automatically adjust security policies and controls based on real-time observations, ensuring that defenses remain effective against evolving threats. Adaptive security measures leverage behavioral insights to customize responses based on the specific behaviors of users and attackers, enhancing overall security posture.

6.8.2.3 Automated Response Systems

Behavioral analysis contributes to the development of automated response systems that can quickly and efficiently mitigate threats.

Automated incident response platforms use behavioral analysis to trigger predefined actions when suspicious behaviors are detected. This approach minimizes the time between threat detection and response, reducing the potential impact of attacks.

Also, integrating behavioral analysis with Security Orchestration, Automation, and Response (SOAR) systems enables more sophisticated and coordinated responses to complex threats [26].

6.8.3 Future Research Directions and Advancements in Behavioral Analysis

The future of behavioral analysis in cybersecurity is promising, with ongoing research and innovations aimed at addressing current challenges and enhancing capabilities. This section explores potential research directions and advancements in behavioral analysis.

6.8.3.1 Enhancing Data Privacy and Ethical Standards

Future research will likely focus on developing techniques that enhance data privacy and ethical standards in behavioral analysis.

Privacy-preserving techniques, such as differential privacy and federated learning, can help protect user data while enabling effective behavioral analysis. These approaches ensure that individual data points remain confidential while still allowing for accurate threat detection [27]. Additionally, establishing ethical frameworks and guidelines for behavioral analysis will be crucial in ensuring responsible and transparent use of this technology.

6.8.3.2 Improving Model Accuracy and Reducing Bias

Advancements in AI and ML will continue to improve the accuracy and fairness of behavioral analysis models.

Research efforts will focus on developing algorithms that can effectively handle diverse and imbalanced datasets, reducing biases and improving detection accuracy across different user groups. Techniques such as transfer learning and multi-task learning can enhance model performance by leveraging knowledge from related tasks and domains [28].

6.8.3.3 Integrating Behavioral Analysis with Emerging Technologies

The integration of behavioral analysis with emerging technologies, such as the Internet of Things (IoT) and 5G networks, will open new avenues for threat detection and mitigation.

Behavioral analytics can provide valuable insights into the security of IoT devices and networks, identifying anomalous behaviors that may indicate compromised devices or unauthorized access. Similarly, the high-speed and low-latency capabilities of 5G networks will enable real-time behavioral analysis on a larger scale, enhancing overall cybersecurity effectiveness.

The future of behavioral analysis in cybersecurity holds great potential, with advancements in AI, ML, behavioral biometrics, quantum computing, and privacy-preserving techniques driving innovation. As cyber threats evolve, the role of behavioral analysis in next-generation cybersecurity will become increasingly important, enabling proactive defense strategies, adaptive security measures, and automated response systems. Ongoing research and development will continue to enhance the capabilities of behavioral analysis, ensuring its effectiveness in safeguarding against emerging threats.

6.9 CONCLUSION

Behavioral analysis is increasingly recognized as a critical component of modern cybersecurity strategies. By focusing on identifying and understanding patterns in user and network behaviors, behavioral analysis can provide early detection of anomalies and potential threats, thereby enhancing the overall security posture of organizations.

The importance of behavioral analysis lies in its ability to detect sophisticated and evolving threats that traditional security measures might miss. Through continuous monitoring and analysis, behavioral analytics can identify subtle deviations from normal behavior, flagging potential malicious activities before they can cause significant harm. This proactive approach enables organizations to respond swiftly and effectively, minimizing the impact of security incidents.

The integration of behavioral analysis with other security measures, such as threat intelligence and traditional security tools, further strengthens an organization's defenses. By combining different data sources and analytical techniques, organizations can gain a comprehensive understanding of the threat landscape and improve their ability to detect and mitigate complex threats.

However, the implementation of behavioral analysis comes with challenges and ethical considerations. Data privacy and ethical concerns must be addressed to ensure the responsible and transparent use of behavioral data. Additionally, biases in AI and ML models need to be mitigated to avoid unfair or inaccurate threat detection.

Despite these challenges, the future of behavioral analysis in cybersecurity looks promising. Emerging technologies such as AI, ML, behavioral biometrics, and quantum computing are driving advancements in this field. These innovations are enhancing the accuracy, efficiency, and scalability of behavioral analysis, making it an indispensable tool in the fight against cyber threats.

Looking ahead, ongoing research and development will continue to improve the capabilities of behavioral analysis. Future directions include enhancing data privacy and ethical standards, reducing biases in models, and integrating behavioral analysis with emerging technologies such as IoT and 5G networks. By staying at the forefront of these advancements, organizations can ensure that their behavioral analysis strategies remain effective in safeguarding against the ever-evolving cyber threat landscape.

In conclusion, behavioral analysis is a powerful and essential component of modern cybersecurity. By leveraging advanced analytical techniques to monitor and understand behaviors, organizations can detect and mitigate threats more effectively, ensuring a robust and resilient security posture in an increasingly complex digital world.

REFERENCES

1. S. Shetty, J.D. Sweeney and L.A. Grieco. 2018. Artificial intelligence and machine learning for cybersecurity: A review. Journal of Cyber Security Technology. 2(1), 1–18.

2. S. Axelsson. 2000. Intrusion detection systems: A survey and taxonomy. Technical Report. Department of Computer Engineering, Chalmers University.
3. N. Moustafa and J. Slay. 2017. The significant feature selection for the detection of malicious activities in IoT networks using machine learning. In 2017 the International Conference on Intelligence and Security Informatics (ISI). 164–166.
4. V. Chandola, A. Banerjee and V. Kumar. 2009. Anomaly detection: A. Survey. ACM Computing Surveys (CSUR. 41(3), 1–58.
5. P. Saxena and R. Thomas. 2018. Behavioral analysis for cybersecurity: UEBA and machine learning. Journal of Information Security and Applications. 41, 44–51.
6. D.E. Denning. 1987. An intrusion-detection model. IEEE Transactions on Software Engineering. 13(2), 222–232.
7. M. Barreno, B. Nelson, R. Sears, A.D. Joseph and J.D. Tygar. 2006. Can machine learning be secure? In Proceedings of the 2006 ACM Symposium on Information, Computer and Communications Security (ASIACCS). 16–25.
8. J.T. Quach and M.L. Machanavajjhala. 2015. LogitBoost: A machine learning algorithm for anomaly detection. In 2015 IEEE International Conference on Big Data (Big Data). 2583–2589.
9. A. Patcha and J.M. Park. 2007. An overview of anomaly detection techniques: Existing solutions and latest technological trends. Computer Networks. 51(12), 3448–3470.
10. A.A. Ghorbani, W. Lu and M. Tavallaee. 2010. Network Intrusion Detection and Prevention: Concepts and Techniques. Springer Science & Business Media.
11. A. Bergholz, J. De Beer, S. Glahn, M.D. Moens, G. Paaß and S. Strobel. 2010. New filtering approaches for phishing email. Journal of Computer Security. 18(1), 7–35.
12. I. Fette, N. Sadeh and A. Tomasic. 2007. Learning to detect phishing emails. In Proceedings of the 16th International Conference on the World Wide Web. 649–656.
13. E. Kolodenker, W. Koch, G. Stringhini and M. Egele. 2017. PayBreak: Defense against cryptographic ransomware. In Proceedings of the 2017 ACM on Asia Conference on Computer and Communications Security. 599–611.
14. A.H. Sung and S. Mukkamala. 2003. The feature selection and intrusion detection problems. In Proceedings of the ninth ACM SIGKDD international conference on Knowledge discovery and data mining. 491–496.
15. F.L. Greitzer and D.A. Frincke. 2010. Combining traditional cyber security audit data with psychosocial data: Towards predictive modeling for insider threat mitigation. In Proceedings of the 43rd Hawaii International Conference on System Sciences. 1–10.
16. S. Hemalatha, M. Mahalakshmi, V. Vignesh, M. Geethalakshmi, D. Balasubramanian and A.A. Jose. 2023. Deep Learning Approaches for Intrusion Detection with Emerging Cybersecurity Challenges. International Conference on Sustainable Communication Networks and Application (ICSCNA). 1522–1529.
17. R. Reshma and A.J. Anand. 2023. Predictive and Comparative Analysis of LENET, ALEXNET and VGG-16 Network Architecture in Smart Behavior Monitoring. Seventh International Conference on Image Information Processing (ICIIP). 450–453.
18. N. Azoury, S. Subrahmanyam and N. Sarkis. 2014. The influence of a data-driven culture on product development and organizational success through the use of business analytics. Journal of Wireless Mobile Networks, Ubiquitous Computing, and Dependable Applications. 15(2). 123–134. https://doi.org/10.58346/JOWUA.2024.I2.009
19. B. Friedman and H. Nissenbaum. 1996. Bias in computer systems. ACM Transactions on Information Systems (TOIS. 14(3), 330–347.
20. M. Armbrust, A. Fox, R. Griffith, A.D. Joseph, R.H. Katz, A. Konwinski and M. Zaharia. 2010. A view of cloud computing. Communications of the ACM. 53(4), 50–58.
21. Y. LeCun, Y. Bengio and G. Hinton. 2015. Deep learning. Nature. 521(7553), 436–444.

22. R.S. Sutton and A.G. Barto. 2018. Reinforcement Learning: An Introduction. MIT Press.
23. A. Montanaro. 2016. Quantum algorithms: An overview. *NPJ Quantum Information*, *2*(1), 1–8. https://doi.org/10.1038/npjqi.2015.23.
24. V. Dunjko and H.J. Briegel. 2018. Machine learning and artificial intelligence in the quantum domain: A review of recent progress. *Reports on Progress in Physics*, *81*(7), 074001. https://doi.org/10.1088/1361-6633/aab406.
25. F. Monrose and A.D. Rubin. 1997. Authentication via keystroke dynamics. In Proceedings of the 4th ACM Conference on Computer and Communications Security. 48–56.
26. J. Lee, Y. Lee and H. Lee. 2017. An integrated approach to proactive threat management. Computers & Security. 70, 35–49.
27. F. Doshi-Velez and B. Kim. 2017. Towards a rigorous science of interpretable machine learning. arXiv preprint arXiv:*1702.08608*.
28. B. McMahan, E. Moore, D. Ramage, S. Hampson and B.A.Y. Arcas. 2017. Communication-efficient learning of deep networks from decentralized data. arXiv preprint arXiv:*1602.05629*.

7 Network Security with Artificial Intelligence

Rachna Rana and Pankaj Bhambri

7.1 INTRODUCTION

"First National Computer Connected Here," is the title of a diminutive sentence in the UCLA learner broadsheet on July 15, 1969 [1]. This paper for a short time describes the effort being completed at UCLA to produce a new system that attaches geographically separated computer networks.

The idea behind the project, sponsored by the Defence Advanced Research Projects Agency (ARPA), was to protect the data flow of military technology configured using a technology called Network Control Protocol (NCP). Since then, the concept of networking has evolved [2].

Currently, various network technologies, such as the Internet, e-commerce, digital goods distribution, and e-mail communication, have become an integral part of everyday life. However, with the increasing reliance on the Internet, the number of cyber threats has also grown at an alarming rate. Some of the most significant security challenges include malware detection, which utilizes signature-based and heuristic search engines to identify potential threats; ransomware, which employs AI-based models to predict and execute attacks while updating itself to evade detection; and distributed denial-of-service (DDoS) attacks, which are analyzed and mitigated using signature-based methods and vulnerability detection techniques. Addressing these threats requires continuous advancements in cybersecurity strategies and threat intelligence frameworks to ensure the safety and integrity of digital systems [3].

The authenticity of its resources is to certify the isolation with guard of the network from threats through Internet of Things (IoT) technology, and phishing services have appropriate control and have a human-based intelligent detection model, human attack can really be touched. This problem arises from the use of virtual private networks (VPNs). Some of the recent threats include peer-to-peer attacks to the cloud, document interception, crypto theft, identity fraud, and more. This is more or less the biggest threat mentioned above [4].

The frequency and sophistication of cyber-attacks is rapidly increasing. From a business perspective, one of the biggest concerns regarding cybersecurity of companies and organizations is lack of strategic planning. This problem goes beyond technical differences. It involves management's lack of understanding of real needs, resulting in an inability to provide appropriate support.

This lack of support holds many organizations back because they are unaware of the need for cybersecurity or are unwilling to invest in it. Of particular concern is the lack of professionals to meet the growing need for cybersecurity expertise. If this

DOI: 10.1201/9781003521020-7

trend continues, it is expected that by 2021, approximately 3.5 million jobs will be opened in cybersecurity field and the cost of global terrorism will be up to 3 trillion dollars [5].

Looking at the current situation, it is easy to see why cybersecurity experts are turning their attention to artificial intelligence (AI) and how AI can help solve some of these problems. For example, machine learning (ML), used in many new AI algorithms, can help detect malware that is difficult to identify and isolate [6].

As malware evolves into traditional security solutions, ML provides an opportunity to learn not only what the malware looks like and behaves, but also how to replace it. In addition, AI systems not only provide detection capabilities, but also perform tasks to correct specific situations, classify situations and threats, freeing experts from repetitive tasks. Some studies estimate that investments in big data and intelligence for science and technology and security products are $96 million in 2021, and will reach $1.088 trillion by 2032 due to the need for trusted data [7].

Despite all the great advances made in the cybersecurity sector in the past few years, especially in the context of AI, it seems necessary to be cautious about the scope of its applications. It is easy to believe that AI is a panacea that can solve all cybersecurity problems, or to blindly believe that AI can overcome all the dangers that modern technology has reached, but we should make it clear that we have not yet achieved the current goal, only some technologies are used that give good results in security applications, and although the system is far from "smart," it is limited to ML and knowing its own level of knowledge as required by AI, supervised machine learning has achieved some great results [8].

Unsupervised machine learning still seems to be the overarching goal in the discipline so far but it still relies on many factors. Enabling human intelligence to obtain content and understand data will eliminate the need for human interaction. Since a domain name system (DNS) server plays a crucial role in managing and resolving domain names, it is essential that the algorithms governing its operation continuously adapt to their environment. These algorithms must function efficiently without needing to reset to a predefined "normal" profile that may overlook evolving threats. In addition, DNS servers often work in coordination with other DNS servers to transfer zone information, ensuring seamless domain resolution. To maintain security and stability, the system must be capable of detecting and flagging malicious traffic while continuing its core functions without interruption [9].

In other words, algorithms should be developed to understand why certain patterns exist behind certain behaviors, rather than blindly learning and assuming model. One increasingly popular technique in this field is to use Bayesian belief networks (BN) to generate experts. BN, also known as causal probability network, is a method that uses probability to represent the relationship between different events, using less energy and resources to deal with additional threats. Big data is growing faster than ever before and ML is crucial to have the capacity to store and analyze this data.

The important thing is to understand the different levels of organization. Therefore, data visualization is one of the areas where ML will play an important role in the future [9].

7.2 ARTIFICIAL INTELLIGENCE AND NETWORK SECURITY

7.2.1 Artificial Intelligence

The goal of the quickly expanding field of AI research is to create systems, ideas, techniques, and technologies that can mimic, enhance, and extend human intellect. It entails analyzing data, finding patterns in it, and drawing conclusions or forecasts from it using computers and algorithms. AI has the potential to revolutionize a wide range of contemporary elements of life, including entertainment, business, and health. The main objective here is to build machines with reaction times comparable to those of human intellect. Natural language processing, medical diagnosis, picture and language recognition, and other areas are few of the outcomes of this field of study. By producing higher-quality data, AI not only advances our understanding of human intellect but also enhances our quality of life.

In the 1970s, many underdeveloped nations extensively studied this sector. However, overcoming the hurdles presented by intelligence proved difficult and progress was slow. AI gained prominence in the 1990s as technology advanced. The development of algorithms and ML has also contributed to skill development. As technology businesses start their research and development efforts, AI has garnered a lot of interest [10].

7.2.2 Network Security

AI is a system which is used for analyzing the network traffic for packets. These might indicate the various kinds of attacks in network. Network security protects our network from different types of attacks like contravention, impingement, and bluff. This comprehensive term encompasses a wide range of solutions, such as software and hardware, rules, guidelines, and configurations for set of connections admittance, system procedure, and largely hazard deterrence. Access control, bug and antivirus software, appliance safety, system analytics, firewalls, organization network encipher, and other mechanism are all part of system safety.

7.2.2.1 Advantages of Network Security

Network security is necessary to protect customer figures and information, preserve the confidentiality of communal information, provide consistent system access and performance, and fight off cyber intrusion. Providing products and services to customers and streamlining business processes are made possible by granting authorized access to systems, apps, and data. A carefully planned safety system explanation reduces transparency expenses while protecting enterprises from costly data breaches and other security disasters.

7.2.2.2 Types of Security Measures for Networks

Figure 7.1 shows the various types of security measures possible for the networks.

7.2.2.2.1 Firewall

Using preset security rules, firewalls regulate both inbound and outbound network traffic. Firewalls keep malicious communications out and are an indispensable

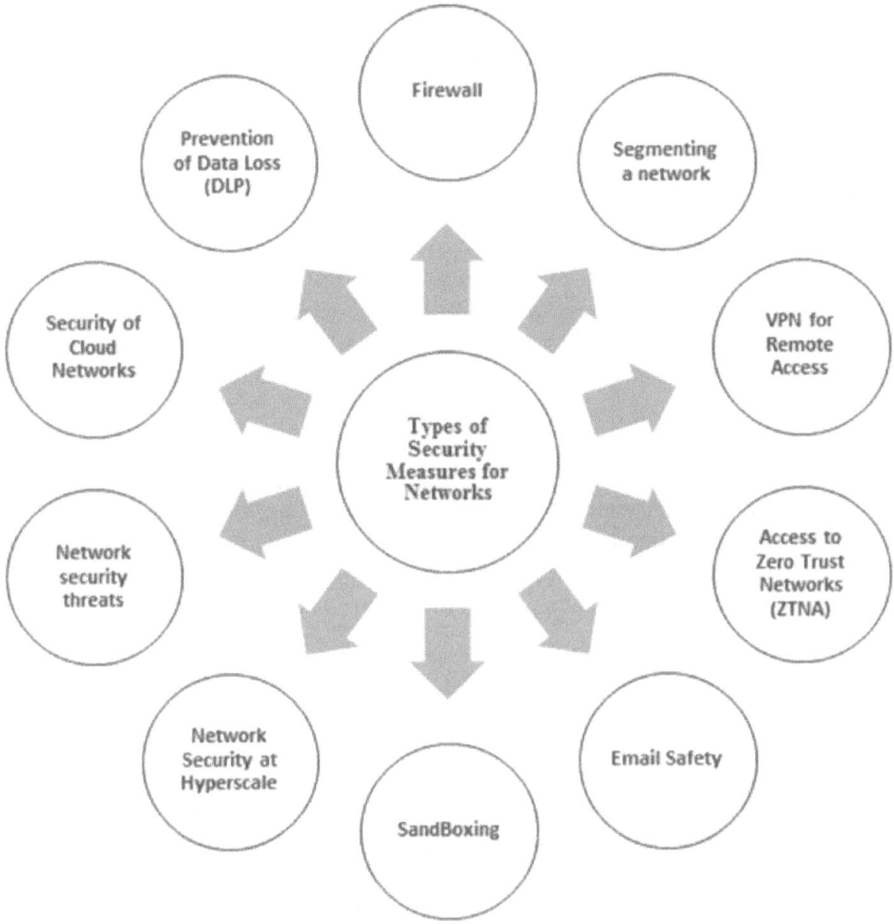

FIGURE 7.1 Types of security measures for networks.

component of regular calculating. In terms of network security, firewalls are essential, especially subsequently Invention Firewalls, which concentrate on overcrowding malware and appliance-layer damages.

7.2.2.2.2 Segmenting a Network

Network segmentation definitively separates assets into groups based on risk, function, or location within an association. As a border entryway unfalteringly divides an association's system from the internet, effectively shielding critical data from outside dangers. Organizations must establish additional internal network borders to significantly strengthen security and access controls.

7.2.2.2.3 VPN for Remote Access

Virtual private network (VPN) enables people who work from home, those on the move, and those accessing the company's network from outside the office to do so safely.

Every user needs to have VPN software installed on their device or access it through a web-based application. To ensure the safety and trustworthiness of important data, there are rigorous checks on the devices used such as the need for multiple forms of verification and the encryption of data.

7.2.2.2.4 Access to Zero-Trust Networks

Under the zero-trust security model, individuals are granted access and permissions strictly necessary for their specific roles. This method diverges from conventional security measures like VPNs, which grant consumers unhindered admission to the premeditated exchange ideas. Software-defined perimeter (SDP) explanations, also referred to as zero-trust network access (ZTNA), offer tailored right of entry to an association's appliance for workers who necessitate it for their occupation tasks.

7.2.2.2.5 Email Protection

Email protection deals with all approach, apparatus, and forces expected at conservation your email financial records and information from outer intimidation. At the same time as the majority email overhaul suppliers presents safety features to defend you, these may not be adequate to discourage hackers from approaching your information.

7.2.2.2.6 Preventing Data Loss

Data loss prevention (DLP) is an essential imitation-security approach that occupies acquaintance and industriousness most excellent preparation to avert receptive information from departure a company. This incorporates confined information like individually specialized information and information related to observance principles for instance PCI DSS, SOX, HIPAA, among others.

7.2.2.2.7 Network Security Threats

Key points to remember: "It includes brute force attacks, Denial of Service (DoS) attacks, and the exploitation of known vulnerabilities that Intrusion Prevention Systems (IPS) technology can identify and block. An exploit is an attack that takes advantage of vulnerability, like a software flaw, to gain control of the system. Attackers often exploit these vulnerabilities before a security patch is available. In these critical situations, an intrusion prevention system can effectively block these attack attempts."

7.2.2.2.8 Sandboxing

Sandboxing is a cyber-security approach that allows you to execute agenda or right of entry information on a host system while mimicking the behavior of end-user operating systems in a secure, contained environment. It monitors files or applications as they are opened for any potentially harmful actions to block threats from entering the network. For instance, it can safely identify and halt malware from infecting users by limiting access to certain file types such as PowerPoint, Microsoft Word, Excel, and PDF.

7.2.2.2.9 Security at Hyper-scale

"Hyper-scale" refers to architecture's ability to adapt as demand increases. This explanation supports quick consumption and leveling up or down to get together varying system safety needs. By powerfully incorporating set of relatives and presenting out belongings in a software- classified system, collection of clarification may make full use of all available hardware assets.

7.2.2.2.10 Security of Cloud Networks

In the modern landscape, applications and workloads are no longer confined to local data centers. Safeguarding your existing data center as applications transition to the cloud necessitates heightened agility and awareness. By leveraging software-defined networking (SDN) and software-defined wide area networking (SD-WAN) solutions in conjunction with firewall-as-a-service (FWaaS) infrastructure, you can effectively fortify both public and private networks.

7.2.2.3 Strong Network Security Will Guard against Viruses

Viruses are malicious files or programs that may be downloaded and distributed by infecting other computer applications with their code. They can also be inactive. Once contaminated, the documentations can extend from one organization to another, may be unfavorable to, or may demolish network information. Figure 7.2 shows a strong network security structure to safe guard against virus attacks.

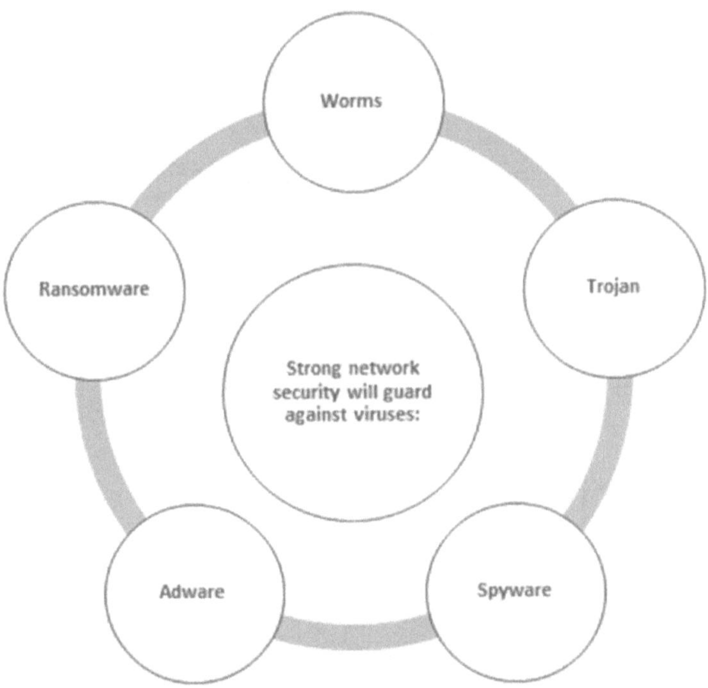

FIGURE 7.2 Strong network security will guard against viruses.

Viruses: They disrupt the computer network by using bandwidth and reducing the equipment's ability to process data efficiently. The virus is a disease that can spread and function on its own, unlike viruses in out body that need the help of a host to spread.

Trojan horse virus: A Trojan horse is a hateful code that camouflages itself as a lawful service but is dangerous and allows unauthorized people to control the computer. Trojan viruses can delete files, open malware on the network, infect others, and steal confidential information.

Spyware: It is a computer virus that collects information about people or groups without their knowledge or consent. We may also share this information with others without your permission.

Adware: This one can redirect your online searches to sites that contain advertisements. It also collects personal information to customize ads based on your past searches and purchases.

Ransomware: Ransomware is a type of Trojan horse malware that encrypts data, making it inaccessible. The aim is to claim compensation by preventing victims from accessing their systems.

7.3 DDOS

7.3.1 Denial of Service Attack

A denial of service (DDoS) attack is a malicious attempt to interrupt traffic to a certain server, service, or network and its surrounds to prevent massive internet usage. A computer virus serves as the foundation for an assault to produce outcomes. Examples of applicable systems include computers and other network services, such as IoT devices.

How does it operate? Attackers can manipulate software. The term "bot" refers to individual devices, and the term "net" refers to a collection of bots, Zombies, Services. One visible sign of an attack is when a website or service suddenly slows down or stops working. However, similar performance problems may require further investigation as they could be caused by various sources, including genuine traffic. Traffic analysis tools can help identify some signs of a DDoS attack [11].

There are unusual traffic patterns, such surges that occur at strange hours of the day or irregular patterns (like every ten minutes). Other variables, particularly with regard to DDoS attacks, will change based on the kind of assault.

7.3.2 Types of DDOS Attacks

To appreciate how dissimilar DDoS attacks effort, it's important to understand how network connections are established. Each layer in the model functions differently, similar to the process of building a house from the ground up.

DDoS attacks can be categorized into three types, even though they usually target a busy device or network. In response to data gathered from the targeted application layer attacks, the attacker may use a series of attack vectors or one or more distinct attack vectors [12].

7.3.2.1 Purpose of Attack

Attackers aim to destroy target resources in order to cause a denial of service, a tactic known as Layer 7 DDoS assaults (relating to Layer 7 of the OSI model). The attack is aimed at the layer responsible for creating web pages on the server and sending them in response to HTTP requests. Making an HTTP request from the client is inexpensive, but response costs on the target server might be substantial since the server frequently loads a large amount of data and performs database queries to build web pages prevention because it might be challenging to determine the harmful impacts of illicit commerce. This assault comes in several complexity levels. Advanced versions will employ various IP addresses, random referrers, and user agents to target random URLs [13].

7.3.2.2 Exhaustion Attacks

Arise from the overuse of network devices like load balancers and firewalls, as well as server resources. Inaccessible next, advance the ball. Workers then get requests for more items without permission, which go unaddressed until they are too sick to accept the package or faint.

7.3.2.3 TCP Handshake

Communication that occurs sporadically in which two computers transmit a sequence of TCP (first connection request) SYN packets meant for incorrect IP addresses in order to establish a network connection. The target's resources are wasted with each contact request that is followed by waiting for the hypothetical final handshake phase, which never happens.

Use all of the bandwidth that is available between the target and the host network to cause congestion. sending a lot of data to the target by employing amplification or other techniques to create a lot of traffic (such utilizing a botnet), saying things like "I want a duplicate of everything, please call me back and repeat my complete order," while in reality the person returning the call was the victim.

With little effort, a long response is created and sent to the victim. You will take delivery of a reply from the server. If your company's website is flooded with satisfied customers after launch, it would be a mistake to cut off all traffic. If the company faces increased traffic from a known attacker, it should work to mitigate the attack [14].

7.3.2.4 Various Formats

There are two types of traffic designs: single-vector attacks and multi-vector attacks. Multi-vector DDoS attacks are those that target more than one protocol at once. An example of such an attack would be DNS amplification, which targets levels 3/4, combined with HTTP flooding, which targets layers 7. Difficult to distinguish from regular traffic - Attackers want to disappear as much as possible to minimize their impact.

Attacks can also be changed to become counterattacks in scenarios with little traffic. Layering will yield the finest outcomes in overcoming the toughest temptations. Proceed in this direction. In its most basic form, when blackhole filtering is employed without further constraints, all valid and harmful network traffic gets routed to the empty path or black holes and lost on the network.

The hotel's Internet Service Provider (ISP) can redirect all website traffic into a black hole to defend against a DDoS assault. Providing the attackers with what they want—an inaccessible network—makes this option suboptimal. Adding volume is another method used to avoid denial of service assaults [15].

7.4 DIGITAL SECURITY AND NETWORK SECURITY

Ensure that only authorized persons have access to computer systems and labs. Prevent personal devices, especially USBs or hard drives, from connecting to the network. Configure your machine for automatic software and operating system updates. Verify the frequent updates of the antivirus software on every PC. The Internet, antivirus programs, SIM cards for smartphones, biometrics, and secure personal gadgets are some examples of these technologies [16].

7.4.1 The Difference between Information Security and Cyber Security

This is not unexpected given that unauthorized access to someone's information, personal, or financial resources is referred to as "cybercrime," emphasizing the importance of cyber security. Digital security differs from cyber security in that it entails safeguarding your online presence (data, identity, and assets). Simultaneously, network security encompasses a wide range of measures to safeguard computers, networks, and other digital devices, as well as the data they contain, against illegal access. Many industry professionals use these two phrases interchangeably; however digital security just protects the words, whereas cyber security covers the entire infrastructure, including all systems, networks, all data.

Showcasing some of the biggest data security breaches over the past ten years is this info graphic from 2019. As if that wasn't frightening enough, this article states that over 7 million data files are hacked every day, and online fraud and abuse surged by 20% in the first three months of 2020 [15].

It's not worldwide news that a stranger finds out you like the original Star Wars trilogy better than the films; it won't jeopardize your financial or personal stability. What kinds of data are therefore in danger? It also has data that can pinpoint your exact position. Identity theft and social engineering frequently exploit personal information.

Furthermore, a hacker possessing your Social Security number (or its equivalent) can create a credit card in your name, which would lower your credit score. Compute your personal payment details. This data consists of PINs, credit and debit card numbers (together with expiration dates), and online banking numbers (transactions and accounts).

When thieves get your online banking credentials, they can transfer funds or make purchases from your account, including purchases of prescription drugs, health insurance, trips to doctors and hospitals, and medical records. Cybercriminals can exploit your health information to order and sell prescription medicines or to create fake insurance claims.

If your digital data is exposed, a lot of things can go wrong. Thankfully, there are several sorts of security available in the digital realms that offer diverse ways of safeguarding. Among these are the following [16].

7.5 ANTI-VIRUS SOFTWARE

Malware and other harmful apps can carry viruses that can corrupt your data and instantly shut down your computer. Not only can a strong antivirus application identify and eliminate these infections, but it can also stop suspicious activities, isolate risks, and stop infection from spreading—even if "the level is high enough." Since firewalls have been around for a while, many cyber security professionals believe they are no longer necessary. Its most sophisticated function, nevertheless, is helpful for barring unauthorized users. In order to restrict access and keep an eye on usage, agents employ authentication procedures and block harmful websites.

Remote monitoring offers flexibility and simplicity, enabling managers to address problems from any location at any time. Vulnerability scanners can help you prepare for attacks by helping you find flaws. Web applications and internal systems can benefit from the deployment of scanners by IT security teams.

7.6 SAFETY VEHICLES

It is very simple (and often used) to target hackers and criminals using this technology, which safeguards the security of your data while it moves across different web sites. The amount of private information that travels over text texts could surprise you. For Android and iOS phones, ChatSecure is a chat program that offers safe encryption; Cryph guards the security of your Mac or Windows online browser. By changing your IP address and enabling anonymous internet browsing, Anonym ox guards against the creation of pseudonyms. It is accessible as a Firefox and Google Chrome add-on. Tor hides every page you visit from advertisers and third-party trackers. Moreover, it removes cookies, cleans your surfing history, and offers many encryption levels. It is free and works with both iOS and Android smart phones. Users of the free, nonprofit signal network may exchange documents, GIFs, music, photos, videos, and text [8].

7.7 IOT AND END POINT SECURITY

Numerous security concerns and laws need to be addressed since the IoT system is susceptible to assaults on all of its tiers. Current IoT research focuses mostly on authentication and control, but with technological improvement, new network protocols such as IPv6 and 5G must be merged to achieve an integrated and competitive IoT architecture. The IoT is mostly developing on a small scale, i.e., in certain sectors or businesses. The way we live now might be drastically altered by the IoT. The most significant challenge in attaining the smart home base is security.

We can demonstrate that the IoT will fundamentally alter society in the future if security concerns like trust, endpoint security, privacy, confidentiality, authentication, access control, global governance, and standards are resolved. To answer current IoT research difficulties, such as models for various devices, alarm and personal development use of control systems, and trust management sites, new technologies for identification, wireless, software, and hardware are required [17].

7.8 GUIDANCE FOR THE FUTURE

Recent years have seen a rapid development of the IoT in fields including pollution monitoring, smart transportation, telemedicine platforms, and logistics tracking. However, addressing IoT-related security vulnerabilities is necessary for IoT to develop and flourish. These are the future research directions that will help safeguard the IoT idea.

7.8.1 Architectural Standards

As of right now, IoT employs many tools, services, and procedures to accomplish various ends. Still, the process of integrating the IoT network must go from a micro to macrolevel in order to accomplish a greater goal, such as connecting several smart buildings to create a smart city. Clear architectural standards for the IoT are now required. These standards should include data models, interfaces, and procedures that can accommodate a broad range of users, materials, languages, and functionalities.

7.8.2 Identity Management

The first step toward achieving identity management in the IoT is the credential exchange between connected devices. These systems are easily vulnerable to man-in-the-middle attacks, which compromise the security of the IoT as a whole. Therefore, self-control or a hub that can keep an eye on the device's connection process via encryption and other mechanisms is required to safeguard the identity thief.

7.8.3 Session Layer

The majority of academics think that starting, ending, and maintaining sessions between two items is not supported by the third layer of the internet of items. It is therefore necessary to have a system that can resolve these issues and make device connection easier. In order to handle the connection, orchestration, and communication of several devices, the decentralized communication system should be utilized as an extra layer in the IoT architecture.

7.8.4 5G Protocol

In order to fully utilize the IoT, IPv4 will never support a large number of uniquely identifying IP goods. For this reason, IPv6, which supports 3.4x1038 devices, is becoming more and more popular.

However, these figures will result in high traffic, which will increase interference and necessitate additional bandwidth. Compared to current technology's 2–1000 Mbps speeds, next-generation communications (5G) should offer rates of 10–800 Gbps, which can handle data from the IoT (4G). Through IPv4/IPv6 core conversion, 5G technology will also handle IPv4 and IPv6. Software-defined networks (SDN), massive MIMO, multiple radio access, and heterogeneous networks (HetNet) will all be possible with the adoption of 5G. But all these technologies have their own security issues.

For instance, HetNets will undergo continuous modification, which will have an immediate impact on the network's authentication procedure, particularly given 5G's low speed requirements. In addition, because of cloud computing's privacy requirements, SDN and cloud computing will see a surge in DDoS assaults. Despite referring to the security and authentication of SDN via the management of security-related user authentication, there should be a broad discussion on the security concerns of 5G and the labeling of all new technologies included in 5G to guarantee IoT security [18–25].

7.9 CONCLUSION

The Information Technology (IT) industry swiftly absorbs buzzwords offered by businesses. Recent years have seen a lot of talk about big data, cloud computing, and AI in various venues; yet, many individuals are not sure what these terms actually represent or how to use them to solve problems effectively. People who are not tech savvy typically respond in one of two ways: either they reject the technology (i.e., new functionality) or, if done right, they label it as cutlery. Often, it takes months or even years for the market to reach its full potential when the dust settles detection of reactivity in real time.

Many attacks manage to evade this procedure, inflict significant harm, and once underway, are unstoppable. Without requiring human assistance, ML can instantaneously identify threats and prevent them before they have a significant negative impact on internet security. Until now, the only area of AI that has proven effective in resolving minor issues is ML. Costs like reduced human intervention in danger detection situations and the development of new technologies and data visualization tools that facilitate the integration of data analytics, data science, and machine learning are the ultimate goals.

REFERENCES

1. M. Abomhara and G. M. Koien, "Security and privacy in the Internet of Things: Current status and open issues," *2014 International Conference on Privacy and Security in Mobile Systems (PRISMS)*, May 2014, https://doi.org/10.1109/prisms.2014.6970594.
2. S. O. Aletan, "Artificial Intelligence Languages and Architectures: Past, Present, and Future," *WORLD SCIENTIFIC eBooks*, pp. 431–484, 1992. https://doi.org/10.1142/9789814354707_0014.
3. K. B. Ali, and F. Zarai, "Efficient MIH/SDN-Based Vertical Handover for 5G HetNets," Jul. 2023, doi: https://doi.org/10.1109/snpd-winter57765.2023.10224059.
4. Z. Chen, "Research on the Application of Intelligent Learning Algorithms in Network Security Situation Awareness and Prediction Methods," *2021 5th Asian Conference on Artificial Intelligence Technology (ACAIT)*, Oct. 2021, doi: https://doi.org/10.1109/acait53529.2021.9731205.
5. G. Currie, and E. Rohren, "Integration of Artificial Intelligence, Machine Learning, and Deep Learning into Clinically Routine Molecular Imaging," *Springer eBooks*, pp. 87–108, Jan. 2022, doi: https://doi.org/10.1007/978-3-031-00119-2_7.
6. E. D. Gennatas, and J. H. Chen, "Artificial Intelligence in Medicine: Past, Present, and Future," *Elsevier eBooks*, pp. 3–18, 2021. https://doi.org/10.1016/b978-0-12-821259-2.00001-6.

7. Shivanand V. Manjaragi, and S. V. Saboji, "An Efficient Handover Authentication Mechanism Using Deep Learning in SDNBased 5G HetNets," *International Journal of Intelligent Engineering and Systems*, vol. 16, no. 6, pp. 753–770, 2023. https://doi.org/10.22266/ijies2023.1231.63.
8. Episode 6: Radiology AI: Past, Present and Future," Nov. 2020. https://doi.org/10.1148/ryai.20201127.podcast.
9. A. F. Khan, and P. Nanda, "Hybrid blockchain-based Authentication Handover and Flow Rule Validation for Secure Software Defined 5G HetNets," *2022 International Wireless Communications and Mobile Computing (IWCMC)*, May 2022. https://doi.org/10.1109/iwcmc55113.2022.9824274.
10. D. Kothari, "A Review of Grey Scale Normalization in Machine Learning and Artificial Intelligence for Bioinformatics Using Convolution Neural Networks," *International Journal for Research in Applied Science and Engineering Technology*, vol. 9, no. 3, pp. 1306–1310, 2021. https://doi.org/10.22214/ijraset.2021.33316.
11. M. M. Mijwil, A. K. Faieq, and A.-H. Al-Mistarehi, "The Significance of Digitalisation and Artificial Intelligence in The Healthcare Sector: A Review," *Asian Journal of Pharmacy, Nursing and Medical Sciences*, vol. 10, no. 3, Nov. 2022. https://doi.org/10.24203/ajpnms.v10i3.7065.
12. Z. Pan and P. Mishra, "The Future of AI-Enabled Cybersecurity," pp. 235–240, Jan. 2023. https://doi.org/10.1007/978-3-031-46479-9_12.
13. J. Qiang, F. Wang, and X. Dang, "Network Security Based on D-S Evidence Theory Optimizing CS-BP Neural Network Situation Assessment," Jun. 2018. https://doi.org/10.1109/cscloud/edgecom.2018.00035.
14. J. Wan, "'Artificial Intelligence, Machine Learning and Deep Learning – Limitations: Privacy and Data Security Issues' Chapter," *Artificial Intelligence in Medicine*, pp. 217–239, 2022. https://doi.org/10.1007/978-981-19-1223-8_10.
15. J. Yang, Y. Yang, L. Zheng, R. Cheng, and S. Lin, "Network Security Situation Assessment Based on Attack Graph Techniques," *Journal of Physics. Conference Series*, vol. 2310, no. 1, pp. 012071–012071, 2022. https://doi.org/10.1088/1742-6596/2310/1/012071.
16. C. Yao, Y. Yang, and K. Yin, "Research on network security situation prediction method based on AM and LSTM hybrid neural network," Sep. 2021, https://doi.org/10.1109/ifeea54171.2021.00072.
17. D. ZHANG, R. ZHENG, Q. WU, and Y. DAI, "Network Security Situation Awareness Model Based on Autonomic Computing," *Journal of Computer Applications*, vol. 33, no. 2, pp. 404–407, 2013. https://doi.org/10.3724/sp.j.1087.2013.00404.
18. X. Zhang, W. Chen, J. Wang, and R. Fang, "Application of Machine Learning Algorithm and Data Evaluation in Computer Network Security Situation Awareness Technology," *Intelligent Decision Technologies*, pp. 1–13, 2023. https://doi.org/10.3233/idt-230238.
19. P. Bhambri, R. Kaur, and S. Kaur, Hosiery Management System: An Automation Software (Vol. 1). Lap Lambert Academic Publishing, 2015. ISBN: 9783659675782.
20. P. Bhambri, and F. Goyal, Development of Phylogenetic Tree Based on Kimura's Method: Based on Un-Weighted Pair Group Method with Arithmetic Mean (UPGMA) and Neighbor Joining (NJ) Scoring Techniques (Vol. 1). Lap Lambert Academic Publishing, 2013. ISBN: 9783659336539.
21. P. Bhambri and V. Paika, Image Recognition Using Neuro-Fuzzy Techniques: Developing a Mamdani's Fuzzy Inference System in MATLAB Using Fuzzy Logic Toolbox (Vol. 1). Lap Lambert Academic Publishing, 2013. ISBN: 9783659460838.
22. P. Bhambri and D. Garg, Enhanced Model for Fusion of Multi-Modality Images: Discrete Wavelet Transformation Using Region Based Fusion Rules (Vol. 1). Lap Lambert Academic Publishing, 2012. ISBN: 9783659208089.

23. P. Bhambri, S. Rani, and A. Khang. (2024). AI-Driven Digital Twin and Resource Optimization in Industry 4.0 Ecosystem. In Intelligent Techniques for Predictive Data Analytics (pp. 47–69). IEEE. https://doi.org/10.1002/9781394227990.ch3
24. P. Bhambri, Study and Implementation of Genetic Intricacies: An Application of Bioinformatics (Vol. 1). Lap Lambert Academic Publishing, 2013. ISBN: 9783659342561.
25. P. Bhambri and P. Bansal, Secondary Structure Prediction of Amino Acids Using GOR Method: On Different Input Formats (Vol. 1). Lap Lambert Academic Publishing, 2013. ISBN: 9783659107306.

8 Endpoint Security and Artificial Intelligence in the Financial Sector

Shaista Alvi

8.1 INTRODUCTION

The financial territory is a key target for cybercriminals owing to the vast sums of money and sensitive personal data it handles. With the growing acceptance of cloud computing, remote work, and internet-connected gadgets, the strike surface for financial institutions has expanded significantly. The current threat landscape cannot be protected by outdated security techniques, so integrating cutting-edge technologies such as artificial intelligence (AI) and machine learning (ML) is essential for efficient endpoint security.

The financial sector has unique characteristics that differentiate its endpoint security requirements from other industries, including strict regulatory compliance (e.g., GDPR, BASEL, SOX), highly sensitive data (e.g., customer financial information, transaction data), a distributed network of endpoints (e.g., ATMs, mobile devices, branch offices), and an increased attack surface due to remote work and cloud adoption [1, 2].

Mobile devices, servers, laptops, desktop computers, routers, and other endpoint devices are all vulnerable to malevolent cyber-attacks and security breaches. Endpoints are susceptible points of entry into any community bank's network since they continue to be the major targets of attackers [2]. The concept of an endpoint in the financial sector is broad and includes not only traditional devices like laptops and desktops but also a wide range of internet-connected devices such as Internet of Things (IoT) devices, automated teller machines (ATMs), and other distributed devices. These endpoints can be exposed to various types of attacks, including zero-day incursions, malware, and file-less attacks, which can compromise the safety of the entire network [1]. Due to the increased attack surfaces and threat vectors associated with the widespread use of mobile devices (e.g., laptops, phones, and tablets), it is imperative to implement stringent endpoint security measures in order to safeguard device access and stop illegal file sharing and program downloads [1, 2].

The number of workers participating in or switching to remote work has increased as a result of COVID-19, exacerbating this danger even more. This trend is probably here to stay [2]. Financial services businesses have implemented cloud and endpoint technology to provide smooth interactions across networks and between client and employee devices. Financial services are vulnerable to cyber-attacks unless suitable security measures are in place. Financial services are recognized as the most

DOI: 10.1201/9781003521020-8

impaired sector, with risks present both outside the organization and internally through employee devices [3]. The business sector is witnessing the second-largest part of pandemic 2019-related cyber-attacks, only behind the health sector, according to the Bank for International Settlements (BIS) [4]. The divergence between the finance, national security, and diplomatic communities is particularly noticeable, and financial authorities confront unique vulnerabilities from cyber assaults [5]. Cyber risks to the financial system are escalating, and the global community must work together to secure it. In 2016, hackers attacked Bangladesh's central bank and exploited vulnerabilities in SWIFT, the international financial system's major electronic payment messaging system, in an attempt to seize $1 billion [6]. Even when the maximum number of transactions was restricted, $101 million disappeared.

The prevailing consensus is that a significant cyberattack represents a risk to financial stability. It's no longer a matter of "if," but rather "when" such an event will occur [7]. The divergence between the finance, national security, and diplomatic communities is particularly noticeable, and financial agencies confront distinct cybersecurity dangers [5]. Financial institutions must safeguard their devices and networks from cyber threats. Endpoint security is an important part of the financial sector's overall cybersecurity strategy. It includes installing and updating software applications to protect against malware, viruses, and other sorts of cyber threats. Figure 8.1 indicates the total cyber incidents in the financial industry worldwide based on the statistics provided by reference [8] and for 2023 it is estimated to gallop more than double over the previous year.

To address these challenges, financial institutions are increasingly turning to AI and ML-powered endpoint security solutions [9]. These technologies can provide automated incident response, real-time threat detection (RTD), and predictive analytics (PA) to ascertain and mitigate digital threats [10]. However, implementation of AI-powered endpoint security in the financial sector faces several challenges, including data privacy and regulatory compliance concerns, integration with legacy systems, and the potential for AI-driven attacks. This research paper aims to explore

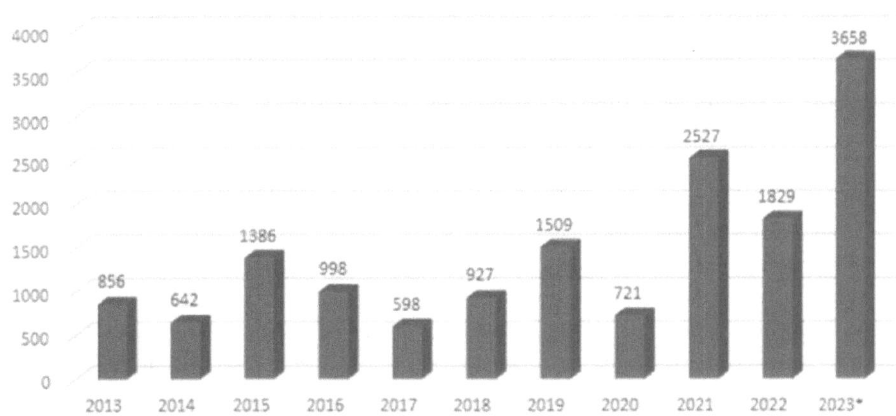

FIGURE 8.1 Cyber incidents in the financial industry worldwide from 2013 to 2023.

AI and ML technologies in endpoint security for the financial sector and focus on the unique characteristics, challenges, and best practices for implementation. The study will review existing research, identify gaps, and recommend future research and practices in this critical area of cybersecurity. The outcomes of this research will contribute to the society on endpoint security, the application of AI and ML technologies in the financial sector.

The structure here onwards is the next section is a literature review that provides a literature survey on the endpoint security and financial sector. The third section elucidates the challenges followed by the fourth section of the future relates to forthcoming developments in end security in the financial sectors. The last is the conclusion section which entails the academic implications and recommendation.

8.2 LITERATURE REVIEW

AI and ML technologies in endpoint security for the financial sector have become more essential in recent years. The unique characteristics, challenges, and best practices for implementation in this domain have been extensively studied. This section of this chapter aims to offer a thorough understanding of the existing academic contributions in this field.

The steep progress of information technology has brought about numerous ways to build enterprise-wide area networks. These include using multiple internal networks, setting up local infrastructure in branch offices, enabling remote office access, supporting mobile offices, and leveraging cloud-based services. However, this varied network structure has led to unclear network boundaries and an increased number of endpoints that depend on network access for business operations. As a result, the risks related to endpoint security have grown significantly [11]. Malware, which encompasses viruses, worms, trojans, and other forms of malicious software, presents a significant danger to the reliability and confidentiality of computer systems and networks.

One of the fundamental features of malware is its capacity to intrude into systems by multiple mechanisms, such as infected email attachments, compromised websites, or exploiting flaws in digital systems [12]. If a system is infected, malware can execute a wide range of destructive behaviors, including stealing sensitive data, monitoring user activity, interrupting system functions, and even giving remote access to the infected machine [13]. Endpoint devices, such as desktop PCs, laptops, and cell phones, are especially vulnerable to malware attacks because of their direct internet connection and regular use for accessing sensitive data and apps. Malware-targeting endpoint devices can jeopardize the security of the entire network because they frequently serve as entry sites for invaders.

As businesses increasingly rely on technology and embrace remote work models, the security perimeter expands beyond the confines of traditional office networks. Every endpoint, regardless of its location or type, becomes a potential target for cybercriminals. Furthermore, research carried out in reference [14] emphasizes the human element in cybersecurity vulnerabilities. A lack of user awareness regarding phishing scams and social engineering tactics can leave endpoints susceptible to infiltration. In addition, unpatched vulnerabilities in software and operating systems create gaps in defenses that attackers can exploit [15].

Existing research has highlighted the growing importance of endpoint security in the financial sector. The economic sector is a key object for digital attacks because of the confidential data involved and the potential income for attackers [12]. Classical endpoint security techniques, such as signature-based detection, have proven ineffective against current and advanced threats that are continually developing and mutating. AI and ML technologies have emerged as dominating tools for improving the protection of endpoints by enabling more advanced threat identification and avoidance strategies [10]. Key advantage of using AI and ML in endpoint security is improved threat detection.

These technologies are capable of examining big data obtained from different resources, such as user behavior, network, and system logs to detect potential dangers. ML techniques learn from a dataset and increase their ability to detect and prevent risks over time, resulting in faster and more accurate threat detection [16]. AI-based behavioral analysis can monitor user behavior and recognize abnormalities that may indicate a possible risk, such as accessing sensitive files at odd times or locations [17]. Another important aspect of AI and ML in endpoint security is advanced threat prevention. These technologies are used to develop proactive threat prevention approaches that can detect and prevent cyber intrusions before they cause damage. AI and ML algorithms can assist security teams in detecting and responding to risk in real time by analyzing trends and abnormalities in user activity and network traffic [18]. AI-powered SOAR (security, orchestration, automation, and response) solutions can connect security tools, integrate disparate security systems, and enable automated responses to select security events [19].

However, AI technology in endpoint security poses distinct issues. One of the main concerns is data privacy and protection. ML models require access to substantial volumes of data [18], raising critical concerns about data privacy and the need to secure this data against potential breaches. Financial businesses embrace dynamic data security processes, such as encryption, data confidentiality, and access controls, to maintain the secrecy and trustworthiness of the data used by their AI/ML models. Model vulnerability is another significant risk associated with AI and ML in endpoint safekeeping. These models are prone to various manipulative attacks, including adversarial attacks and model poisoning, which can compromise the integrity of the financial decisions made and expose institutions to financial losses and reputational harm [20].

To mitigate these risks, financial institutions must implement secure model development and maintenance practices, such as model auditing, secure deployment, and safe inference and model serving. The "black box" nature of certain ML models also poses challenges in maintaining algorithmic transparency and compliance with regulatory requirements. Financial institutions must strive towards implementing explainable AI (XAI) frameworks to enhance the interpretability and accountability of their ML-driven decisions, ensuring transparency and building customer trust [21].

8.3 CHALLENGES IN FINANCIAL SECTOR IN IMPLEMENTING END POINT SECURITY

Implementing endpoint security in the financial sector poses a complex set of challenges, primarily due to the unique needs of the industry and the ever-evolving landscape of cybersecurity threats. This sector, being heavily regulated and dealing with sensitive financial data, demands stringent security measures. However, the fast-paced nature of technological advancements and the increasing sophistication of cyber threats create a constantly shifting environment that financial institutions must navigate. Here are some of the primary challenges identified in recent research:

8.3.1 Complexity of Endpoint Devices

In the financial sector, the complexity and diversity of endpoint devices present significant challenges for implementing comprehensive security measures. The variety of devices includes traditional PCs, laptops, handheld devices such as smartphones and tablets, and an increasing number of Internet of Things (IoT) devices. Each of these device types can run on different operating systems, such as Windows, macOS, iOS, Android, and various Linux distributions. In addition, these devices often operate on different software versions, each with unique security vulnerabilities and requirements. This diversity complicates the standardization of security measures across an organization, making it challenging to implement uniform security policies.

One particular area of concern is the Bring Your Own Device (BYOD) environment [22], which is increasingly common in the financial sector. BYOD policies allow employees to use their personal devices for work purposes, offering convenience and flexibility but also introducing significant security risks. Personal devices are often less secure than corporate-issued ones, as they may not have the same level of security controls, such as encryption, antivirus software, and regular updates. This disparity makes it difficult to ensure that all devices accessing the corporate network adhere to the same security standards, increasing the risk of data breaches and other cyber threats.

To address these challenges, financial institutions must adopt comprehensive device management solutions that can provide visibility and enforce security protocols across all endpoints, regardless of the device type or operating system. Consequently, organizations must have highly skilled personnel [23] to adopt comprehensive device management solutions to ensure visibility and enforce security protocols across all endpoints.

8.3.2 User Awareness and Behavior

User awareness and behavior are critical factors in maintaining cybersecurity within any organization, especially in the financial sector, where sensitive data and assets are at constant risk of cyber threats. Despite technological advancements in security infrastructure, human error remains one of the most significant exposures [22]. Personnel often lack awareness of the security hazards associated with their actions, such as clicking on malicious links, visiting dangerous websites, or installing unauthorized software. This lack of awareness not only increases the likelihood of security incidents but can also lead to resistance against security measures that are perceived as hindrances to productivity.

One of the primary issues is that employees may not fully understand the potential consequences of their actions in a cybersecurity context. For instance, clicking on a phishing email can lead to malware infections or data breaches, while using weak or reused passwords can make accounts vulnerable to unauthorized access. The gap in understanding often stems from a lack of regular and comprehensive security training. Many organizations provide only cursory training sessions during onboarding, which are insufficient for keeping employees updated on evolving cyber threats and best practices.

To address these challenges, organizations need to implement a robust and ongoing security training program. This program should be designed to educate employees on the importance of cybersecurity, the specific threats they might encounter, and their individual responsibilities in maintaining a secure environment. Effective training should cover various topics, including recognizing phishing attempts, understanding the importance of strong passwords, and the dangers [22] of using public Wi-Fi for accessing sensitive information. Interactive training modules, simulations of real-world scenarios, and regular assessments can help reinforce learning and ensure that employees retain the information.

Moreover, training should not be a one-time event but rather an ongoing process. The cybersecurity landscape is continuously evolving, with new threats and attack vectors emerging regularly. Regularly updating the training content and conducting refresher courses can help keep employees informed about the latest threats and security practices. In addition, creating a culture of security awareness within the organization is crucial. This can be achieved by integrating cybersecurity discussions into regular team meetings, sharing updates on recent security incidents, and celebrating positive security behaviors among employees.

Another critical aspect of enhancing user awareness and behavior is addressing the perception that security measures are obstacles to productivity. Employees may view security protocols, such as multi-factor authentication (MFA) or restrictions on software installations, as inconvenient or time-consuming. This perception can lead to resistance or even attempts to circumvent these measures, thereby undermining the organization's overall security posture. To mitigate this issue, it's essential to implement user-friendly security measures that balance security needs with usability.

For instance, single sign-on (SSO) solutions can streamline the authentication process by allowing users to access multiple applications with a single set of credentials [24], reducing the need to remember multiple passwords. Similarly, employing adaptive authentication methods that assess the risk level of each login attempt can provide additional security without imposing unnecessary hurdles on users. In addition, providing secure, easy-to-use alternatives to potentially risky behaviors, such as using company-approved cloud storage services instead of unauthorized personal accounts, can help ensure compliance with security policies.

Engaging employees in the security process can also enhance their understanding and adherence to security protocols. Organizations can establish security ambassador programs where selected employees receive advanced training and serve as liaisons between the security team and other staff. These ambassadors can help raise awareness, provide support, and answer questions, making security more approachable and less intimidating. Furthermore, encouraging employees to report suspicious activities or potential security issues can create a sense of shared responsibility and vigilance.

Feedback mechanisms are another valuable tool in fostering a positive security culture. Regularly soliciting feedback from employees about the effectiveness and usability of security measures can provide insights into potential areas for improvement. This feedback can help security teams refine their strategies, making security policies more practical and less disruptive to daily operations. In addition, recognizing and rewarding good security practices can motivate employees to be more vigilant and proactive in maintaining cybersecurity.

8.3.3 Evolving Threat Landscape

In the financial sector, the evolving threat landscape poses substantial challenges that require constant vigilance and adaptation. As a major target for cybercriminals, financial institutions face an array of sophisticated threats, including ransomware attacks, phishing schemes, and data breaches. The frequency and severity of these threats are markedly higher compared to other industries, necessitating a proactive and dynamic approach to cybersecurity.

Ransomware attacks, in particular, have become increasingly prevalent and damaging in the financial sector. Cybercriminals employ ransomware to encrypt critical data and demand substantial ransoms for its release. These attacks can disrupt operations, lead to significant financial losses, and damage an institution's reputation. Financial institutions are attractive targets due to their high-value data and the urgency with which they need to recover from attacks. As ransomware techniques evolve, with attackers using more advanced encryption methods and leveraging double extortion tactics—where they not only encrypt data but also threaten to leak it—financial institutions must continuously enhance their defenses and response strategies.

Phishing attacks also present a major threat to financial institutions. These attacks involve deceptive emails or messages designed to trick recipients into divulging sensitive information, such as login credentials or financial data [10]. Sophisticated phishing campaigns can be highly convincing, often mimicking trusted sources or leveraging current events to increase their effectiveness. The rise of spear phishing, where attackers target specific individuals or departments within an organization, further complicates the challenge. Financial institutions must therefore invest in advanced phishing detection and training programs to equip employees with the skills to recognize and respond to such threats.

Data breaches, another significant concern, involve unauthorized access to sensitive information, potentially exposing customer data, financial records, and proprietary information. The financial sector's stringent regulatory environment adds another layer of complexity to managing data breaches. Regulations such as the General Data Protection Regulation (GDPR) and the Payment Card Industry Data Security Standard (PCI DSS) [25] impose rigorous requirements for data protection and breach notification. Compliance with these regulations is critical, as breaches can result in severe financial penalties and damage to customer trust [26]. Financial institutions must implement robust security measures, including data encryption, access controls, and regular security audits, to safeguard against unauthorized access and ensure regulatory compliance.

The rapid advancement of cyber threats requires financial institutions to be agile and proactive in their security efforts. Traditional security measures are often insufficient to counter new and emerging threats. As cybercriminals develop more sophisticated techniques, financial institutions must continuously update their security infrastructure and adopt cutting-edge technologies to stay ahead of potential threats. This includes investing in advanced threat detection and response systems, such as security information and event management (SIEM) platforms and endpoint detection and response (EDR) solutions, which provide real-time visibility into network activities and enable rapid response to potential incidents.

One of the key challenges in adapting to the evolving threat landscape is managing the balance between resource allocation and security needs. Financial institutions must continually assess their security posture, identify vulnerabilities, and allocate resources effectively to address emerging threats. This ongoing need for adaptation can strain financial and human resources, complicating security management. To mitigate these challenges, institutions often turn to external partners, such as cybersecurity firms and threat intelligence providers, to gain access to specialized expertise and advanced technologies. Collaboration with these partners can enhance threat detection capabilities and provide valuable insights into emerging threats and attack trends.

In addition, integrating threat intelligence into security operations is crucial for staying ahead of cybercriminals. Threat intelligence provides valuable information about the tactics, techniques, and procedures used by attackers, allowing financial institutions to anticipate and prepare for potential threats. By leveraging threat intelligence feeds and analysis, organizations can enhance their ability to detect and respond to threats in a timely manner. However, effectively utilizing threat intelligence requires advanced analytics and expertise, further highlighting the need for skilled cybersecurity professionals and robust security infrastructure.

The evolving threat landscape also emphasizes the importance of continuous monitoring and assessment. Financial institutions must implement comprehensive monitoring solutions that provide visibility into network activities, user behaviors, and system configurations. Regular vulnerability assessments and penetration testing are essential for identifying weaknesses in security defenses and validating the effectiveness of security measures. By maintaining a proactive approach to monitoring and assessment, institutions can detect and address potential vulnerabilities before they are exploited by cybercriminals. This ongoing need for adaptation can strain resources and complicate security management [9].

8.3.4 Regulatory Compliance

The financial sector operates within a framework of stringent regulatory oversight that governs data protection and cybersecurity practices. Compliance with these regulations is not only a legal obligation but also a critical component of maintaining trust and credibility with clients. Implementing endpoint security measures that align with these regulatory requirements presents a multifaceted challenge, particularly in the face of rapid technological advancements and the increasing sophistication of cyber threats.

Financial institutions are subject to a variety of regulations that mandate specific security measures to protect sensitive data. Key regulations include the General Data Protection Regulation (GDPR) in the European Union, which sets forth requirements for data protection and privacy, and the Payment Card Industry Data Security Standard (PCI DSS), which establishes security standards for organizations handling payment card information. In addition, there are industry-specific regulations, such as the Gramm-Leach-Bliley Act (GLBA) in the United States, which mandates safeguarding customer financial information.

Each of these regulations imposes detailed requirements for data protection, including encryption, access controls, data retention, and breach notification procedures. For instance, GDPR mandates that organizations implement technical and organizational measures to ensure a level of security appropriate to the risk. This includes encryption of personal data, regular testing of security measures, and ensuring that personal data is processed securely. Similarly, PCI DSS requires that organizations implement robust security measures, such as firewall configurations, secure password policies, and regular security testing to protect cardholder data.

The challenge of aligning endpoint security measures with these regulatory requirements is compounded by the rapid pace of technological change and the evolving nature of cyber threats. As new technologies emerge, such as cloud computing and mobile devices, and as cyber threats become more sophisticated, regulatory standards may lag behind the latest developments. Financial institutions must continuously adapt their security strategies to address these changes while ensuring compliance with existing regulations.

One significant challenge is the need to implement endpoint security measures that are both effective and compliant. This involves deploying advanced security technologies, such as endpoint detection and response (EDR) systems, security information and event management (SIEM) platforms, and data encryption solutions [26]. These technologies must be configured to meet regulatory requirements and provide robust protection against cyber threats. However, the complexity of managing and integrating these technologies can be daunting, particularly for institutions with limited resources or expertise.

In addition, financial institutions must ensure that their endpoint security strategies are consistently applied across a diverse array of devices and environments. This includes managing security for traditional endpoints, such as desktops and laptops, as well as newer technologies like mobile devices and Internet of Things (IoT) devices. Each type of endpoint may have different security requirements and vulnerabilities, making it essential to implement comprehensive security policies and controls that address the specific needs of each device type.

Compliance with regulatory requirements also involves maintaining thorough documentation and conducting regular audits. Financial institutions must document their security policies, procedures, and practices to demonstrate compliance with regulations. This documentation should include details on how security measures are implemented, how data is protected, and how breaches are handled. Regular audits, both internal and external, are necessary to assess the effectiveness of security measures and identify areas for improvement. These audits help ensure that security practices are in line with regulatory standards and provide evidence of compliance during regulatory inspections.

Another aspect of regulatory compliance is the need for ongoing employee training and awareness. Employees must be educated about the regulatory requirements that apply to their roles and the importance of adhering to security policies and procedures. Regular training programs should cover topics such as data protection, incident reporting, and secure handling of sensitive information. By fostering a culture of compliance and security awareness, financial institutions can reduce the risk of accidental breaches and ensure that employees understand their role in maintaining regulatory compliance.

The process of achieving and maintaining compliance can be resource-intensive, requiring significant investments in technology, personnel, and processes. Financial institutions must allocate resources effectively to balance the need for robust security with the demands of regulatory compliance. This may involve hiring specialized compliance and cybersecurity professionals, investing in advanced security technologies, and developing comprehensive security policies and procedures.

In addition to managing regulatory compliance, financial institutions must also stay informed about changes in regulatory standards and emerging best practices. Regulations are subject to periodic updates and revisions, and organizations must adapt their security practices to reflect these changes. Establishments must ensure that their endpoint safekeeping strategies align with compliance mandates while effectively safeguarding sensitive financial data [9].

8.4 FUTURE DEVELOPMENT IN ENDPOINT SECURITY IN FINANCIAL SECTOR

Future developments in endpoint security within the financial sector are expected to focus on several key trends and technologies that enhance protection against increasingly sophisticated cyber threats. As per Figure 8.2 based on reference [27], over the following five years, it is predicted to increase at an annual pace of 12.93%, reaching a market volume of around USD 26.9 billion. Furthermore, the expected revenue for the Endpoint Security market in 2024 is roughly 14.32 billion US dollars. Most significant anticipated advancements are discussed in subsequent subsections

8.4.1 AI and ML Integration

Combination of AI and ML in endpoint protection solutions is poised to revolutionize how financial institutions identify and respond to dangers. AI can analyze gigantic volumes of data to recognize anomaly patterns symptomatic of digital threats in real time. This proactive approach enables financial firms to enhance their threat detection capabilities and respond more swiftly to potential breaches, thereby reducing the risk of significant financial loss or data compromise [9].

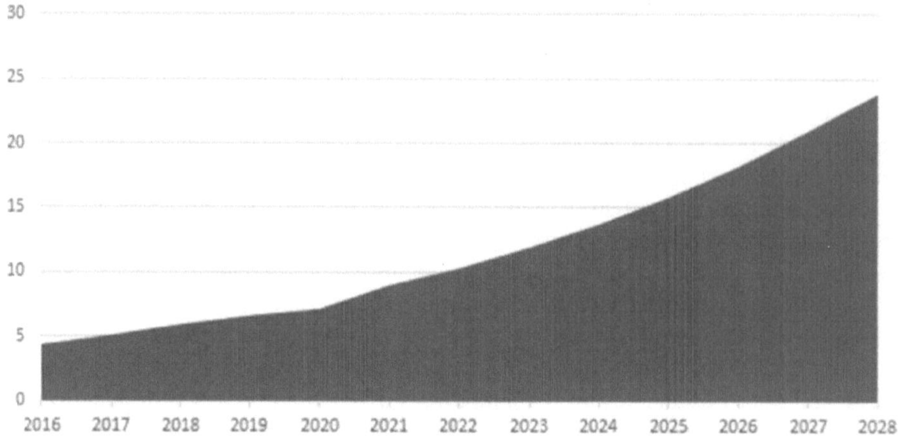

FIGURE 8.2 Endpoint security revenue in billion.

8.4.2 Extended Detection and Response

Adopting extended detection and response (XDR) [28] solutions are becoming increasingly vital for financial institutions. XDR provides comprehensive visibility across various endpoints and networks, integrating data from multiple security tools to improve incident detection and response times. This holistic approach allows security teams to better understand and manage threats, particularly in complex IT environments typical of financial organizations [29].

8.4.3 Focus on Compliance and Regulatory Requirements

As regulatory frameworks evolve, financial institutions need to ensure that their endpoint security strategies align with compliance mandates. This takes account of implementing robust protection measures that meet industry-specific guidelines, such as GDPR. Continuous risk evaluations and changes to security procedures will be required to maintain certification and successfully protect client data [30].

The future of endpoint security in the financial sector will be characterized by integrating AI and advanced analytics, adopting XDR solutions, automating security processes, using blockchain technology, and a strong focus on regulatory compliance. These developments will collectively enhance the ability of financial institutions to safeguard sensitive data and counter to embryonic cyber threats effectively.

8.5 CONCLUSION

The cutting-edge technologies in endpoint security for the financial sector are vital constituent in combating cyber threats. Even though the advanced technologies offer significant advantages in terms of improved threat detection and prevention, they nevertheless, introduce unique challenges related to data privacy, model vulnerability, and algorithmic transparency. Existing research has highlighted the growing importance of endpoint security in the financial sector, with studies showing that a significant portion of cyber-attacks target endpoint devices. Researchers have explored the use of EDR solutions, which leverage AI with ML to detect and respond to risk instantaneously. Implementing AI-powered endpoint security in the financial sector faces several challenges, including data privacy and regulatory compliance concerns. With the unique characteristics, challenges and limitations of AI-powered endpoint security, financial institutions can enhance their cybersecurity posture, ensure regulatory compliance, and safeguard their critical assets and customer data.

To address the challenges and effectively leverage AI-powered endpoint security in the financial sector, several key recommendations are proposed. First, developing robust governance frameworks is essential to ensure the ethical use of AI. This includes forming policies and guidelines that govern the application and management of AI. Second, investing in upskilling security personnel is crucial so they can understand and manage AI-powered solutions effectively. This includes training programs and workshops to enhance their technical skills and knowledge. Third, prioritizing the integration of AI-powered endpoint security with existing security tools and enterprise systems is vital for creating a cohesive and comprehensive security strategy. Fourth, implementing comprehensive data protection and privacy measures is necessary to comply with regulatory requirements and protect sensitive information. Lastly, continuously inspecting and assessing the execution and security of AI-powered solutions is essential to ascertain and mitigate emerging risk. This involves regular evaluations and updates to ensure that the AI systems are functioning optimally and addressing new challenges as they arise.

Therefore, by adopting a comprehensive approach financial institutions can effectively leverage AI and ML to enhance their endpoint security and protect their sensitive data and assets.

REFERENCES

1. C. Adams, "Endpoint Security 101 for Financial Services." Accessed: Jul. 17, 2024. Available: https://www.venturelynkrm.com/blog/endpoint-security-101-for-financial-services.
2. A. Hussain, W. Mark, and A. Toins, "Endpoint Security: On the Frontline of Cyber Risk," Community Banking Connections. Accessed: Jul. 17, 2024. [Online]. Available: https://www.communitybankingconnections.org/articles/2021/i3/endpoint-security-on-the-frontline-of-cyber-risk.
3. Check Point Software, "Cyber Security for Financial Services," Check Point Software. Accessed: Jul. 17, 2024. [Online]. Available: https://www.checkpoint.com/industry/financial-services/.
4. M. Peihani, "Regulation of Cyber Risk in the Banking System: A Canadian Case Study," *Journal of Financial Regulation*, vol. 8, no. 2, pp. 139–161, 2022. doi: 10.1093/jfr/fjac006.
5. T. Maurer, and A. Nelson, "Cyber Threats to the Financial System Are Growing, and the Global Community Must Cooperate to Protect It," *IMF F&D*, 03, pp. 1-4, 2021, [Online]. Available: https://www.imf.org/external/pubs/ft/fandd/2021/03/pdf/global-cyber-threat-to-financial-systems-maurer.pdf
6. O. Gulyás, and G. Kiss, "Impact of Cyber-Attacks on the Financial Institutions," *Procedia Computer Science*, vol. 219, pp. 84–90, 2023. doi: 10.1016/j.procs.2023.01.267.

7. I. Leroy, and I. Zolotaryova, "Insights for Economic Security: Recovery Strategies from Cyber-Attacks," in *2023 13th International Conference on Dependable Systems, Services and Technologies (DESSERT)*, Oct. 2023, pp. 1–7. doi: 10.1109/DESSERT61349.2023.10416490.
8. Statista, "Cyber incidents in financial industry worldwide 2022," Statista. Accessed: Jul. 25, 2024. [Online]. Available: https://www.statista.com/statistics/1310985/number-of-cyber-incidents-in-financial-industry-worldwide/
9. M. Gibbard, "Robust Endpoint Security Strategies for Finance Firms," The Phishing Report. Accessed: Jul. 23, 2024. [Online]. Available: https://thephishingreport.net/robust-endpoint-security-strategies-for-finance-firms/
10. I. H. Sarker, "AI-Driven Cybersecurity and Threat Intelligence: Cyber Automation, Intelligent Decision-Making and Explainability | SpringerLink." Accessed: Jul. 25, 2024. [Online]. Available: https://link.springer.com/book/10.1007/978-3-031-54497-2
11. R. Ward, and B. Beyer, "Beyondcorp: a New Approach to Enterprise Security," *Login:: the Magazine of USENIX & SAGE*, vol. 39, no. 6, pp. 6–11, 2014.
12. S. S. Goswami, S. Mondal, R. Haider, J. Nayak and A. Sil., "Exploring the Impact of Artificial Intelligence Integration on Cybersecurity: A Comprehensive Analysis," 2(2), pp. 73-93, May 2024, doi: 10.56578/jii020202.
13. N. Scaife, H. Carter, P. Traynor, and K. R. B. Butler, "CryptoLock (and Drop It): Stopping Ransomware Attacks on User Data," in *2016 IEEE 36th International Conference on Distributed Computing Systems (ICDCS)*, Jun. 2016, pp. 303–312. doi: 10.1109/ICDCS.2016.46.
14. N. A. Odeh, D. Eleyan, and A. Eleyan, "A Survey of Social Engineering Attacks: Detection and Prevention Tools". *Vol.*, no. 18, 2021.
15. O. Temizkan, R. L. Kumar, S. Park, and C. Subramaniam, "Patch Release Behaviors of Software Vendors in Response to Vulnerabilities: An Empirical Analysis," *Journal of Management Information Systems*, vol. 28, no. 4, pp. 305–338, 2012. doi: 10.2753/MIS0742-1222280411.
16. M. Alam, J. Ferreira, J. Fonseca, Eds., *Intelligent Transportation Systems*, vol. 52. in Studies in Systems, Decision and Control, vol. 52. Cham: Springer International Publishing, 2016. doi: 10.1007/978-3-319-28183-4.
17. M. M. Hossain, M. Fotouhi and R. Hasan, "Towards an Analysis of Security Issues, Challenges, and Open Problems in the Internet of Things," 2015 IEEE World Congress on Services, New York, NY, USA, 2015, pp. 21–28, doi: 10.1109/SERVICES.2015.12.
18. A. A. Aburomman, and M. B. I. Reaz, "A Survey of Intrusion Detection Systems Based on Ensemble and Hybrid Classifiers," *Computers & Security*, vol. 65, pp. 135–152, 2017. doi: 10.1016/j.cose.2016.11.004.
19. IBM, "What is SOAR (security orchestration, automation and response)? | IBM." Accessed: Jul. 22, 2024. [Online]. Available: https://www.ibm.com/topics/security-orchestration-automation-response
20. G. Olaoye, and F. Williams, "Cybersecurity and AI-based threat detection in financial systems," Mar. 2024.
21. N. Rane, S. Choudhary, and J. Rane, "Explainable Artificial Intelligence (XAI) Approaches for Transparency and Accountability in Financial Decision-Making," *SSRN Electronic Journal*, pp. 1-17, 2023. doi: 10.2139/ssrn.4640316.
22. J. Edwards Dr, "Mobile Security," in Mastering Cybersecurity: Strategies, Technologies, and Best Practices, Dr. J. Edwards, Ed., Berkeley, CA: Apress, 2024, pp. 173–221. doi: 10.1007/979-8-8688-0297-3_7.
23. Check Point Software, "The Top 7 Enterprise Endpoint Security Challenges," Check Point Software. Accessed: Jul. 23, 2024. [Online]. Available: https://www.checkpoint.com/cyber-hub/cyber-security/the-top-7-enterprise-endpoint-security-challenges/

24. U. Upadhyay *et al.*, "Mitigating Risks in the Cloud-Based Metaverse Access Control Strategies and Techniques," *International Journal of Cloud Applications and Computing (IJCAC)*, vol. 14, no. 1, pp. 1–30, 2024, doi: 10.4018/IJCAC.334364.
25. M. M. Husin, S. Aziz, M. M. Husin, and S. Aziz, "Navigating Fintech Disruptions: Safeguarding Data Security in the Digital Era," https://services.igi-global.com/resolvedoi/resolve.aspx?doi=10.4018/979-8-3693-3633-5.ch007. Accessed: Jul. 29, 2024. [Online]. Available: https://www.igi-global.com/gateway/chapter/www.igi-global.com/gateway/chapter/351513
26. J. Edwards Dr, *Mastering Cybersecurity: Strategies, Technologies, and Best Practices.* Berkeley, CA: Apress, 2024. doi: 10.1007/979-8-8688-0297-3.
27. Statista, "Endpoint Security - Worldwide | Statista Market Forecast," Statista. Accessed: Jul. 25, 2024. Available: https://www.statista.com/outlook/tmo/cybersecurity/cyber-solutions/endpoint-security/worldwide.
28. L. Dekel, I. Leybovich, P. Zilberman, and R. Puzis, "MABAT: A Multi-Armed Bandit Approach for Threat-Hunting," *IEEE Transactions on Information Forensics and Security*, vol. 18, pp. 477–490, 2023. doi: 10.1109/TIFS.2022.3215010.
29. A. M. Freed, "Securing the Financial Sector Now and Into the Future with XDR." Accessed: Jul. 23, 2024. [Online]. Available: https://www.cybereason.com/blog/securing-the-financial-sector-now-and-into-the-future-with-xdr.
30. Safecore, "Cyber security in the financial sector: scenario, risks and future challenges | Safecore." Accessed: Jul. 23, 2024. Available: https://safecore.io/en/industry/cyber-security-in-the-financial-sector-the-scenario-future-risks-and-challenges/.

9 Cloud Security and Artificial Intelligence

Mansi Sharma, David Raymond, Induni Weerarathna, Praveen Kumar, and Abeny Ramadan Chadar

9.1 INTRODUCTION

Cloud computing was invented by internet service providers (ISPs) to support the maximum number of users and elastic services using minimal resources. In just a few years, emerging cloud computing has become the most popular technology. The evolution of cloud computing has progressed from an internal IT system to a public service, from a cost-saving tool to a revenue generator, and from an ISP to a telecom, beginning with the publication of core papers by Google in 2003, the commercialization of Amazon EC2 in 2006, and the service offering of AT&T Synaptic Hosting. This chapter presents the concept, history, advantages and disadvantages of cloud computing, as well as the value chain and standardization efforts [1]. The way cloud computing is used has completely transformed how data is stored, managed, and processed by both individuals and organizations. Cloud computing provides flexible and versatile resources through the internet, reducing the need for substantial investments in on-site hardware. Cloud computing uses the internet to deliver computing services, allowing users to access and use data storage, processing power, and software applications as required. As with utilities like electricity or water, cloud service providers (CSPs) usually supply these services and charge depending on consumption [2].

Security in the cloud involves a collection of rules, restrictions, protocols, and tools aimed at safeguarding data, applications, and infrastructure linked to cloud computing. It deals with different security issues, including unauthorized access, data loss, data breaches, and adherence to regulatory mandates. Cloud security encompasses a wide array of methods and tactics, such as data encryption, identity and access management (IAM), threat detection and response, network security, and compliance management [3, 4]. In today's computerized world, maintaining solid cloud security is greatly imperative because it plays a crucial part in defending touchy information, maintaining administrative compliance, and diminishing cyber dangers.

With organizations progressively utilizing the cloud to store and handle huge volumes of sensitive information, it is vital to execute strong security measures to anticipate unauthorized access and information breaches. It is basic to follow to lawful and regulatory commitments such as GDPR and HIPAA to maintain a strategic distance from confronting critical fines and legal results. Successful cloud security measures are basic for guaranteeing continuous trade operations by minimizing downtime and

DOI: 10.1201/9781003521020-9

disturbances caused by cyber-attacks or information loss [10]. Also, it provides strong belief with clients and partners illustrating a devotion to securing data. Besides, contributing in cloud security can result in significant fetched savings by avoiding financial repercussions related with security occurrences. Keeping up assurance is significant as foundation scales quickly, requiring adaptable security measures. Defending computerized resources, maintaining organizational notoriety, and keeping up operational strength makes cloud security crucial [5].

9.2 ROLE OF AI IN ENHANCING CLOUD SECURITY

The use of artificial intelligence (AI) is revolutionizing cloud security through the introduction of sophisticated capabilities that significantly enhance cybersecurity posture overall, threat detection, and response. AI plays a critical role in the complex digital world of today, as businesses are depending more and more on cloud services to store, analyze, and manage vast amounts of sensitive data. The ability of AI to handle and evaluate massive volumes of data in real time is its primary contribution to cloud security. Because of their size and complexity, traditional techniques for threat identification and security monitoring sometimes find it difficult to keep up with cloud settings [6]. AI-powered machine learning techniques are very good at finding patterns and abnormalities in data to detect potential security risks. They can look at endless sums of information, including as network traffic, user behavior, and logs, to distinguish anomalous action or takeoffs from the standard. Besides, AI moves forward the accuracy and viability of risk location by ceaselessly learning from modern information.

As AI models retain and assess more information over time, they get superior at recognizing both known and obscure dangers. By being proactive and quickly recognizing and tending to cyberattacks, businesses may decrease potential harm and the impact of security episodes. AI is fundamental for automating and upgrading security operations in cloud situations, not fair for threat detection [7]. AI-driven arrangements have the potential to computerize routine tasks like powerlessness evaluations, fix administration, and system monitoring. This may move forward operational proficiency and ensure that security measures are actualized reliably over complicated cloud frameworks. AI-powered analytics offer comprehensive experiences into security occasions and events by joining different data sources and distinguishing the fundamental causes of security breaches. This makes a difference in businesses which distinguishes the root causes of defense-related vulnerabilities and deficiencies, enabling them to create more effective cybersecurity arrangements and utilize assets wisely to lower risks [8].

We currently access and use computing resources in a completely new way because of cloud computing. The foundation of cloud services is a sophisticated, yet well-planned architecture that works behind the scenes. An architecture for cloud computing may be seen as a tiered model, where each layer contributes in a different way to the overall smooth operation of the cloud [9]. The frontend represents the user's perspective. It includes laptops, desktop computers, tablets, and smartphones as well as other gadgets used to communicate with cloud services. The user interface (UI) that enables consumers to access and make use of cloud services or apps is housed on these devices. The backend is the central component of the cloud, unseen by the user but essential to service delivery. And it contains various sub-layers. The cloud's physical base is made up of cloud infrastructure layers. It consists of a sizable pool of virtualized resources that are all under the cloud provider's management, including servers, storage, and networking hardware. The resources are distributed and expanded dynamically in response to user requirements, guaranteeing maximum efficiency. The software environment that cloud apps run in is managed by cloud runtime layer. It consists of virtualization software (e.g., KVM or Hyper-V) that builds containers and virtual machines (VMs) for effectively separating and running applications. Large volumes of user, application, and system data can be stored using the resources provided by storage layer. Different types of storage solutions can be distinguished, including file storage, object storage, and block storage [10].

There are various services which are the backbone of cloud architecture as shown in Figure 9.1 such as software as a service (SaaS), infrastructure as a service (IaaS), and platform as a service (PaaS). Access to software programs distributed via the internet is made possible by the SaaS strategy. Users may simply access the program using a web browser or a specialized client application; they do not need to install or maintain it. The whole application, including security and upgrades, is managed by the cloud provider. Users may access virtualized computer resources like servers, storage, and networking whenever they need them with IaaS paradigm. Because they have complete control over these resources, users may set up and oversee their own cloud-based IT infrastructure. PaaS paradigm provides a cloud application

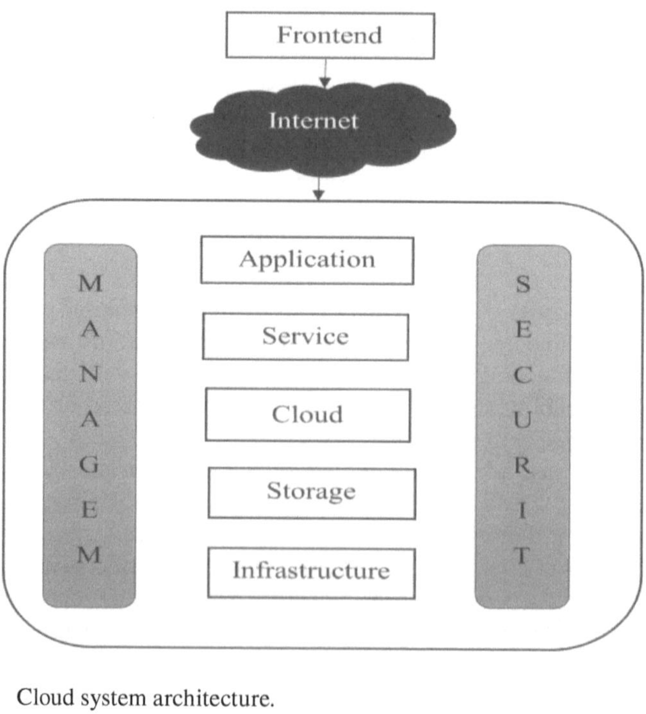

FIGURE 9.1 Cloud system architecture.

development and deployment platform. By giving users access to vital resources including operating systems, databases, middleware, and development tools, PaaS frees users up to concentrate on creating and implementing applications rather than maintaining the underlying infrastructure [11]. The smooth communication between these levels is what makes cloud computing so magical. The request is sent over the network to the cloud provider's backend when a user engages with a cloud application via the frontend. On the basis of the request, the management layer then allocates resources from the cloud infrastructure. The program is executed in a virtual environment created by the cloud runtime, which uses storage to access data. The chosen service model (IaaS, PaaS, or SaaS) provides the functionality that is asked for. The security layer guarantees system protection and data integrity throughout this procedure [12].

There are several benefits to a well-designed cloud computing infrastructure. The capacity to simply scale up or down cloud resources in response to demand removes the requirement for a one-time hardware and software investment. A wide range of alternatives are provided by cloud services, enabling consumers to select the service model that best meets their requirements. Since users only pay for the resources they use, costly software licensing and hardware upkeep are not required, CSPs guarantee less downtime and data loss by providing high availability and disaster recovery procedures. Compared to many on-premises implementations, cloud providers offer a more secure environment since they substantially invest in security measures [13].

Table 9.1 shows the comparison between traditional and AI-driven security in cloud as AI-driven security in cloud is efficient for the large scaled data and for preventing data from the cyber threat. Today, a huge amount of confidential data is at risk. It can be affected by malware and ransomware types of cyberattacks. These attacks can lead to data leaks. To ensure better security, an appropriate security solution is needed. Table 9.1 shows different types of security solutions. This helps in making a decision about choosing the best security solution. A good security solution enhances data confidentiality.

9.3 CLOUD COMPUTING FUNDAMENTALS

A model for providing on-demand internet access to computer resources is called cloud computing. Imagine having unlimited access to apps, storage, and processing power without having to worry about maintaining the physical infrastructure [20, 21]. There are various characteristic of cloud computing to be formed in real time.

9.3.1 Virtualization and Abstraction

Virtual resources are separated from physical computer resources, such as servers, storage, and networks. It is possible to dynamically provide and manage these virtual resources without relying on the underlying hardware. Because of this, customers may obtain processing power without being concerned about the infrastructure [22].

TABLE 9.1
Comparison of Traditional Security vs. AI-Driven Security in Cloud Environments

Sr. no	Traditional Security	AI-Driven Security
1	Rule-based security depends on signature detection and pre-established security policies to recognize and stop attacks.	Large-scale data is analyzed by machine learning algorithms, which then use the results to find trends, anticipate dangers, and automate security actions [14].
2	Effective against known threats with proven signatures; well-known and recognizable to security specialists; provides precise control over security rules, enabling customized access control.	Continually picks up new skills and adjusts to changing viruses and attack methods. frees up security staff to work on strategic objectives by automating security activities [15].
3	May provide a large number of false positives, which can cause security flaws and alert fatigue. is unable to expand to accommodate the enormous volume of data produced in cloud systems.	It needs training data and continuous observation to guarantee efficacy and accuracy. may be difficult to manage and implement, needing certain expertise [16].
4	May be difficult to manage and implement, needing certain expertise. In certain situations, the restricted explainability of AI judgments might impede openness and confidence. vulnerable to bias in the training set, which might cause erroneous threat identification.	Detecting and preventing threats proactively. Automating repetitive security procedures [17].
5	lists of access controls for user authorization. Use firewalls to stop unwanted traffic.	UEBA (user and entity behavior analytics) is used to find unusual activities. Utilizing network traffic analysis (NTA), one may spot questionable network activity [18].
6	While traditional security will always be important, its efficacy will depend more and more on how well it integrates with AI-driven solutions.	It is anticipated that AI-driven security, which provides automated security management and sophisticated threat detection, would become increasingly significant [19].

9.3.2 Pooling of Resources and Multi-Tenancy

Multiple users (also known as tenants) can be served simultaneously via cloud infrastructure. Tenants are given dynamic allocations of shared resources, such as CPU, memory, and storage, according to their usage. Cloud providers benefit from economies of scale and efficient utilization of resources [23].

9.3.3 Self-Service on Demand

Users have the ability to provision and manage cloud resources through an API or self-service interface, granting them significant power and flexibility while eliminating the need for human IT intervention [24].

9.3.4 Wide-Ranging Network Connectivity

Through an API or self-service interface, users can provision and manage cloud resources, giving them considerable power and flexibility while removing the necessity for human IT intervention [25].

9.3.5 Service Metrics

Cloud providers track and measure the number of resources used by each tenant. A pay-per-use approach that bills consumers based on their actual use is encouraged, as is cost efficiency [26].

9.3.6 Quick Elasticity

Cloud resources may be quickly scaled up or down to meet demand fluctuations. As a result, customers may adapt to changing workloads without having to overprovision resources or make significant upfront investments [27].

9.3.7 Elevated Accessibility and Dependability:

Cloud providers design their infrastructure with high availability and fault tolerance in mind. Redundancy techniques ensure that services will continue to run even if certain hardware components malfunction [28].

9.4 CLOUD DEPLOYMENT MODELS

9.4.1 Private Cloud

Businesses that want more control and protection over their data might choose the private cloud option. Private clouds are inaccessible to the general public, in contrast to public clouds. Because of this exclusivity, businesses are able to customize the cloud to meet their own requirements, taking care of issues like security and bandwidth constraints that may occur with public cloud services. When compared to using pre-existing resources in a public cloud, building and maintaining private clouds might be more expensive, even if they offer greater freedom regarding ownership, operation, and administration (internal or external). Furthermore, private cloud management calls for specialized IT knowledge that may not be easily found inside an organization [21].

By the examination of private clouds [28] it can be said that they come in two varieties: on-premise private cloud, which is also referred to as a "internal cloud," is housed in the data center of an enterprise. It offers safety and a more consistent approach, but its size and scalability are frequently constrained. In addition, with this paradigm, the capital and operating expenditures for the physical resources would fall on the IT department of a company. The ideal applications for on-premise private clouds are those that need total control over the infrastructure and security configuration [29]. Externally hosted private cloud, an external cloud computing provider hosts this private cloud architecture. The service provider offers a private cloud environment that is exclusive and fully guaranteed. Organizations who would rather not use a public cloud infrastructure are advised to adopt this format because of there

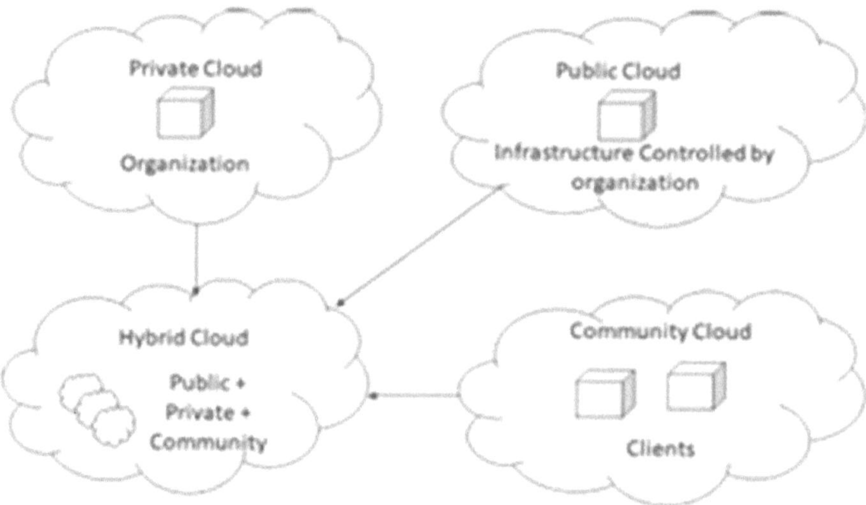

FIGURE 9.2 Deployment model of cloud computing mainly public, private, hybrid (combination of public and private), and community cloud.

are many challenges associated with sharing of physical resources. Figure 9.2 shows how cloud will work as public, private, and hybrid.

9.4.2 Public Cloud

Anyone with an internet connection may easily access a wide range of features and services offered by public cloud computing. It is therefore a well-liked option for companies of all kinds. The pay-as-you-go concept allows customers to only pay for the resources they really utilize, in contrast to typical IT systems. This results in considerable financial savings. With almost infinite processing and storage capacity, public clouds enable enterprises to simply scale their requirements up or down as needed. Additionally, the cloud provider manages all infrastructure, upgrades, and maintenance, freeing up the IT team of the organization to concentrate on core business operations. A public cloud is an appealing alternative for companies searching for a quick and simple solution because it is straightforward to set up and requires little initial expenditure. Although security is a top priority, public cloud providers protect customer data with strong security methods including access limits and authentication. Prominent public cloud services include Microsoft Azure, IBM Blue Cloud, Google App Engine, Amazon EC2, and IBM Blue Cloud. All things considered, the public cloud provides easily scaled, reasonably priced, and easily accessible means for companies to utilize computer resources and services over the internet [21, 30].

9.4.3 Hybrid Cloud

A distinct method of using cloud computing is provided by the communal cloud. It serves a certain set of organizations with comparable needs and demands, such as local government agencies. By sharing resources like processing power and storage, this strategy enables these organizations to possibly save money in comparison to individual public cloud subscriptions. The community offers flexibility based on resources and experience, with the option to operate the infrastructure in-house or contract it out to a third party supplier. However, cost dispersion and shared administration are not without challenges. The security needs of distinctive educate may contrast, and guaranteeing framework compatibility inside the community cloud may give challenges. Generally speaking, the community cloud gives organizations arranged to address security issues with a adjust between cost, control, and particular necessities [31].

9.4.4 Community Cloud

The community cloud is made for participation and offers shared framework to businesses with comparable requirements. Think about research institutions and colleges trading apps and capacity. This strategy diminishes IT costs by pooling assets and maybe avoids the overhead of numerous public cloud memberships. Depending on their level of ability, the educators may select to handle things themselves or enter into a contract with a cloud benefit supplier, who offers adaptability. There is a trade-off, though. Indeed, if it's less expensive, it's crucial to require under consideration any differences in security necessities over institutions and make beyond any doubt that different frameworks within the community cloud work together [32].

9.5 SECURITY CHALLENGES IN CLOUD ENVIRONMENTS

Cloud security necessitates ongoing caution. Although cloud systems provide benefits, there are a number of security dangers associated with them. Misconfigurations, inadequate access control, and a lack of general system visibility are the most frequent ones. Vulnerabilities can also be caused by unregulated cloud services, insider attacks, data breaches, and insecure APIs. Lastly, material that is not encrypted is susceptible to interception. Organizations may safeguard their cloud environment by being aware of these typical dangers [4]. Globally, new technology suppliers and consumers continue to struggle with security issues. The recent Cambridge Analytical data breach, which revealed that over 86 million Facebook users' personal information had been improperly and unapproved utilized, is proof of certain security flaws in most modern technology and I.T. platforms. The use and dissemination of cloud computing technology are being impeded by security concerns. This is because privacy concerns are making a lot of consumers quickly lose faith in the cloud [33]. The security risks associated with cloud computing might prevent consumers from reaping the rewards of this innovative technology. Cloud dangers do not exclude users in the educational sector. These dangers frequently come from the network mediums that the client uses to access the cloud service as well as the cloud infrastructure itself [34].

According to studies, misconfigurations are the most common vulnerability in cloud security. According to the NSA's cloud security study, this is frequently caused by development teams that don't properly comprehend security best practices or don't do enough peer review. Serious repercussions may ensue, from unapproved access to total system failures. The research paper in reference [35] presents the findings of a Delphi survey that focuses on the most significant concerns that businesses have when deciding whether or not to utilize cloud computing. A Delphi panel comprising 34 experts with varying domain backgrounds participated. Divided into three subpanels, the panelists were IT and cloud computing professionals who represented a diverse range of clients, suppliers, and academics. Three steps made up the Delphi process: ideation, refining, and rating. In the first step, the panelists selected 55 topics of concern. These were then assessed, categorized, and arranged into 10 groups: security, strategy, legal and ethical, IT governance, migration, culture, business, availability, impact, and awareness. After ranking the top 18 issues in each subpanel, a moderate intra-panel consensus was reached. The experts were also questioned 16 more times in order to gain a better grasp of the problems and the reasons for the importance of some problems over others [35]. A substantial amount of research on Google Scholar confirms that data leaks are still a top worry in cloud security. Attackers use a variety of strategies, such as phishing and cloud storage flaws, to obtain private data. In order to reduce the danger of data breaches, scholarly studies highlight the necessity of strong data encryption solutions, both in transit and at rest [36]. Table 9.2 shows the list of cloud security vulnerabilities with its potential impact and how data will be affected from cyber-attack and what kind of problem may arise will be shown.

TABLE 9.2
List of Common Cloud Security Vulnerabilities with Descriptions and Potential Impacts

Sr. No.	Vulnerability	Description	Potential Impact
1	Data breaches	Attackers using flaws, phishing, or other techniques to take private data.	Sensitive data loss, financial losses, reputational harm, and legal ramifications [37].
2	Lack of encryption	Unencrypted data in transit or at rest, which leaves it open to misuse and interception.	Data breaches, sensitive information exposure, and noncompliance with regulations [38].
3	Shadow IT	Improper usage of cloud services that is not under the authority of the IT department and may not follow security guidelines.	Greater attack surface, possible data breaches, and challenges implementing security regulations [39].
4	Misconfigurations	Improper configuration of cloud resources, making them vulnerable to hacking or broken down.	Unauthorized access, data breaches [40].
5	Poor access management	Users with too many rights, insufficient MFA, or weak passwords giving them more access than they require to cloud resources.	Enhanced danger of data breaches, privilege escalation, and illegal access [41].
6	Lack of visibility	Cloud activity is difficult to observe because of its opaque and complicated structures.	Inadequate identification and handling of security risks [42].

9.6 AI-Driven METHODS FOR ENHANCING CLOUD SECURITY

AI-driven risk detection tackles a significant development in cybersecurity by addressing the ability of AI to support defense systems against more sophisticated threats. Through the integration of intricate AI algorithms, computer-based intelligence-driven systems adeptly analyze vast quantities of data on a regular basis, swiftly identifying and mitigating possible cyber threats. This approach takes a proactive attitude and finds patterns, abnormalities, and irregularities in system logs, network traffic, and user activities [43].

Cloud security is undergoing a revolution because of the AI, which provides proactive techniques for stopping, identifying, and handling cyberattacks. AI is particularly good at sifting through large volumes of data to find trends and abnormalities that human security measures might overlook. This enables AI to anticipate and thwart a range of threats, including as phishing attempts, malware, DDoS assaults, and zero-day vulnerabilities. Figure 9.3 shows observing user behavior and network traffic patterns. AI may also detect insider threats and advanced persistent threats (APTs).

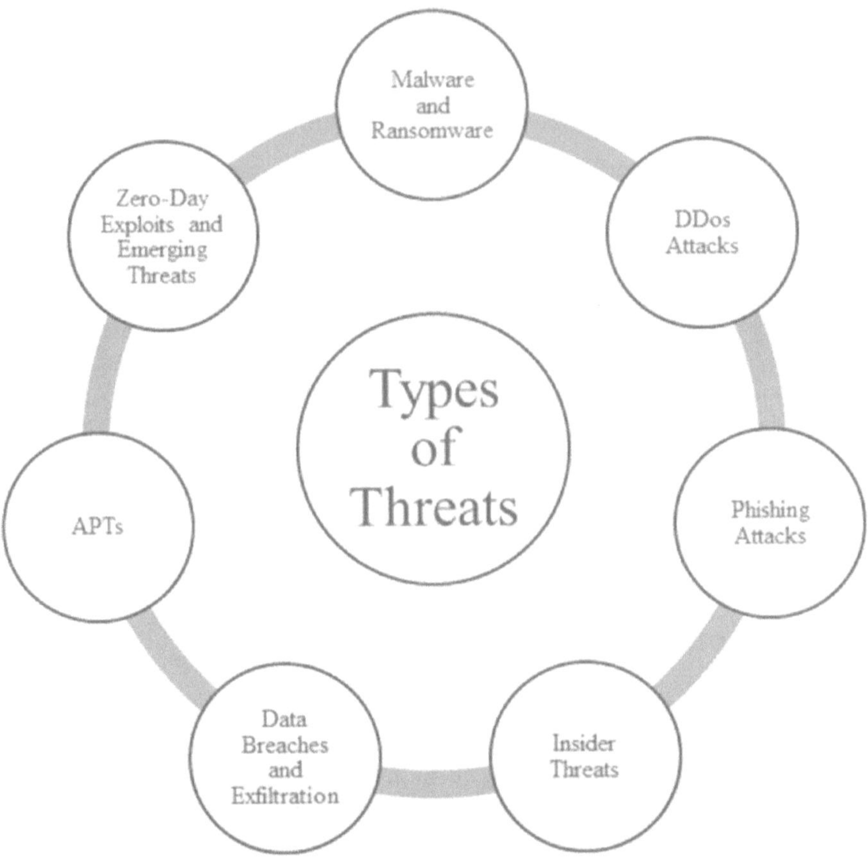

FIGURE 9.3 Types of threat addressed by artificial intelligence.

Moreover, AI can automatically mitigate the effects of security incidents by quarantining infected devices or screening hostile communications. Businesses can keep ahead of the changing threat landscape and dramatically improve their cloud security posture by utilizing AI [44].

9.7 THREAT PREVENTION USING AI

9.7.1 Predictive Analytics for Threat Prevention

Consider a security framework that effectively predicts dangers instead of just reacting to them. Predictive analytics are empowered by AI accurately. Through a careful examination of past assault information and risk insights, AI can recognize designs and uncover the strategies utilized by recognized risk actors. It may at that point utilize this data to estimate upcoming assaults, empowering companies to require preventative activity. AI is competent of assessing helplessness reports and prioritizing fixing agreeing to the possibility of an misuse. By doing this, the attack surface is reduced and major vulnerabilities are addressed first. AI can recommend tighter access limits for critical data and systems based on user behavior and historical security breaches [45, 46].

9.7.2 Real-time Monitoring and Anomaly Detection

Novel attacks can circumvent signature-based detection, a common feature of traditional security methods. AI, on the other hand, employs a different approach that combines real-time monitoring with anomaly identification. AI monitors network data, searching for anomalies that deviate from typical user behavior or patterns. Unusual attempts to log in, questionable data transfers, or abrupt spikes in network traffic might all be indicators of this. Like it can with network traffic, AI can also monitor irregularities in user behavior. Unusual access times, attempts to access data without permission, or a sudden increase in activity from a certain user account might all be relevant variables [47]. Table 9.3 states the AI-based algorithms and their applications used for preventing cyber threats, like malwares, phishing attacks, on confidential data. The subset of AI will be used to prevent data from the Malware, phishing attack, etc. social media threat will be detected using the NLP and different type of Insider threat detection will be done through the UEBA.

9.8 THREAT DETECTION WITH AI

9.8.1 Machine Learning Models for Identifying Malicious Activities

Network intrusion detection systems, or NIDSs, have been developed. Their challenging mission is to gather data in order to create an intelligent NIDS that is capable of accurately detecting both known and unknown assaults. This research suggests an anomaly detection method for IICSs based on deep learning models that may use data gathered from TCP/IP packets to train and validate in order to overcome this difficulty. It consists of a series of training steps carried out with a

TABLE 9.3
Examples of AI Algorithms Used in Threat Prevention with Their Specific Applications

Sr. No	Algorithms	Applications	Description
1	Supervised machine learning	Malware and phishing detection	It examines a large quantity of data in order to find patterns linked to phishing attempts or known malware [48].
2	Unsupervised machine learning	Real-time network traffic monitoring	It examines network traffic patterns in order to identify any unusual activity. finds abnormalities, such abrupt traffic surges or strange data transfers, that might point to a cyber-attack [49].
3	Natural language processing	Social media threat detection	It examines postings on social media and online discussions to find threats, such planned assaults or attempts to attract new members. aids in keeping companies ahead of any internet dangers [50].
4	User and entity behavior analytics	Insider threat detection	It examines user behavior to spot oddities that could point to a compromised account or malevolent intent. aids in preventing insider threats from contractors or employees [51].

deep feedforward neural network architecture and deep auto-encoder, assessed on two popular network datasets: NSL-KDD and UNSW-NB15. Figure 9.4 shows that the ADS proposed deployment model, experimental findings outperform eight previously published strategies in terms of detection rate and false positive rate, this methodology might be utilized in real IICS environments. Organizations may save a lot of money by automating security activities and enhancing threat detection. Through proactive threat identification and mitigation, firms may steer clear of the financial consequences associated with data loss, downtime, and security breaches. Lower operating expenses result from the decreased requirement for human intervention in repeated processes [50–52].

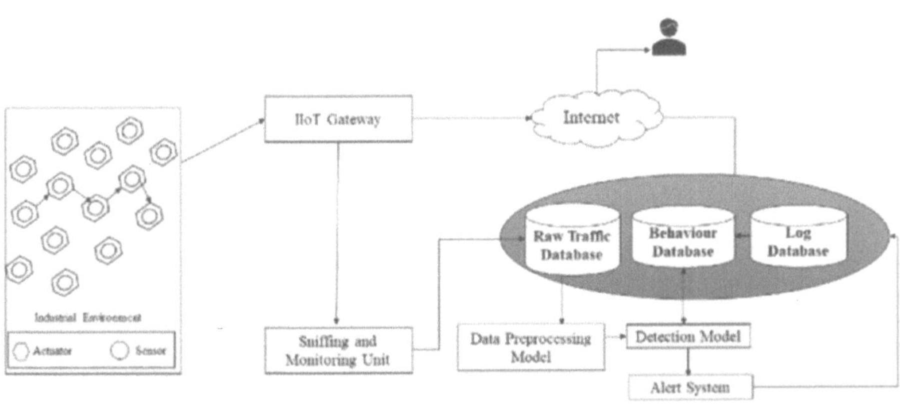

FIGURE 9.4 Suggest ADS deployment structure.

9.9 AI IN INCIDENT RESPONSE

A methodical strategy is necessary for digital forensics investigations to guarantee that evidence is gathered and examined efficiently. Following ACPO rules, this procedure consists of four main components. First, using best procedures, investigators acquire evidence at the site. The gathered data is then safely saved on portable devices in the form of digital files or logs. After that, scientists examine the data using clustering algorithms. In order to do this, fresh data must be found, compared to the information already in existence, and then arranged for more study. Clustering facilitates the discovery of latent relationships and patterns in the data. Ultimately, researchers examine the clustering model's output to comprehend the connections among various data sets. This all-encompassing strategy guarantees the integrity of the evidence and offers insightful information for developing a compelling case [53, 54]. Figure 9.5 shows the proposed model by the authors Hasan et.al which shows the learning process of model. Table 9.4 shows the comparison of response times and effectiveness before and after AI implementation. There are various challenges will be faced by before implementation of AI in response times and effectiveness like before manually work will be done after implementing AI it suggest that automation system in response times. alert also examine by manually now seems to be automated. And fast detection will be done.

9.10 AI APPLICATIONS IN DIFFERENT CLOUD DEPLOYMENT MODELS

An examination of over 500 academic publications on proactive cloud and predictive technologies scheduling of resources. Next, because the first critical step in developing a prediction model is identifying relevant and comprehensive datasets, we offer some statistics on the most popular cloud datasets that were found through this investigation. In the event that no comprehensive datasets are available, we once more offer a few often used cloud benchmarks for workload trace development. It is crucial to define a prediction model's determining objective since doing so leads

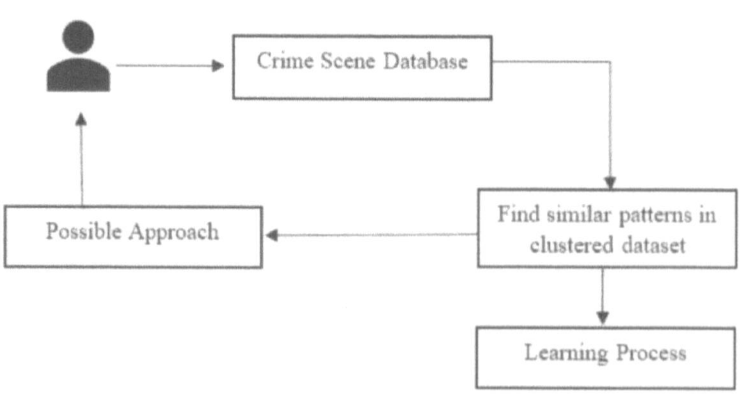

FIGURE 9.5 AI in incident response.

TABLE 9.4
Comparison of Response Times and Effectiveness Before and after AI Implementation

Sr. No.	Feature	Before AI Implementation	After AI Implementation
1	Response time	The process of manually analyzing warnings and occurrences may be laborious and slow. An excessive number of notifications may be too much for security professionals to handle at once.	Real-time data analysis using AI makes it possible to identify and rank dangers more quickly. Quicker reaction times save downtime and possible harm [55].
2	Workload	Security staff are overburdened by the manual examination of every alert, routine duties take up time that might be used for preventive security measures. Excessive stress can result in burnout and mistakes made by people.	Routine chores are automated by AI, freeing up security staff for important occurrences [56].
3	Threat prediction	Imitated capacity to foresee upcoming assaults; depends on security knowledge and past data analysis	To forecast possible attack vectors, we may examine past data and threat information. This allows us to spot trends and warning signs of new dangers [57].
4	Scalability	Human resources restricted by team composition and experience, challenges in growing response operations in the event of large-scale assaults.	AI is scalable to manage high data and alert volumes. It allows for a successful reaction even in the face of intricate or pervasive threats [58].
5	Compliance	Upholding security rules may be difficult and time-consuming.	By automating data analysis and reporting, AI can aid with compliance, guarantees accurate and timely reporting of security issues [59].

to crucial scheduling decisions. Figure 9.6 shows a percentage classification of the cloud systems' most anticipated elements [60].

The data center lab discussed in [59] have tested a hybrid AI application. In this case, a customer can choose to effectively move sensitive data into a cloud provider's proprietary large language model (LLM) by using data lakes in addition to using their data center for safely handling sensitive data. The cloud-based LLM's seamless interaction with the client's data enables modifications, improving the system's capacity to produce results that are more pertinent and accurate. The LLM can be reintegrated into the on-premise data center when it has been adjusted to match the client's demands. With this usage, utilizing AI for exercises like automated decision-making, predictive analysis, and content generation is made less complex. It does this whereas securing the protection of information and utilizing the complete potential and versatility of cloud AI technology. The upgraded adaptability and capabilities of

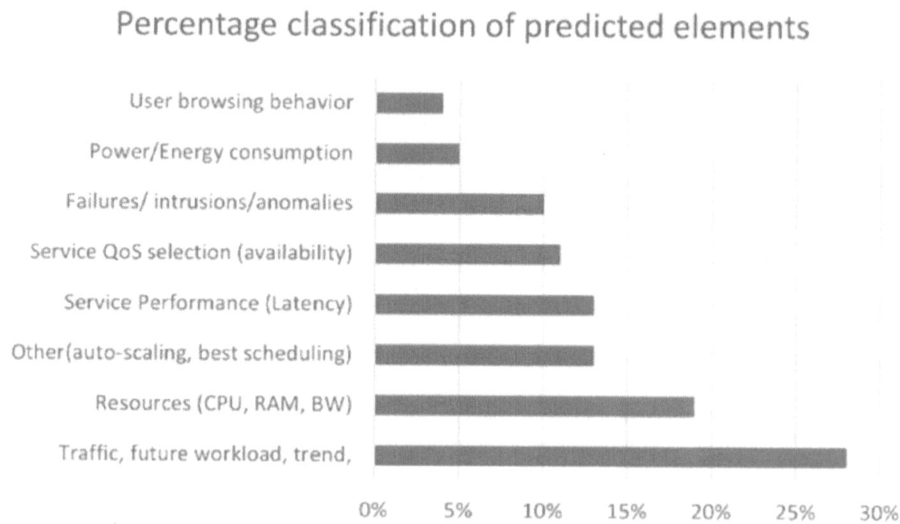

FIGURE 9.6 Cloud presented element categorization as a percentage.

cloud-based AI arrangements are combined with the security and control that come with on-premise foundation in this hybrid architecture. By utilizing the benefits of both on-premise and cloud innovations, businesses may effectively consolidate AI through the utilization of this strategy. By finding an adjustment between encouraging innovation and maintaining data integrity, businesses can stay competitive in the rapidly changing computerized scene of current [57]. Figure 9.7 explores the hybrid AI usage and propriety of LLM.

9.11 CHALLENGES AND FUTURE DIRECTIONS AND TRENDS

Although there are still certain obstacles to be solved, AI has the potential to completely transform cloud security. Since AI models need to be trained on enormous amounts of data, data privacy may provide a significant obstacle. Ensuring the confidentiality and security of sensitive data is essential for user confidence and compliance. Moreover, the decision-making forms of AI models may become dark since to their complexity. It is pivotal to comprehend how AI recognizes and handles dangers in security situations. Moreover, unfriendly actors may utilize AI through adversarial assaults, tricking it with wrong information to set up wrong cautions or ignore genuine dangers. AI and cloud security have shinning futures ahead of them. Contributing in AI-powered arrangements and receiving these patterns may help enterprises make a cloud environment that's more economical, reliable, and secure. This cooperative strategy, in which AI and humans collaborate, enables firms to remain ahead of the constantly changing panorama of cyber threats. Utilizing AI's capacity is now essential for guaranteeing the security and prosperity of businesses in the digital era, as cloud use keeps rising. A trained staff with knowledge of both security and AI is also essential to the success of AI security solutions [61].

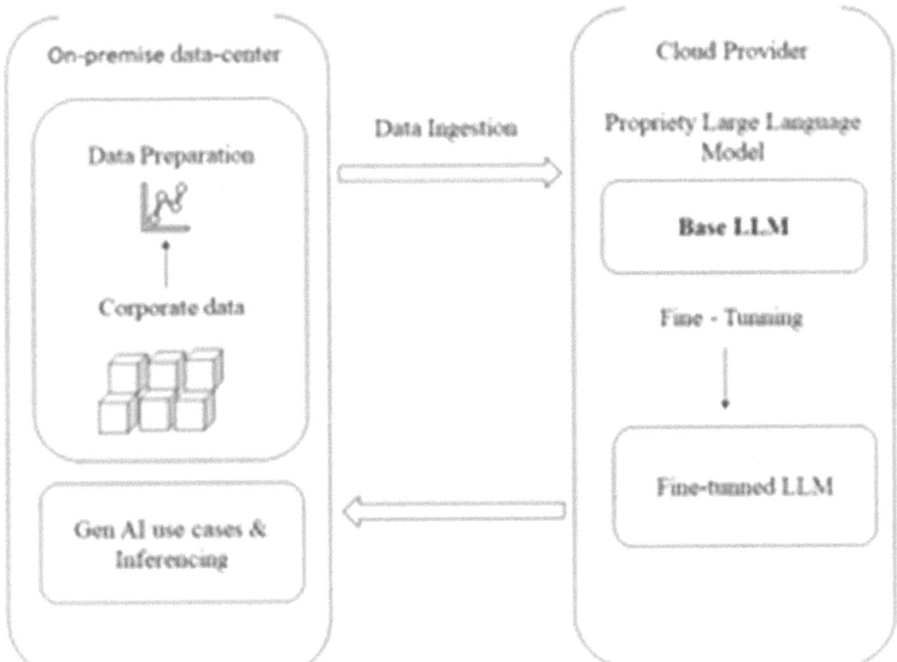

FIGURE 9.7 Hybrid AI usage.

It is anticipated that humans and AI will work together more often. While humans will still be in charge of making final decisions and providing oversight, AI will undertake the labor-intensive task of danger identification and analysis. There will be an emphasis on developing "Explainable AI" models that are more transparent so that security experts can understand the AI's logic. Improving cloud security solutions will need federated learning, a method for training AI models on decentralized datasets without sacrificing privacy. AI-powered security automation will automate monotonous processes like incident response, vulnerability scanning, and log analysis, freeing up human time for strategic work. Lastly, massive security data streams will be continually analyzed by AI, allowing for the anticipation and defense against cyberattacks [62]. Data privacy presents one of the main security problems for AI. Extensive data is needed for AI models to be trained effectively. Federated Learning provides an answer. With the use of this method, dispersed datasets from many places may be used to train AI models without compromising data privacy. Without explicitly sharing the data, each network member trains the model using their own local data. This collaborative approach promotes the development of more resilient AI security solutions for the cloud while adhering to data privacy laws [61]. The increasing reliance on cloud computing for apps, processing power, and storage necessitates a solid security posture. Traditional security measures are often insufficient due to the sophistication and ongoing evolution of cyber threats. Usually where cloud security-revolutionizing AI comes into its own as a powerful companion. Cloud frameworks may be upgraded for more prominent security and flexibility inside businesses by joining AI and human abilities [4].

AI and cloud security have a bright future ahead of them, with numerous important trends and directions already apparent. An significant development is the growing use of AI to predictive threat intelligence, which helps businesses detect and mitigate security issues before they manifest. Since it takes less time and effort to handle security concerns, AI-driven incident response process automation is also becoming more and more popular. Cyberattack damage may be reduced and recovery periods shortened via automation. Furthermore, by providing immutable records of transactions and events, combining AI with cutting-edge technologies like blockchain may improve cloud security frameworks and boost security. The creation of explainable AI (XAI), which seeks to make AI models more transparent and understandable so that security experts can understand them [63].

In the future, AI and cloud security could enhance regulatory compliance. Automating data classification and maintaining clear audit trails, AI can speed up compliance processes, ensuring that companies meet their regulatory obligations while reducing the burden on compliance staff.

Security automation driven by AI is another fascinating development. AI will automate repetitive processes that need a large number of human resources, such incident response, vulnerability scanning, and log analysis. Security experts may now devote more of their important time to more strategic tasks like threat hunting and security planning. AI is capable of real-time analysis of enormous volumes of security data from many sources, such as system logs, network traffic, and user activity. This makes it possible to find minute irregularities that more conventional security technologies would overlook. AI drastically cuts down on the time required to detect and address threats by automating the first detection and analysis stage. This minimizes downtime and harm [64].

9.12 CONCLUSION

Nowadays, strong security measures are required due to the growing dependence on cloud services. The integration of AI into cloud security is a revolutionary development in the protection of digital infrastructures. In order to combat the constantly changing world of cyber threats, comprehensive security standards are crucial as enterprises depend more and more on cloud services for computing, storage, and applications. The basic ideas of cloud computing have been covered in this chapter, along with a discussion of the several security concerns that can occur in cloud systems. It has been shown through an investigation of AI-driven tactics how these cutting-edge technologies may greatly improve cloud security postures, including attack prevention, detection, and response methods. This chapter looked at the relationship that exists between AI and cloud security. It discussed the security threats associated with cloud computing as well as its fundamentals. AI-powered technologies, such machine learning and natural language processing, provide enormous promise to enhance cloud security through proactive threat prevention, detection, and response. The chapter covered the real-world uses of AI in deployment models for private, public, and hybrid clouds in more depth. Additionally, it provided instances from the actual world of how AI may improve regulatory compliance, automate incident response, and bolster threat intelligence. By embracing innovation and making wise investments in AI-driven security technologies, businesses can guarantee a route to sustained success and strengthen their digital resilience in the face of escalating cyber threats.

REFERENCES

1. Coutinho, Emanuel Ferreira, Flávio Rubens de Carvalho Sousa, Paulo Antonio Leal Rego, Danielo Gonçalves Gomes, and José Neuman de Souza. "Elasticity in Cloud Computing: A Survey." *Annals of Telecommunications-Annales des Télécommunications* 70 (2015): 289–309.
2. Al-Dhuraibi, Yahya, Fawaz Paraiso, Nabil Djarallah, and Philippe Merle. "Elasticity in Cloud Computing: State of the Art and Research Challenges." *IEEE Transactions on Services Computing* 11, no. 2 (2017): 430–447.
3. Birje, Mahantesh N, Praveen S Challagidad, R. H. Goudar, and Manisha T Tapale. "Cloud Computing Review: Concepts, Technology, Challenges and Security." *International Journal of Cloud Computing* 6, no. 1 (2017): 32–57.
4. Vaishnavi, R., J. Anand, and R. Janarthanan, "Efficient Security for Desktop Data Grid using Cryptographic Protocol", *IEEE International Conference on Control, Automation, Communication and Energy Conservation, Kongu Engineering College, Erode*, Vol. 1, pp. 305–311, 4-6 June 2009.
5. Kim, Gene, Jez Humble, Patrick Debois, John Willis, and Nicole Forsgren. The DevOps Handbook: How to Create World-Class Agility, Reliability, & Security in Technology Organizations. *IT Revolution*, 2021.
6. El Kafhali, Said, Iman El Mir, and Mohamed Hanini. "Security Threats, Defense Mechanisms, Challenges, and Future Directions in Cloud Computing." *Archives of Computational Methods in Engineering* 29, no. 1 (2022): 223–246.
7. Liu, Qiang, Pan Li, Wentao Zhao, Wei Cai, Shui Yu, and Victor C. M Leung. "A Survey on Security Threats and Defensive Techniques of Machine Learning: A Data-Driven View." *IEEE Access* 6 (2018): 12103–12117.
8. Stutz, Dalmo, Joaquim T de Assis, Asif A Laghari, Abdullah A Khan, Nikolaos Andreopoulos, Andrey Terziev, Anand Deshpande, Dhanashree Kulkarni, and Edwiges G. H Grata. "Enhancing Security in Cloud Computing Using Artificial Intelligence (AI)." *Journal* (2024): 179–220.
9. Namasudra, S. "Cloud Computing: A New Era." *Journal of Future and Applied Sciences* 10, no. 2 (2018).
10. Mansouri, Yaser, Adel Nadjaran Toosi, and Rajkumar Buyya. 2017. Data Storage Management in Cloud Environments: Taxonomy, Survey, and Future Directions. *ACM Comput. Surv.* 50, 6, Article 91 (November 2018), 51 pages.
11. Groom, Frank M. The Basics of Cloud Computing. In *Enterprise Cloud Computing for Non-Engineers*. 2018, Auerbach Publications: 1–42.
12. Nayyar, Aman. *Handbook of Cloud Computing: Basic to Advance Research on the Concepts and Design of Cloud Computing*. 2019, BPB Publications.
13. Hurwitz, Judith S, and Daniel Kirsch. *Cloud Computing for Dummies*. 2020, John Wiley & Sons.
14. Muthumanickam, Kannan. *A Behavior-Based Kernel Level Authentication Mechanism for Protecting System Services Against Malicious Code Attacks*." 2016 Department of Computer Science and Engineering, PEC, Pondicherry University.
15. Diogenes, Yvan, and Emre Ozkaya. *Cybersecurity–Attack and Defense Strategies: Counter Modern Threats and Employ State-of-the-Art Tools and Techniques to Protect Your Organization Against Cybercriminals*. 2019, Packt Publishing Ltd.
16. Bauer, Erik. *Lean Computing for the Cloud*. 2016, John Wiley & Sons.

17. Jarvis, Allen, James Johnson, and Praveen Anand. *Successful Management of Cloud Computing and DevOps*. 2022, Quality Press.
18. Seppänen, Mikko. "Methods for Managed Deployment of User Behavior Analytics to SIEM Product." 2021.
19. Rangaraju, Sandeep. "Secure by Intelligence: Enhancing Products with AI-Driven Security Measures." *International Journal of Security and Engineering* 9, no. 3 (2023): 36–41.
20. Gorelik, Evgeny. *Cloud Computing Models*. 2013, Massachusetts Institute of Technology.
21. Diaby, Thierno, Boubacar Bah, and Jean Ibrahime Rad. "Cloud Computing: A Review of the Concepts and Deployment Models." *International Journal of IT and Computer Science* 9, no. 6 (2017): 50–58.
22. Papagianni, Chrysa, Aris Leivadeas, Symeon Papavassiliou, Vasilis Maglaris, Cristina Cervello-Pastor, and Alvaro Monje. "On the Optimal Allocation of Virtual Resources in Cloud Computing Networks." *Computer Networks* 62, no. 6 (2013): 1060–1071.
23. Ngo, Canh, Yuri Demchenko, and Cees de Laat. "Multi-Tenant Attribute-Based Access Control for Cloud Infrastructure Services." *Journal of Cloud Computing* 27 (2016): 65–84.
24. Raj, Pethuru, Anupama Raman, Pethuru Raj, and Anupama Raman. "Multi-Cloud Management: Technologies, Tools, and Techniques." In *Software-Defined Cloud Centers*. 2018, Springer, 219–240.
25. Christensen, James H. "Using RESTful Web-Services and Cloud Computing to Create Next Generation Mobile Applications." in *Proceedings of the 24th ACM SIGPLAN Conference Companion on Object Oriented Programming Systems Languages and Applications*. 2009.
26. Espadas, Javier, Arturo Molina, Guillermo Jiménez, Martín Molina, Raúl Ramírez, and David Concha. "A Tenant-Based Resource Allocation Model for Scaling Software-as-a-Service Applications Over Cloud Computing Infrastructures." *Journal of Cloud Computing* 29, no. 1 (2013): 273–286.
27. Delimitrou, Christos, and Christos Kozyrakis. "Hcloud: Resource-Efficient Provisioning in Shared Cloud Systems." in *Proceedings of the Twenty-First International Conference on Architectural Support for Programming Languages and Operating Systems*. 2016.
28. Cheraghlou, Mehdi Nazari, Ahmad Khadem-Zadeh, and Majid Haghparast. "A Survey of Fault Tolerance Architecture in Cloud Computing." *International Journal of Cloud Computing* 61 (2016): 81–92.
29. Mezgár, István, and Ulrich J Rauschecker. "The Challenge of Networked Enterprises for Cloud Computing Interoperability." *Computers and Industrial Engineering* 65, no. 4 (2014): 657–674.
30. Grossman, Robert L. "The Case for Cloud Computing." *International Journal of Cloud Computing* 11, no. 2 (2009): 23–27.
31. Rittinghouse, John W, and James F Ransome. *Cloud Computing: Implementation, Management, and Security*. 2017: CRC Press.
32. Thakur, Neeti, Dhananjay Bisen, Vikas Rohit, and Neelesh Gupta. "Review on Cloud Computing: Issues, Services and Models." *International Journal of Computer Science* 91, no. 9 (2014).
33. Okai, Safiya, and Mueen Uddin, Amad Arshad, Raed Alsaqour, and Asadullah Shah. "Cloud Computing Adoption Model for Universities to Increase ICT Proficiency." *International Journal of Computer Applications* 4, no. 3 (2014): 2158244014546461.
34. Harfoushi, Osama, Bader Alfawwaz, Nazeeh A Ghatasheh, Ruba Obiedat, M. Mua'ad, and Hossam Faris. "Data Security Issues and Challenges in Cloud Computing: A Conceptual Analysis and Review." *International Journal of Computer Science* 6, no. 1, (2014): 1–7.

35. El-Gazzar, Rania, Eli Hustad, and Dag H Olsen. "Understanding Cloud Computing Adoption Issues: A Delphi Study Approach." *Procedia Computer Science* 118 (2016): 64–84.
36. Tabrizchi, Hossein, and Mohammad Javad Kuchaki Rafsanjani. "A Survey on Security Challenges in Cloud Computing: Issues, Threats, and Solutions." *The Journal of Supercomputing* 76, no. 12 (2020): 9493–9532.
37. Gupta, B.B., N.A. Arachchilage, and K.E. Psannis. "Defending Against Phishing Attacks: Taxonomy of Methods, Current Issues and Future Directions." *Security and Privacy* 67 (2018): 247–267.
38. Morris, Stephen A. *The Misuse of Encryption and the Risks Posed to National Security.* 2017, Utica College.
39. Islam, Tanveer, D. Manivannan, and Sherali Zeadally. "A Classification and Characterization of Security Threats in Cloud Computing." *International Journal of Network and General Computing* 7, no. 1 (2016): 268–285.
40. Wikina, Samuel B. "What Caused the Breach? An Examination of Use of Information Technology and Health Data Breaches." *Proceedings of the 11th Annual International Conference on Health Information Management* (2014, Fall).
41. Howlader, Md. M. R., "User Attribute Aware Multi-Factor Authentication Framework for Cloud-Based Systems." 2018.
42. Shedden, Paul, et al. "Asset Identification in Information Security Risk Assessment: A Business Practice Approach." *Journal of Information Security* 39, no. 1 (2016): 15.
43. Chandana, P., and Chandrashekhar Gulzar. "Securing Cyberspace: A Comprehensive Journey through AI's Impact on Cyber Security." *Journal of Privacy and Security* 44, no. 2 (2023).
44. Yaseen, Ali J. "AI-Driven Threat Detection and Response: A Paradigm Shift in Cybersecurity." *International Journal of Information Security and Cybersecurity* 7, no. 12 (2023): 25–43.
45. Xiao, Ping. "Malware Cyber Threat Intelligence System for Internet of Things (IoT) Using Machine Learning." *Journal of Communication and Mobility* (2024): 53–90.
46. Reshma, R., and A. J Anand. "Predictive and Comparative Analysis of LENET, ALEXNET, and VGG-16 Network Architecture in Smart Behavior Monitoring." in *2023 Seventh International Conference on Image Information Processing (ICIIP).* 2023. IEEE.
47. Zeadally, Sherali, Erwin Adi, Zubair Baig, and Imran A Khan. "Harnessing Artificial Intelligence Capabilities to Improve Cybersecurity." *International Journal of Intelligent Engineering* 8 (2020): 23817–23837.
48. Thomas, Kevin, Fang Li, Amir Zand, Jason Barrett, Joseph Ranieri, Luca Invernizzi, Yury Markov, Oana Comanescu, Veena Eranti, Adam Moscicki, and David Margolis. "Data Breaches, Phishing, or Malware? Understanding the Risks of Stolen Credentials." in *Proceedings of the 2017 ACM SIGSAC Conference on Computer and Communications Security.* 2017.
49. Kalutarage, Harsha K, Siraj A Shaikh, Indika P Wickramasinghe, Qin Zhou, and Anne E James. "Detecting Stealthy Attacks: Efficient Monitoring of Suspicious Activities on Computer Networks." *Journal of Security and Networks* 47 (2015): 327–344.
50. Fire, Michael, Roy Goldschmidt, and Yuval Elovici. "Online Social Networks: Threats and Solutions." *Journal of Cyber Security* 16, no. 4 (2014): 2019–2036.
51. Eterovic-Soric, Brett, Kim-Kwang Raymond Choo, Helen Ashman, and Sameera Mubarak. "Stalking the Stalkers–Detecting and Deterring Stalking Behaviours Using Technology: A Review." *International Journal of Cybersecurity* 70 (2017): 278–289.
52. Al-Hawawreh, Muna, Nour Moustafa, and Elena Sitnikova. "Identification of Malicious Activities in the Industrial Internet of Things Based on Deep Learning Models." *Computers in Industry* 41 (2018): 1–11.

53. Dunn, Heather, Laurie Quinn, Susan J Corbridge, Kamal Eldeirawi, Mary Kapella, and Eileen G Collins. "Cluster Analysis in Nursing Research: An Introduction, Historical Perspective, and Future Directions." *Journal of Nursing Research* 40, no. 11 (2018): 1658–1676.
54. Hemalatha, S., M. Mahalakshmi, V. Vignesh, M. Geethalakshmi, D. Balasubramanian, and Anand A Jose. "Deep Learning Approaches for Intrusion Detection with Emerging Cybersecurity Challenges." in *2023 International Conference on Sustainable Communication Networks and Application (ICSCNA)*. 2023. IEEE.
55. Shi, Q., and M. J Abdel-Aty. "Big Data Applications in Real-Time Traffic Operation and Safety Monitoring and Improvement on Urban Expressways." *Transportation Research Part C: Emerging Technologies* 58 (2015): 380–394.
56. Phillips, D. "Robots in the Library: Gauging Attitudes Towards Developments in Robotics and AI, and the Potential Implications for Library Services." 2017.
57. Bakdash, Jonathan Z, Steve Hutchinson, Erin G Zaroukian, Laura R Marusich, Saravanan Thirumuruganathan, Charmaine Sample, Blaine Hoffman, and Gautam Das. "Malware in the Future? Forecasting of Analyst Detection of Cyber Events." *Journal of Cybersecurity* 4, no. 1 (2018).
58. Sontan, A. D, and S. V Samuel. "The Intersection of Artificial Intelligence and Cybersecurity: Challenges and Opportunities." *World Journal of Applied Research and Reviews,* 21, no. 2 (2024): 1720–1736.
59. Kingston, J. J. "Using Artificial Intelligence to Support Compliance with the General Data Protection Regulation." *AI & Law* 25, no. 4 (2017): 429–443.
60. Ikhlasse, Hamzaoui, Duthil Benjamin, Courboulay Vincent, and Medromi Hicham. "An Overall Statistical Analysis of AI Tools Deployed in Cloud Computing and Networking Systems." in *2020 5th International Conference on Cloud Computing and Artificial Intelligence: Technologies and Applications (CloudTech)*. 2020. IEEE.
61. Subramanian, N., and A Jeyaraj. "Recent Security Challenges in Cloud Computing." *Journal of Computer Engineering* 71 (2018): 28–42.
62. Girasa, R., "Applications of AI and Projections of AI Impact. In *AI as a Disruptive Technology*. 2020, Palgrave Macmillan, Cham, 23–67.
63. Pan, Y., and Zhang, L. "Roles of Artificial Intelligence in Construction Engineering and Management: A Critical Review and Future Trends." *Automation in Construction* 122 (2021): 103517.
64. Hassan, M., L. A.-R Aziz, and Y. J Andriansyah. "The Role of Artificial Intelligence in Modern Banking: An Exploration of AI-Driven Approaches for Enhanced Fraud Prevention, Risk Management, and Regulatory Compliance." *Research on Cybersecurity in Banking and Applications* 6, no. 1 (2023): 110–132.

10 Adversarial Attacks on AI Security Systems

Investigating the Vulnerability of AI-Powered Security Solutions

Shaista Alvi

10.1 INTRODUCTION

The cutting-edge artificial intelligence (AI) technology disrupts various sectors including the financial sector as digitalization has enabled new product evolution [1]. AI is a broad phrase that refers to the creation of computer systems capable of doing tasks that normally require human intelligence, such as learning, problem-solving, decision-making, and natural language processing (NLP). AI has advanced greatly since its inception in the 1950s, with advances in fields like machine learning, deep learning, and NLP. The origins of AI may be traced back to the 1950s, when pioneering researchers such as A. Turing, J. McCarthy, and M. Minsky established the groundwork for the subject [2]. AI has progressed significantly throughout time, with notable milestones including the invention of professional systems in the 1970s and the renaissance of interest in the late 1980s and 1990s due to technological developments in computer power and information storage [3].

AI systems can be broadly classified into two types: First based on rule-based systems use established rules and logical reasoning to solve problems, whereas second, machine learning algorithms that learn from data to generate predictions and judgements. In the early days of AI research, the emphasis was on creating rule-based systems, in which computers were programmed with a set of rules and logical reasoning to solve certain issues. This method, known as "symbolic AI," had several triumphs, including the construction of chess-playing programs that could outperform human players. However, the limitations of this method became clear as it struggled to deal with the complexity and ambiguity of real-world problems.

The 1970s and 1980s saw a revolution in the field of AI, with the creation of machine learning, a subfield that focused on allowing computers to study from data and improve their performance. This trend was prompted by the availability of greater datasets and the increasing processing capability of computers. ML algorithms, such as neural networks and decision trees, were able to solve more complicated issues by recognizing patterns and making data-driven predictions.

DOI: 10.1201/9781003521020-10

The 1990s and early 2000s witnessed a surge in the development of AI applications, with the rise of expert systems, NLP, and computer vision [4]. Expert systems, for example, were employed in a variety of industries to offer expert advice and decision-making assistance. NLP allowed computers to interpret and generate human language, resulting in the creation of chatbots and virtual assistants. Computer vision, on the other hand, enabled machines to interpret and analyze visual input, paving the door for applications such as picture identification and driverless cars [5].

The present era of machine learning has been defined by significant breakthroughs in deep learning, a powerful technology that uses artificial neural networks to interpret and learn from massive volumes of data [6]. Deep learning has enabled substantial improvements in image identification, NLP, and speech recognition, outperforming humans on numerous tasks. One of the most notable uses of deep learning is computer vision, where machine learning algorithms can now correctly identify objects, faces, and scenes in photos and movies. This technology has been widely used in a variety of applications, including driverless vehicles and medical imaging analysis.

NLP is another area in which machine learning has advanced significantly. Using deep learning techniques, machines can now interpret and synthesize human language with astonishing precision, enabling applications such as language translation, text summarization, and conversational chatbots.

One of the significant achievements in machine learning during this period was the rediscovery of backpropagation, a strong method that enabled neural networks to learn complicated patterns in data [7]. Neural networks, inspired by the structure and function of the human brain, were able to solve a variety of challenges, including image identification and NLP. The increase in the development of machine learning applications, with the rise of expert systems, support vector machines, and recurrent neural networks [8]. Expert systems, for example, were used in various industries to provide expert-level advice and decision-making support. Support vector machines, on the other hand, demonstrated remarkable performance in tasks like spam filtering and medical diagnosis. In machine learning, there are numerous strategies such as reinforcement learning supervised learning, and, unsupervised learning, each having its own strengths and uses [9].

NLP is another field in which AI has advanced significantly [10]. By leveraging deep learning algorithms, these systems can now comprehend and generate human language with remarkable accuracy, enabling applications such as language translation, text summarization, and conversational chatbots [8]. AI breakthroughs have also resulted in the development of intelligent personal assistants, capable of understanding and responding to voice commands, scheduling appointments, and doing a variety of other functions. These AI-powered assistants have become a vital part of many people's daily life, indicating the increasing integration of AI into our everyday.

Another continuing research subject is the investigation of reinforcement learning, a paradigm in which AI agents learn by interacting with their surroundings and receiving feedback in the form of rewards or punishments [11]. Reinforcement learning has shown promise in areas like game-playing, robotics, and decision-making, as it allows agents to learn complex behaviors through trial and error. Along with these improvements, researchers are contending with ongoing obstructions in the field of AI and machine learning. Issues such as interpretability, robustness, and safety have become increasingly important as these technologies become more integrated into real-world applications [12].

As AL and ML continue to advance, the research landscape has gotten more dynamic and diversified. Researchers across the globe are pushing the boundaries of what is possible, exploring new algorithms, architectures, and approaches to tackle complex problems and unlock the full potential of these transformative technologies.

The advancement of generative adversarial networks (GANs) has also received substantial interest, as these models have shown the ability to create realistic synthetic data ranging from images to text [13]. This has opened up new possibilities in areas like content creation, data augmentation, and even the generation of fake media, raising concerns on the ethical implications of such technologies.

The research community continues to prioritize the development of systems that are transparent, reliable, and ethical [14]. Furthermore, the ethical implications of AI and machine learning, including concerns about algorithmic bias, privacy, and the impact on employment, have sparked a growing body of research aimed at developing ethical and governance structures to ensure that these technologies are developed and deployed flawlessly [15].

While AI has made great progress, there are still hurdles and ethical concerns in the sector. Rapid advances in AI and deep learning have transformed many industries, including security and safety-critical applications. However, the widespread use of AI systems has made them vulnerable to adversarial attacks, in which fraudulently constructed inputs drive these systems to act in unexpected and potentially destructive ways. The purpose of this introduction is to discuss the growing threat of adversarial assaults on AI-powered security systems, as well as the urgent necessity to address this issue.

The growing dependence on AI in cybersecurity has resulted in both benefits and new concerns. AI-based systems have shown possibilities in improving cyber threat detection and mitigation by utilizing their abilities to analyze vast volumes of data, discover trends, and adapt to shifting attack techniques [16]. However, the very nature of AI systems' susceptibility to adversarial attacks poses a significant concern [17]. Attackers can design carefully crafted inputs, known as adversarial examples, that can mislead AI models into making erroneous decisions, potentially compromising the integrity of the security system [18].

Adversarial instances are inputs that are nebulous from valid ones to the human, but they can cause AI algorithms to make inaccurate predictions or classifications [19]. This vulnerability of AI systems has far-reaching implications, as adversarial attacks can be used to bypass security measures, evade detection, and even manipulate the decision-making processes of critical AI-based security applications [16].

There is an emerging concern in today's landscape, where AI, such as deep learning, can be incorporated to sophisticated models [20]. AI models confront a variety of digital risks that interfere with their sampling, learning, and decision-making skills. With the advancement of AI, the threat of digital assaults, cybercrimes, and virus attacks has increased rapidly. Traditional attack methods have developed, prompting attackers to adopt innovative strategies [20].

This chapter discusses the vulnerabilities of AI systems and the potential adversarial attacks they face. It explores traditional approaches and how AI can enhance the multilayered defense mechanisms used in organizations today. The chapter concludes with discussions about the future possibilities of AI in the face of these emerging threats.

The chapter flow here onwards is the next section is a literature review that provides a literature survey on the adversarial attack on AI. The third section is about the forms of adversarial attack followed by the fourth section which elucidates the defensive contemporary strategy for combatting adversarial attacks. After that, section 5 articulates challenges faced by AI in this context which is followed by a conclusion with the future potential of AI in the face of these evolving threats.

10.2 LITERATURE REVIEW

Adversarial incidents are deliberate alterations of input data that lead AI systems to produce inaccurate or unwanted outputs, weakening the confidence and reliability of these technologies [20, 21]. The increasing reliance on these technologies has also exposed a concerning vulnerability – the susceptibility of AI/ML models to adversarial attacks. The significance of combating adversarial attacks cannot be emphasized, especially in safety-critical applications where model failures can be disastrous—allowing attackers to impersonate legitimate users. Research published in leading journals has shed light on the intricacies of these attacks and the urgent need for robust countermeasures.

A seminal study investigated the incorporation of hidden Trojan models directly into neural networks, a form of adversarial attack that can be particularly challenging to detect [22]. The researchers demonstrated the feasibility of these attacks and the potential for widespread impact, highlighting the critical need for advanced detection and mitigation techniques. Building on this foundation, a comprehensive framework for defending against such covert attacks proposed architectural modifications to enhance the resilience of neural networks [23]. This work underscores the importance of proactive measures in securing AI systems against evolving threats.

Furthermore, a study delved into the use of clean, unmodified data for deceptive purposes, offering countermeasures to defend against these more subtle yet equally dangerous adversarial attacks [24]. This research emphasizes the multifaceted nature of the adversarial attack landscape and the need for a diverse arsenal of defensive strategies. Recognizing the urgency of automating the detection of adversarial threats, researchers have proposed innovative anomaly detection techniques [25]. These methods leverage the power of AI itself to identify maliciously trained models, paving the way for more proactive and scalable security measures.

A study delved into the development of a novel defense mechanism called Feature Denoising, which aims to remove adversarial perturbations from input images [26]. The researchers demonstrated the efficacy of their approach in boosting the robustness of deep neural networks against a wide range of adversarial attacks, demonstrating its practical utility.

Another study examined the sensitivity of medical image analysis AI models to adversarial attacks [27]. The study highlighted the need for robust defenses in safety-critical domains like healthcare, where the consequences of model failures can be severe. The authors proposed a framework for evaluating the robustness of medical AI systems and called for the development of standardized testing protocols to ensure the reliability of these technologies. Another study focusses on the development of a defense mechanism known as Thermometer Encoding, which intends to advance the resilience of neural networks to adversarial threats. The researchers demonstrated that their approach can effectively fight against numerous attacks while maintaining the underlying model's performance on clean data.

Additional research investigated the use of ensemble methods to protect against adversarial attacks [28]. The researchers showed that by combining multiple models, each trained with a different defense mechanism, they could achieve a higher level of robustness compared to individual defenses. Alongside academic efforts, corporate and governmental bodies have drawn extensively from the insights provided by these studies to standardize security measures across various applications of neural networks, underscoring the widespread recognition of the adversarial attack challenge [29]. The research community's collaborative work emphasizes the crucial need of dealing with adversarial threats in order to pursue responsible AI development.

10.3 TYPES OF ADVERSARIAL ATTACKS

As AI systems progress and spread, they become more vulnerable to adversarial attacks. These assaults involve the deliberate modification of input data to cause AI systems to deliver inaccurate or unwanted results, weakening the credibility and reliability of these technologies [20]. The importance of understanding and mitigating adversarial attacks cannot be overstated. As AI systems are increasingly incorporated into essential infrastructure and decision-making processes, the potential repercussions of model failures can be severe, ranging from financial losses to dangers to human safety. Researchers and practitioners have recognized the urgency of this challenge, with a growing body of literature dedicated to exploring the various types of adversarial attacks and developing effective countermeasures. Adversarial assaults on AI systems are roughly classified into three types: evasion attacks, poisoning attacks, and model extraction attacks [30].

10.3.1 Evasion Attacks

Evasion assaults involve the manipulation of input data during the deployment phase to induce an already trained model to make false predictions [20]. These assaults are the most prevalent and well-studied type of adversarial attack, with numerous techniques having been proposed to generate adversarial examples [13]. One of the most influential studies in this area was conducted by [31], who developed a powerful optimization-based attack that can bypass many existing defenses. Evasion attempts have been observed in a variety of applications, including malware detection [32]. Evasion attacks entail manipulating input data to avoid detection or categorization by the AI model [33]. These attacks can be particularly pernicious because the changes made to the input data are often unnoticeable to human observers but can cause the model to erroneously classify it. To illustrate, an attacker could subtly manipulate the image of a stop sign, making it unrecognizable to a self-driving car's computer vision system. Evasion attacks can have severe consequences in real-world applications, as they can lead to critical failures in safety-critical systems or enable malicious actors to bypass security measures. Researchers proposed different defense techniques, to improve the metrics of AI models against evasion attempts [26, 34].

10.3.2 Poisoning Attacks

PA targets the training phase of machine learning models, attempting to inject erroneous data into the pilot, causing the model to develop unfitting patterns. These attacks can be particularly insidious, as they can lead to model failures even when clean data is used during deployment PA involves the introduction of corrupted data into the pilot dataset, which can cause the AI model to study inappropriate links and make faulty predictions [35]. These assaults can be particularly challenging to detect, as the impact of the poisoned data may not be immediately apparent, and the model's poor performance could be attributed to other factors, such as overfitting or natural noise in the data. PA can have far-reaching consequences, as they can undermine the reliability of systems in critical domains like healthcare, finance, and transportation. Researchers have invented various techniques to detect and mitigate poisoning attacks, such as robust optimization and anomaly detection [36, 37]. The use of GANs to generate poisoning examples that are indistinguishable from clean data [24]. The researchers demonstrated that their approach can significantly degrade the performance of machine learning across the tasks. As AI systems become more reliant on crowdsourced or online data, the possibility of poisoning attacks is expected to grow, emphasizing the importance of strong data sanitization and anomaly detection approaches [20].

10.3.3 Model Extraction Attacks

MEA purpose is to acquire the functionality of a model by probing it with carefully designed queries. These attacks can be particularly problematic when the model being targeted is a commercial service or when the model itself contains sensitive or proprietary information [38]. One study indicated that model extraction assaults can be used to steal the functionality of cutting-edge image classification models, having implications for the protection of intellectual property in the AI business [39]. MEA can also be used to bypass access control mechanisms and gain unauthorized access to sensitive data or resources. Researchers found that model extraction attacks can be used to purloin the performance of biometric authentication systems, potentially allowing attackers to impersonate legal users.

10.3.4 Backdoor Attacks

Backdoor attacks involve the introduction of a hidden trigger or pattern into the AI model during the pilot stage, which can cause the model to produce a specific, corrupted output when the trigger is present in the inserted data. These attacks can be particularly insidious, as the model may perform well on standard test cases but exhibit malicious behavior only when the trigger is present.

Backdoor attacks can have severe consequences, as they can enable adversaries to compromise the integrity of AI systems without being detected. Researchers have explored techniques, such as neural cleanse and spectral signatures, to detect and mitigate these attacks [22, 40]. The research community has made significant progress in developing defense mechanisms to mitigate the impact of adversarial assaults on AI systems.

10.4 PROTECTING AGAINST ADVERSARIAL ATTACKS

As adversarial assaults continue to grow, academia and practitioners have proposed a range of defensive strategies to protect AI systems.

10.4.1 Adversarial Training

One of the most effective defenses is AT, which involves training the AI model on both clean and adversarial data to improve its robustness. AT has proven to be effective in enhancing the resistance of models against a wide range of attacks, demonstrating its usefulness across a variety of tasks and datasets [41].

10.4.2 Input Transformation

Input processing techniques such as noise injection or inserted transformation can help mitigate the impact of adversarial attacks [42]. These methods aim to remove or reduce the adversarial perturbations before they reach the AI model while preserving the essential features of the input.

10.4.3 Ensemble Methods

Combining multiple AI models with diverse architectures and training techniques, known as ensemble methods, can increase the overall robustness of the system [28]. Ensemble approaches, by exploiting the complimentary capabilities of various models, can make it difficult for attackers to uncover a single adversarial case that can deceive all the models at once. Ensemble methods are effective in defending against both evasion and PA, with studies demonstrating their effectiveness in a range of security-critical applications [35].

10.4.4 Robust Model Design

Techniques like gradient obfuscation and feature squeezing can be used to make the AI model more robust and resistant to adversarial threats [43, 44]. These methods make it more difficult for attackers to craft actual adversarial examples by modifying the model's internal structure or the input data.

10.4.5 Anomaly Detection

Continuous surveillance and anomaly detection systems monitor activities in real-time, identifying and responding to potential threats by analyzing patterns, trends, and deviations from expected behavior. These systems help detect adversarial attacks early, enabling proactive security measures to mitigate risks and protect critical infrastructure [45]. These systems continuously monitor the model's inputs and outputs, identifying unexpected patterns that could indicate an ongoing attack. AI systems face significant challenges due to adversarial attacks, which exploit the vulnerabilities and limitations inherent in machine learning models.

10.5 VULNERABILITIES IN AI-POWERED SECURITY SYSTEMS

AI systems face significant challenges due to adversarial attacks, which exploit the vulnerabilities and limitations inherent in machine learning models. Some of the key challenges are discussed in subsequent sections.

10.5.1 Inherent Model Vulnerabilities

Modern neural networks, which form the backbone of many AI systems, have inherent vulnerabilities that make them susceptible to adversarial attacks. Their linear behavior in high-dimensional spaces allows small input changes to create drastically altered outputs, making them prone to misclassification [43].

10.5.2 Evolving Attack Strategies

Attackers continuously refine and redefine their strategies to maximize the impact of adversarial attacks. As AI developers bolster their defenses, adversaries innovate by harnessing newer algorithms, incorporating AI in their attacks, and exploiting missed system vulnerabilities [30]. This perpetual evolution of attack strategies poses a significant challenge to AI systems.

10.5.3 Lack of Awareness and Standardization

Many AI users and developers lack awareness about the rising prominence of adversarial attacks, leading to vulnerabilities in deployed models. The lack of standardized testing methodologies and standards makes it difficult to measure the resilience of AI systems against these threats.

10.5.4 Transferability of Attacks

Adversarial instances designed for one model can frequently trick other models with comparable structures, even when they differ in training data and architecture [28]. This transferability of attacks exacerbates the challenge of protecting against vulnerabilities, as a single attack can potentially compromise multiple AI systems.

10.5.5 Difficulty in Detection and Mitigation

The subtle nature of adversarial perturbations and the constant evolution of attack techniques make it challenging to detect and mitigate these attacks effectively.

Existing defense mechanisms often fail to provide comprehensive protection against the wide range of adversarial threats [42].

To address these challenges, researchers and practitioners must work collaboratively to develop robust defense strategies, establish ethical frameworks, and promote awareness about the risks of adversarial attacks. Prioritizing security and resilience in AI development assurances the legitimacy and dependability of these disruptive technologies.

10.6 CONCLUSION AND FUTURE RESEARCH

Adversarial attacks pose a substantial risk to the widespread implementation and deployment of AI systems. As AI becomes more integrated into essential structures and decision-making procedures, model failures can have major consequences, ranging from financial losses to threats to human safety. By understanding the vulnerabilities of AI models and exploring innovative countermeasures, we can work towards creating more secure and trustworthy AI technologies that can fulfil their transformative potential without compromising safety and reliability. Researchers and practitioners have recognized the urgency of this challenge, with a growing body of literature dedicated to exploring various types of adversarial attacks and developing effective countermeasures. Studies published in top-ranked journals have explored various aspects of the problem, from developing novel defense mechanisms to investigating the fundamental limits of adversarial robustness. While significant progress has been made in defending against adversarial attacks, much work remains to be done. However, the field remains an active area of research, with many open challenges and opportunities for further exploration. As AI systems become more complex and powerful, new attack vectors are likely to emerge, requiring ongoing research and development to stay ahead of the curve. By collaborating across disciplines and industries, we can build a future in which AI systems are secure, dependable, and trustworthy, for the benefit of civilization as a whole.

Academics, industry, and politicians are collaborating to create guidelines for ethics and governance structures to ensure that AI systems are developed and executed responsibly. This includes resolving concerns about algorithmic bias, privacy, and transparency in order to develop more safe and trustworthy AI solutions.

REFERENCES

1. S. Alvi, "Technology Based Uop Green Bond Reshaping the Issuance," *Academy of Marketing Studies Journal*, vol. 25, no. 3S, pp. 1–11, 2021.
2. J. McCarthy, M. L. Minsky, N. Rochester, and C. E. Shannon, "A Proposal for the Dartmouth Summer Research Project on Artificial Intelligence, August 31, 1955," *AI Magazine*, vol. 27, no. 4, Art. no. 4, Dec. 2006, https://doi.org/10.1609/aimag.v27i4.1904.
3. K. Abhishek, "Introduction to artificial intelligence," Simple Talk. Accessed: Jul. 16, 2024. [Online]. Available: https://www.red-gate.com/simple-talk/development/data-science-development/introduction-to-artificial-intelligence/

4. S. J. Russell, and P. Norvig, *Artificial Intelligence : a Modern Approach*. Pearson, 2016. Accessed: Jul. 16, 2024. [Online]. Available: https://thuvienso.hoasen.edu.vn/handle/123456789/8967
5. Faster Roi," FasterCapital. Accessed: Aug. 05, 2024. [Online]. Available: https://fastercapital.com/keyword/faster-roi.html
6. What is Deep Learning?," Market Business News. Accessed: Aug. 05, 2024. [Online]. Available: https://marketbusinessnews.com/financial-glossary/deep-learning/
7. F. Jiang *et al.*, "Artificial Intelligence in Healthcare: Past, Present and Future," *Stroke Vasc Neurol*, vol. 2, no. 4, Dec. 2017, https://doi.org/10.1136/svn-2017-000101.
8. O. Zawacki-Richter, V. I. Marín, M. Bond, and F. Gouverneur, "Systematic Review of Research on Artificial Intelligence Applications in Higher Education – Where Are the Educators?," *International Journal of Educational Technology in Higher Education*, vol. 16, no. 1, p. 39, 2019. https://doi.org/10.1186/s41239-019-0171-0.
9. N. Saini, "Research Paper on Artificial Intelligence and Its Applications," vol. 8, no. 4, 2023.
10. MathAware AI: Created by AI, Enriched by Human Expertise & Charme. Backed Up by Facts. | MathAware: AI Generators, Reviews & Research!" Accessed: Aug. 05, 2024. [Online]. Available: https://www.mathaware.org/
11. R. S. Sutton, and A. G. Barto, *Reinforcement Learning, Second Edition: An Introduction*. MIT Press, 2018.
12. Z. C. Lipton, "The Mythos of Model Interpretability," Mar. 06, 2017, *arXiv*: arXiv:1606.03490. https://doi.org/10.48550/arXiv.1606.03490.
13. I. Goodfellow *et al.*, "Generative Adversarial Nets," in *Advances in Neural Information Processing Systems*, Curran Associates, Inc., 2014. Accessed: Jul. 16, 2024. [Online].
14. T. Sipola, J. Alatalo, M. Wolfmayr, T. Kokkonen, Eds., *Artificial Intelligence for Security: Enhancing Protection in a Changing World*. Cham: Springer Nature Switzerland, 2024. https://doi.org/10.1007/978-3-031-57452-8.
15. A. Jobin, M. Ienca, and E. Vayena, "The Global Landscape of AI Ethics Guidelines," *Nature Machine Intelligence*, vol. 1, no. 9, pp. 389–399, 2019. https://doi.org/10.1038/s42256-019-0088-2.
16. I. Stoica *et al.*, "A Berkeley View of Systems Challenges for AI," Dec. 15, 2017, *arXiv*: arXiv:1712.05855. https://doi.org/10.48550/arXiv.1712.05855.
17. K. Ren, T. Zheng, Z. Qin, and X. Liu, "Adversarial Attacks and Defenses in Deep Learning," *Engineering*, vol. 6, no. 3, pp. 346–360, 2020. https://doi.org/10.1016/j.eng.2019.12.012.
18. N. Siroya, and P. M. Mandot, *Role of AI in Cyber Security*. John Wiley & Sons, 2021.
19. J. Zhang, and C. Li, "Adversarial Examples: Opportunities and Challenges," *IEEE Trans. Neural Netw. Learning Syst*, pp. 1–16, 2019. https://doi.org/10.1109/TNNLS.2019.2933524.
20. N. Bhargava, R. Bhargava, P. S. Rathore, and R. Agrawal, *Artificial Intelligence and Data Mining Approaches in Security Frameworks*. John Wiley & Sons, 2021.
21. Palo Alto Networks, "What Is Adversarial AI in Machine Learning?," Palo Alto Networks. Accessed: Jul. 16, 2024. Available: https://www.paloaltonetworks.com/cyberpedia/what-are-adversarial-attacks-on-AI-Machine-Learning.
22. Y. Liu *et al.*, "Trojaning Attack on Neural Networks," in *Proceedings 2018 Network and Distributed System Security Symposium*, San Diego, CA: Internet Society, 2018. https://doi.org/10.14722/ndss.2018.23291.
23. K. Eykholt *et al.*, "Robust Physical-World Attacks on Deep Learning Visual Classification," in *2018 IEEE/CVF Conference on Computer Vision and Pattern Recognition*, Jun. 2018, pp. 1625–1634. https://doi.org/10.1109/CVPR.2018.00175.

24. A. Shafahi *et al.*, "Poison Frogs! Targeted Clean-Label Poisoning Attacks on Neural Networks," Nov. 10, 2018, *arXiv*: arXiv:1804.00792. https://doi.org/10.48550/arXiv.1804.00792.
25. T. Gu, B. Dolan-Gavitt, and S. Garg, "BadNets: Identifying Vulnerabilities in the Machine Learning Model Supply Chain," Mar. 11, 2019, *arXiv*: arXiv:1708.06733. https://doi.org/10.48550/arXiv.1708.06733.
26. C. Xie, Y. Wu, L. van der Maaten, A. Yuille, and K. He, "Feature Denoising for Improving Adversarial Robustness," Mar. 25, 2019, *arXiv*: arXiv:1812.03411. https://doi.org/10.48550/arXiv.1812.03411.
27. S. G. Finlayson, J. D. Bowers, J. Ito, J. L. Zittrain, A. L. Beam, and I. S. Kohane, "Adversarial Attacks on Medical Machine Learning," *Science*, vol. 363, no. 6433, pp. 1287–1289, 2019. https://doi.org/10.1126/science.aaw4399.
28. Y. Liu, X. Chen, C. Liu, and D. X. Song, "Delving into Transferable Adversarial Examples and Black-box Attacks," *ArXiv*, vol. abs/1611.02770, 2016. Available: https://api.semanticscholar.org/CorpusID:17707860.
29. L. Lyu, H. Yu, and Q. Yang, "Threats to Federated Learning: A Survey," Mar. 04, 2020, *arXiv*: arXiv:2003.02133. https://doi.org/10.48550/arXiv.2003.02133.
30. N. Papernot, P. McDaniel, S. Jha, M. Fredrikson, Z. B. Celik, and A. Swami, "The Limitations of Deep Learning in Adversarial Settings," in *2016 IEEE European Symposium on Security and Privacy (EuroS&P)*, 2016, pp. 372–387. https://doi.org/10.1109/EuroSP.2016.36.
31. N. Carlini, and D. Wagner, "Towards Evaluating the Robustness of Neural Networks," in *2017 IEEE Symposium on Security and Privacy (SP)*, 2017, pp. 39–57. https://doi.org/10.1109/SP.2017.49.
32. K. Grosse, N. Papernot, P. Manoharan, M. Backes, and P. McDaniel, "Adversarial Examples for Malware Detection," in *Computer Security – ESORICS 2017*, vol. 10493, S. N. Foley, D. Gollmann, and E. Snekkenes, Eds., in Lecture Notes in Computer Science, vol. 10493., Cham: Springer International Publishing, 2017, pp. 62–79. https://doi.org/10.1007/978-3-319-66399-9_4.
33. C. Szegedy *et al.*, "Intriguing properties of neural networks: 2nd International Conference on Learning Representations, ICLR 2014," Jan. 2014. Accessed: Jul. 16, 2024. Available: http://www.scopus.com/inward/record.url?scp=85070854365&partnerID=8YFLogxK
34. A. Madry, A. Makelov, L. Schmidt, D. Tsipras, and A. Vladu, "Towards Deep Learning Models Resistant to Adversarial Attacks," Sep. 04, 2019, *arXiv*: arXiv:1706.06083. Accessed: Jul. 16, 2024. [Online]. Available: http://arxiv.org/abs/1706.06083
35. B. Biggio, B. Nelson, and P. Laskov, "Poisoning Attacks against Support Vector Machines," 2012.
36. J. Steinhardt, P. W. W. Koh, and P. S. Liang, "Certified Defenses for Data Poisoning Attacks," in *Advances in Neural Information Processing Systems*, Curran Associates, Inc., 2017. Accessed: Jul. 16, 2024. [Online]. Available: https://proceedings.neurips.cc/paper_files/paper/2017/hash/9d7311ba459f9e45ed746755a32dcd11-Abstract.html
37. M. Jagielski, A. Oprea, B. Biggio, C. Liu, C. Nita-Rotaru, and B. Li, "Manipulating Machine Learning: Poisoning Attacks and Countermeasures for Regression Learning," in *2018 IEEE Symposium on Security and Privacy (SP)*, May 2018, pp. 19–35. https://doi.org/10.1109/SP.2018.00057.
38. F. Tramèr, F. Zhang, A. Juels, M. K. Reiter, and T. Ristenpart, "Stealing machine learning models via prediction APIs," in *Proceedings of the 25th USENIX Conference on Security Symposium*, in SEC'16. USA: USENIX Association, Aug. 2016, pp. 601–618.
39. T. Orekondy, B. Schiele, and M. Fritz, "Knockoff Nets: Stealing Functionality of Black-Box Models," in *2019 IEEE/CVF Conference on Computer Vision and Pattern Recognition (CVPR)*, 2019, pp. 4949–4958. https://doi.org/10.1109/CVPR.2019.00509.

40. B. Tran, J. Li, and A. Madry, "Spectral Signatures in Backdoor Attacks," Nov. 01, 2018, *arXiv*: arXiv:1811.00636. https://doi.org/10.48550/arXiv.1811.00636.
41. H. Zhang, H. Chen, Z. Song, D. Boning, I. S. Dhillon, and C.-J. Hsieh, "The Limitations of Adversarial Training and the Blind-Spot Attack," Jan. 15, 2019, *arXiv*: arXiv:1901.04684. https://doi.org/10.48550/arXiv.1901.04684.
42. C. Guo, M. Rana, M. Cisse, and L. van der Maaten, "Countering Adversarial Images using Input Transformations," Jan. 25, 2018, *arXiv*: arXiv:1711.00117. https://doi.org/10.48550/arXiv.1711.00117.
43. J. Buckman, A. Roy, C. Raffel, and I. Goodfellow, "Thermometer Encoding: One Hot Way to Resist Adversarial Examples," 2018.
44. W. Xu, D. Evans, and Y. Qi, "Feature Squeezing: Detecting Adversarial Examples in Deep Neural Networks," in *Proceedings 2018 Network and Distributed System Security Symposium*, San Diego, CA: Internet Society, 2018. https://doi.org/10.14722/ndss.2018.23198.
45. A. I. Lumenova, "Adversarial Attacks in ML: Detection & Defense Strategies," Lumenova AI. Accessed: Jul. 16, 2024. Available: https://www.lumenova.ai/blog/adversarial-attacks-ml-detection-defense-strategies/

11 Ethical Considerations and Privacy in AI-Powered Security

Gowtham H, Nandha Gopal J, and A. Jose Anand

11.1 INTRODUCTION

Artificial intelligence (AI)-powered finance involves the application of technology for AI and methods within the financial sector to improve different aspects of financial operations, decision-making, and customer service. AI systems in this context are built to process large volumes of data, recognize trends, make forecasts, and automate tasks that were traditionally done by humans. AI-powered finance covers a broad spectrum of applications across various segments of the financial industry. Some typical examples include:

- *Evaluation and management of risks*: AI algorithms are capable of analyzing historical data and market trends to more effectively assess and manage financial risks. This encompasses areas such as portfolio optimization, fraud detection, and credit risk assessment.
- *Financial trading and investment strategies*: To make well-informed investment decisions, AI algorithms can analyze news, market data, and social media sentiment. These algorithms can also automate trading strategies, executing trades rapidly and with minimal human involvement.
- *Customer support and tailored experiences*: AI-driven chatbots and virtual assistants offer personalized customer support by answering queries and assisting with financial planning. Using natural language processing (NLP), these systems can accurately understand and respond to customer inquiries.
- *Adherence to regulations and compliance reporting*: AI systems can aid in compliance by monitoring transactions, identifying suspicious activities, and ensuring that regulatory requirements are met. They can also automate the regulatory reporting process which reduces manual effort and enhances accuracy.
- *Identifying and preventing fraud*: AI algorithms have the capability to examine enormous volumes of transactional data to look for irregularities and possible fraud patterns, and prevent fraudulent activities in real time.
- *Predictive analytics and economic forecasting*: AI methods, such as machine learning, enable financial institutions to generate precise forecasts, predict market trends, and make data-driven decisions related to pricing, investments, and risk management.

DOI: 10.1201/9781003521020-11

The implementation of AI in finance provides numerous benefits such as better risk management, increased productivity, increased accuracy, cost savings, and improved client experiences. But it also brings with it difficulties and worries about data protection, security, and ethics, as previously discussed. Overall, AI-powered finance signifies a major advancement in the financial industry, transforming traditional processes and allowing financial institutions to utilize data-driven insights for more informed decision-making and better customer service [1].

11.2 SIGNIFICANCE OF ETHICS

Significance of ethics, security, and data privacy in AI-powered finance ethics, security, and data protection is essential in AI-powered banking for the following reasons:

- *Protecting individual privacy*: In AI-powered finance, financial institutions gather and handle large quantities of personal and sensitive information from their customers. To ensure data privacy, people must have control over how their information is gathered, saved, and used, courtesy of privacy rights. This helps to build trust between customers and financial institutions, supporting long-term relationships and preserving a positive reputation.
- *Mitigating data breaches and cyber threats*: Financial sector is a desirable target for online thieves since financial data is valuable and sensitive. Both individuals and businesses may suffer large financial losses as well as reputational harm from a data breach and potential legal issues. Thus, it is essential to enforce strict data security policies in order to protect against cyber threats, unauthorized access, and breaches.
- *Ensuring fairness and avoiding bias*: AI algorithms in finance rely on historical data to make decisions and predictions. If this data includes biases or discriminatory patterns, it can lead to unfair outcomes. Ethical standards require that financial AI systems are intended to be equitable, transparent, and unbiased, ensuring that all individuals are treated fairly regardless of their demographic characteristics.
- *Ensuring transparency and clarity*: AI algorithms in finance can be quite complex, which makes it difficult to discern the factors influencing their decision-making. Transparent and explainable AI systems are crucial for ensuring accountability and regulatory compliance, as well as for enabling individuals to understand and challenge decisions that impact them. In addition, transparent AI systems assist financial institutions in identifying and correcting errors or biases in their models.
- *Ensuring adherence to regulatory standards*: Financial industry is governed by numerous regulations and legal frameworks concerning data privacy, security, and ethics. Significant fines and legal ramifications may result if these rules are broken. Adhering to these requirements reflects a commitment to responsible data management and ethical practices.
- *Maintaining customer confidence and loyalty*: Data building and maintaining consumer trust requires careful attention to privacy, security, and ethics. People are more likely to interact with financial organizations and

divulge their data when they have faith that their personal information is handled sensibly and safely. Building and maintaining trust is essential for retaining customers and creating favorable word-of-mouth referrals.

Financial organizations may lower risks and increase client trust in AI-powered finance by making data privacy, security, and ethics their first priority, ensuring regulatory compliance, and supporting the creation of a fair and reliable financial ecosystem. This approach also protects individuals' rights, encourages responsible AI practices, and helps preserve the integrity and reputation of the financial industry as a whole [2].

11.2.1 Data Protection in AI-Driven Financial Systems

The handling of sensitive and personal consumer data by financial organizations makes data privacy an essential component of AI-powered finance. Sustaining trust and complying with regulatory requirements includes maintaining the confidentiality and integrity of people's data while safeguarding their right to privacy. Important things to think about when it comes to data privacy [3] in AI-powered finance are:

- *Getting consent and restricting data use*: Before collecting and using a person's personal information, financial institutions must obtain the person's informed consent. They must guarantee that the data is utilized exclusively for the stated purpose and provide a clear explanation of the reason behind the collection. Any future use of the data must either have legal or acceptable grounds or additional consent.
- *Retention and data collection minimization*: Financial institutions should simply gather and keep the absolute minimum of information required to fulfil their stated objectives. Reducing the amount of needless data collected can help lower the possibility of misuse or unauthorized access. In order to prevent the unnecessary keeping of personal data, it is also important to set explicit policies regarding data retention.
- *Protective safeguards*: To prevent unwanted access, loss, or destruction of personal data, strong security measures are necessary. This entails putting procedures in place including access controls, encryption, safe storage, and regular security assessments. Information breaches may have detrimental effects on a person's reputation and legal standing, as well as financial institutions.
- *Techniques for data pseudonymization and anonymization*: Financial institutions can employ strategies like anonymization and pseudonymization to enhance privacy protection. In order to prevent data from being linked back to particular people, personally identifiable information must be removed or encrypted. This process is anonymization. Contrarily, pseudonymization substitutes direct identifiers for pseudonyms, enabling the usage of data for specific objectives while maintaining the privacy of individuals.
- *Data exchange and external service suppliers*: Financial institutions should set up strong data protection agreements before exchanging data with

outside suppliers. These agreements must clearly outline each party's duties and responsibilities as well as the security and privacy measures that have been put in place for the shared data.
- *Adhering to regulatory standards*: Financial institutions operating across various jurisdictions must abide by any data protection regulations, including the General Data Protection Regulation (GDPR) in the European Union and the California Consumer Privacy Act (CCPA) in the United States. Honoring the rights of data subjects, such as the ones to access, rectification, erasure, and processing restriction, is part of compliance.
- *Evaluations of privacy risks*: Before deploying financial institutions should do privacy impact assessments (PIAs) for AI systems that handle personal data. PIAs assist in identifying and mitigating privacy concerns associated with AI use, guaranteeing that the required protections are in place to people's right to privacy.

Financial institutions may build consumer trust, comply with legal requirements, and lower the risk of data breaches and unauthorized access by putting data protection first in AI-powered finance. This strategy improves the credibility and standing of financial institutions in the sector while also protecting individuals' privacy [4].

11.2.2 Securing Data in AI-Driven Financial Systems

Given that financial institutions handle vast volumes of sensitive financial and personal data, data security is a crucial component of AI-powered finance. Protecting this data against loss, alteration, and unauthorized access is crucial to maintaining client confidence, adhering to legal requirements, and the integrity of financial systems [5]. Key considerations related to data security in AI-powered finance include:

- *Access management*: Financial institutions should establish robust controls on access to sensitive data to ensure that only those who are authorized can access it. Setting up strong authentication protocols, role-based access controls (RBAC), and regular privilege evaluations and revocations are necessary to achieve this. Limiting data access to those who have a legitimate need to know helps reduce the possibility of unwanted access or misuse.
- *Data encryption*: Encryption is necessary to safeguard data during both transmission and storage. To protect sensitive data, financial institutions should employ strong encryption methods that, even in the event that they are intercepted, render the data unreadable without the right decryption keys. This covers encrypting information that is saved on portable devices, transferred over networks, and kept in databases.
- *Prevention of data loss*: Financial institutions should deploy measures for data loss prevention (DLP) to identify and stop sensitive data from being disclosed or sent without authorization. DLP systems are able to keep an eye on data transfers, spot possible security breaches, and enforce rules to keep data safe across various channels, including e-mail, file transfers, and cloud storage.

- *Robust infrastructure security*: Installing firewalls, installing intrusion detection and prevention systems (IDPSs), and updating software and systems on a regular basis are all important steps financial institutions should take to provide a safe infrastructure for processing and storing data. Network segmentation can also aid in the isolation of systems and sensitive data, lessening the possible consequences of a security breach.
- *Incident management and surveillance*: Strong incident response procedures are essential for financial institutions to properly identify, handle, and recover from security issues. This involves proactive monitoring of systems and networks for suspicious activities, prompt incident reporting, and a well-defined process for managing and remediating incidents.
- *Security for vendors and external partners*: Financial institutions frequently collaborate with outside suppliers and service providers, thus it is critical to confirm that these suppliers have sufficient security measures in place to safeguard the data they manage. This involves conducting due diligence assessments, including security audits and evaluations of their data handling practices, and establishing appropriate contractual agreements to address data security requirements.
- *The purpose of staff education and awareness*: It is to inform staff members about best practices for data security, the risks associated with data breaches, and the importance of following established security policies and procedures, financial institutions should give priority to employee awareness and training programs. Regular training sessions and awareness campaigns should be organized for staff members for them to be security-conscious.

Financial institutions can lower the risk of data breaches, unauthorized access, and data manipulation by putting strong data security measures in place. This method helps to preserve consumers' confidence and trust while protecting sensitive financial data, regulators, and stakeholders within the AI-powered finance ecosystem [6].

11.2.3 Protective Strategies and Protocols for Data Security

In order to safeguard data in finance powered by AI, financial organizations must to adopt a comprehensive set of security measures and protocols. Key security measures to consider include:

- *Data encryption techniques*: Make use of robust encryption technologies to protect data while it's being moved or stored. Ensure that sensitive data is encrypted when it is being transmitted across networks and that it is encrypted when it is stored in files and databases. This makes it more likely that compromised data will stay unreadable in the absence of the necessary decryption keys.
- *Access management systems*: Implement strong access restrictions to limit authorized personnel's access to sensitive data. Utilize multi-factor authentication (MFA) and other robust authentication methods to verify user identities. Use RBAC to assign access privileges in line with job roles and

responsibilities. Review access privileges on a regular basis and remove personnel who are no longer in need of them.

- *Protected network architecture*: By using IDPSs, firewalls, and verifying that routers, switches, and other network components and equipment are configured securely, you can maintain a secure network infrastructure. By separating critical data and systems, you can reduce the possible impact of a compromise by using network segmentation.
- *Frequent patch management and security updates*: Maintain software functionality use the most recent security fixes to keep apps and systems current. Update security patches frequently to fix known flaws and protect against new threats. Establish a thorough patch management procedure to guarantee that patches are applied timely and consistently across all systems within the organization.
- *Robust password guidelines*: Enforce stringent password regulations that demand staff members to use complicated passwords and to update them often. Steer clear of default or simple-to-guess passwords. To safely store and manage passwords, think about putting password management software or password vaults into place.
- *Initiatives for employee awareness and education*: Organize regular security awareness training sessions to inform staff members on phishing scams, social engineering tactics, security best practices, and safe data handling. Make sure staff members are trained to identify and report possible security incidents, and that they are aware of their roles and duties in data protection.
- *Creating and maintaining an incident response*: Plan is the incident management strategy a procedure that outlines what should be done in the event of a data breach or security incident. Procedures for promptly detecting, containing, investigating, and correcting issues should be included in the plan. Test and update the strategy frequently to make sure it remains effective and up-to-date.
- *Plan for data backup and recovery*: Put in place an extensive a data backup plan to guarantee that, in the event of a security breach, system failure, or corrupted data, data may be recovered. To reduce downtime and data loss, regularly test backups to ensure their integrity and create a disaster recovery strategy during major incidents.
- *Security for suppliers and external partners*: Assess the safety procedures prior to doing business with third-party vendors and service providers. Make sure there are contracts in place that outline obligations and expectations for security. Assess vendors' security postures on a regular basis and carry out audits to confirm that security standards are being followed.
- *Compliance with regulatory standards*: Remain up to date on pertinent legislation pertaining to privacy and data protection, including the CCPA, GDPR, and other laws that apply to your industry. Make sure that these legal and regulatory standards are followed, and put in place the proper measures to safeguard data as needed.

Sensitive data in AI-powered finance must be protected by implementing a multi-layered, comprehensive data security strategy. Financial institutions should regularly do security audits and penetration tests, monitor and assess their security posture, and keep up with the latest developments in data security best practices and new threats [7].

11.2.4 The Moral Consequences of AI-Powered Financial Systems

Financial institutions need to take into account a number of ethical issues related to AI-powered finance in order to guarantee the ethical and equitable application of this technology. Important moral considerations consist of:

- *Equity and impartiality*: Financial institutions have to make sure that biases based on socioeconomic class, gender, or race are not reinforced or amplified by AI systems. To minimize prejudice and guarantee equitable results for every person, AI systems must be carefully designed and trained. It is also critical to conduct routine audits and monitoring of AI systems to identify and address any potential biases that may develop over time.
- *Clarity and understandability*: Financial AI systems should be open and transparent, offering comprehensible explanations for their decisions and recommendations. It needs to be possible for users to comprehend how AI models operate and what variables affect their choices. This openness promotes confidence, enables users to confirm and dispute results, and aids in the detection of any biases or mistakes.
- *Safeguarding privacy*: Financial organizations are expected to manage personal information in a way that respects individuals' right to privacy. Complying with applicable data protection standards and obtaining informed consent are essential steps for collecting, storing, and processing data. AI systems ought to be created with the least amount of personal data collection and usage while still accomplishing the goals for which it was created.
- *Responsibility and oversight*: Financial institutions must take responsibility for the choices and acts made by AI systems. Clearly defined responsibilities should be established, with appropriate oversight and governance structures in place. This involves defining responsibility for any damage or negative impact brought about by AI systems and setting up mechanisms for redress and dispute resolution.
- *Data management and protection*: Robust data governance procedures must to be used to ensure the privacy of, precision, and integrity of the data utilized by AI systems. This include maintaining data security, assuring data quality, and adhering to laws controlling data usage. Prioritizing data security procedures is vital for financial institutions to prevent any unauthorized access, breaches, or abuse.
- *Human supervision and involvement*: AI systems should be designed to complement human experts instead of completely replacing them. In order to guarantee that AI outputs are precise, dependable, and compliant with ethical

norms, human supervision and involvement are essential. When required, humans ought to be able to examine, challenge, and overrule AI choices.

- *Systemic risks and financial consequences*: Financial institutions should evaluate the broader systemic risks and potential economic impacts of AI adoption in finance. This includes assessing risks such as market manipulation, concentration of power, and effects on employment. Developing mitigation strategies to address these risks is essential to ensure that AI-powered finance benefits the society as a whole.
- *Ongoing surveillance and assessment*: It is necessary to regularly monitor and assess AI systems in order to appraise their efficacy, morality, and overall performance. When deploying and using AI technologies, financial institutions should be proactive in identifying and resolving any unintended repercussions, biases, or ethical issues that may arise.

By proactively addressing these ethical considerations, financial institutions can foster trust, fairness, and accountability in AI-powered finance. This approach not only helps mitigate risks but also ensures that the application of AI technologies is done in a way that upholds social values and enhances the well-being of individuals and communities [8].

11.2.5 AI-Driven Finance: Harmonizing Data Privacy, Security, and Ethical Practices

In AI-powered finance, achieving a balance between data privacy, security, and ethics is essential to guaranteeing responsible and trustworthy use of AI while safeguarding individual rights and interests. Here are some considerations for achieving this balance:

- *Designing with privacy in mind*: Design and create AI systems with privacy considerations in mind from the beginning. Implement measures such as data minimization, purpose limitation, and obtaining user consent to ensure that personal data is collected, processed, and stored with privacy in mind. Conduct PIAs to identify and mitigate privacy risks associated with AI systems.
- *Robust data protection*: Put strong data security procedures in place to safeguard sensitive financial and personal data. This includes using encryption, access controls, secure infrastructure, regular security updates, and having incident response plans in place. By safeguarding data against unauthorized access, breaches, or misuse, financial institutions can protect individuals' privacy while maintaining the integrity of their systems.
- *Ethical principles and norms*: Clearly define ethical principles and guidelines for the use of AI in finance. These guidelines should cover issues such as fairness, transparency, explainability, and accountability. They should provide a framework for responsible AI development, deployment, and use, ensuring that AI systems operate in alignment with ethical principles and societal values.

- *Ethical frameworks and criteria*: Provide unambiguous ethical standards and standards for AI use in finance. These guidelines should address key issues such as fairness, transparency, explainability, and accountability. They should offer a framework for responsible AI development, deployment, and use, ensuring that AI systems function in a way that aligns with ethical principles and societal values.
- *User autonomy and knowledgeable consent*: Empower users by clearly informing them about how their data will be used in AI-powered finance. Obtain informed consent for data collection and processing, and give individuals control over their data, including the ability to revoke consent or request data deletion. Transparent communication and user-friendly interfaces can build trust and enable individuals to make informed decisions about their data.
- *Routine audits and oversight*: Perform routine AI audits and monitoring systems to ensure compliance with privacy, security, and ethical standards. This includes ongoing evaluation of algorithmic biases, data quality, and system performance. Regular reviews help identify and address potential risks, unintended consequences, or ethical concerns, enabling timely corrective actions.
- *Partnerships with regulatory bodies and industry*: Financial institutions should actively collaborate with regulators, industry organizations, and experts to develop best practices, standards, and regulations for AI-powered finance. This engagement ensures that ethical, security, and privacy concerns are effectively addressed and incorporated into regulatory frameworks and industry guidelines [9].
- *Ongoing learning and professional development*: Develop an ethical, security, and privacy-conscious culture among staff members who deal with AI-powered finance. To guarantee that staff members are aware of the importance of data privacy, security, and ethical issues, provide frequent training sessions and educational initiatives. Employees that receive this training are better able to recognize hazards, make educated decisions, and follow best practices.
- *Open communication and transparency*: Engage in transparent and open dialogue with the public, customers, and stakeholders about the application of AI in finance. Clearly explain how AI systems are used, their advantages, and the security measures in place to privacy and security. Solicit feedback and address concerns to build trust and promote responsible AI practices [10].

Balancing data privacy, security, and ethics in AI-powered finance necessitates a multidimensional approach that encompasses legal requirements, industry standards, and societal expectations. By integrating privacy, security, and moral issues financial institutions may encourage ethical AI practices and guarantee that the advantages of AI are realized while preserving individual rights and interests at every step of AI development and implementation [11].

11.3 METHODOLOGY

11.3.1 Method of Research

In order to gather and compile scholarly articles, academic publications, conference papers, and pertinent literature from reliable databases, this literature evaluation follows a methodical procedure. In order to capture the most recent developments in privacy-preserving methods within AI-powered cybersecurity, it focusses on articles from the past ten years [12].

11.3.2 Search Methodology

A variety of academic all databases, pertinent institutional libraries, were searched thoroughly using this approach. To find pertinent material, a variety of keyword combinations were employed, including "privacy-preserving techniques," "AI in cyber security," "differential privacy," "homomorphic encryption," "secure multi-party computation," "federated learning," and "cyber security challenges" [13].

11.3.3 Qualification Standards

The papers that made up this evaluation of the literature had to meet certain requirements: they had to be published in all journals, academic publications, or conferences that dealt with privacy-preserving methods in AI-powered cybersecurity. Furthermore, a key inclusion criterion was relevance to subjects including cybersecurity possibilities and problems, AI integration, privacy protection, and issues [14].

11.3.4 Guidelines for Exclusion

Publications without enough rigor and relevance, or those that did not explicitly address the junction of AI-powered cybersecurity and privacy-preserving strategies, were not taken into consideration. Opinion pieces, editorials, and blog posts—all non-peer-reviewed—were also excluded [15].

11.3.5 Gathering and Examining Data

The process of extracting data from the chosen literature required a careful reading, analysis, and extraction of the most important concepts, techniques, conclusions, and constraints. Then, utilizing AI to power cybersecurity, this extracted data was synthesized and grouped into themes that addressed privacy-preserving strategies, obstacles, and possibilities.

11.3.6 In-Depth Evaluation

To assess the benefits, drawbacks, and implications of the examined literature, a critical analysis was done. This procedure yielded a thorough grasp of the existing situation, pointing up inconsistencies, gaps, and places in need of more investigation.

11.4 FUTURE SCOPE

11.4.1 Advances in Hybrid Privacy Security Techniques

A review of the literature indicates that there is room for more investigation into creating and applying hybrid privacy-preserving techniques. The drawbacks of individual approaches might be addressed by combining techniques like differential privacy with homomorphic encryption providing a more reliable and well-rounded approach to data privacy in AI-powered cybersecurity.

11.4.2 Improved Computational Performance

One major difficulty that still has to be addressed is the computational overhead that comes with privacy-preserving approaches. Subsequent investigations have to concentrate on creating novel approaches or enhancements to lessen this computational load while upholding strong privacy measures. More effective implementations might be made possible by taking advantage of developments in remote computing, hardware acceleration, or creative algorithmic enhancements [16].

11.4.3 Cross-Disciplinary Partnerships

Experts in the fields of data privacy, AI, cryptography, cybersecurity, and law must work together due to the interdisciplinary nature of privacy-preserving solutions in cybersecurity. By incorporating these many viewpoints and skill sets, complete frameworks that successfully strike a compromise between privacy and utility and guarantee adherence to changing regulatory norms may be developed.

11.4.4 Legal and Moral Repercussions

It is imperative to conduct more research on the moral implications of privacy-preserving measures. It is critical to comprehend the potential biases, societal effects, and ethical issues related to using AI-powered cybersecurity solutions. Moreover, constant oversight and conformity to developing regulatory frameworks, like GDPR and new data privacy regulations, are necessary to guarantee that AI-powered cybersecurity solutions are morally and legally sound.

11.4.5 Training and Knowledge

It is essential to raise awareness and educate stakeholders, including developers, legislators, and end users. A more informed approach to adopting and implementing preserving technologies can be fostered by initiatives that emphasize the value of privacy, AI ethics, and responsible data handling in cybersecurity scenarios.

11.5 RESULTS AND CONVERSATION

11.5.1 Utilizing AI-Powered Internet of Things in Integrated Smart Health Care Systems

The healthcare sector is rapidly adopting Internet of Things (IoT) powered by AI (AIIoT) to improve medical services and devices. Patients who require ongoing care, individuals with chronic illnesses, and the elderly can all benefit from this integration. Sodhro projects that by 2025, spending on AIIoT solutions for healthcare will total one trillion US dollars. This investment could significantly advance the accessibility, personalization, and timeliness of healthcare services. For example, Figure 11.1 demonstrates a smart healthcare system showcasing these advancements.

In recent years, hospitals have embraced AI-driven IoT technology more and more, including them into cloud-based systems, electronic medical records, and patient rooms. Experts predict that digital healthcare will revolutionize the sector by lowering prices and increasing access to diagnosis, treatment, and preventive care. Handling patients at high risk remains a significant challenge in controlling healthcare expenses. Chronic disease management accounts for approximately 30% of healthcare spending in the United States, with substantial costs attributed to conditions such as heart disease, diabetes, and asthma [17].

AI-driven IoT devices, particularly in healthcare, have the potential to transform patient monitoring and management. By integrating these devices with cloud storage solutions like Amazon AWS, healthcare providers can leverage real-time data analysis to improve patient outcomes and reduce costs. This is further improved by the Internet of Medical Things (IoMT), which links medical devices over Wi-Fi, enabling seamless data transfer and integration with health IT systems. This connectivity allows for continuous monitoring of high-risk patients, timely interventions, and a reduction in unnecessary expenses.

Research shows that numerous domains in health care can significantly benefit from the integration of AI-driven IoT technology. One such domain is elder care, where monitoring the movements and health of elderly individuals in hospitals and nursing facilities is essential. This includes the use of various bedside devices, such

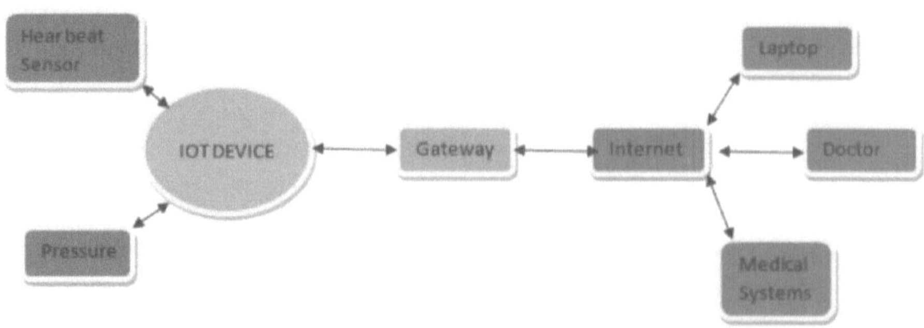

FIGURE 11.1 An illustration of a smart healthcare system.

as EKG monitors. The field has seen continuous growth, with current advancements emerging globally in AI-powered IoT applications.

Patients and providers stand to benefit greatly from the increasing use of AI-driven IoT in healthcare. AI-powered IoT enables remote monitoring and communication to improve medical treatments. Furthermore, wearable technology and mobile medical apps are included in the AI-driven IoT for health care enable patients to record their own health information. This trend is largely driven by the data revolution, which allows individuals to improve their well-being by utilizing connected devices like wearables, tablets, and smartphones [10].

The analysis of data gathered from making better decisions is aided by data from portable devices, diagnostic information from imaging equipment, and electronic medical records. Patients are able to manage their personal health more actively because to this skill. Such tailored, data-driven health assessments will be available soon is expected to become a standard practice, providing patients with customized strategies to combat illnesses. The insights generated from this data empower individuals to learn how to improve their well-being and motivate them to take charge of their health. Researchers suggest that in the field of clinical decision support software, a new industry is developing which is closely tied to AI-driven IoT. This sector is expanding, enhancing the role of connected devices by integrating them more directly into clinical decision-making processes [18].

11.5.2 Security Concerns with AI-powered IoT

Increasing number of devices are being connected to the internet as a result of the fast growth of AI-driven IoT, which makes them more susceptible to security risks. In late 2018, for example, two security researchers discovered that more than 68,000 medical systems were accessible online, more than 12,000 of which belonged to a single healthcare provider. One of the main issues with this finding was that these devices were connected through computers running on the outdated Windows XP operating system, known for its numerous exploitable vulnerabilities. The researchers used to find these systems, use Shodan, a search engine built to find IoT devices. Due to their hard-coded login passwords and vulnerability to brute-force assaults, these devices are especially vulnerable to hacking. Using straightforward Shodan searches, the researchers were able to find a number of medical equipment throughout their analysis, including nuclear medicine systems, anesthesia machines, and infusion devices. They claimed that more than 50,000 Secure Shell (SSH) authentications on these fictitious medical devices had been accomplished by attackers, who had also successfully installed malware payloads. Furthermore, the majority of the time the attackers left the compromised PCs as part of their botnets, seemingly oblivious to the extent of their compromise [19].

11.5.3 IoT-Based Healthcare Systems' Privacy Risks

As there are so many unknowns about how the AI-driven IoT will affect society, it is difficult to quantify the possible risks that could arise. However, the results of mass data collecting from social media, mobile networks, and smartphone sensors are already apparent. The effects of IoT powered by AI on privacy can be better understood by examining how similar technologies have historically impacted privacy. This comparison suggests that even if individual data transmissions from endpoint devices do not immediately raise privacy concerns, the aggregation of data from multiple sources can lead to significant privacy issues. In addition, certain features of AI-driven IoT make privacy threats particularly complex. Data collection often occurs passively, intrusively, and pervasively, which means users may be unaware that their activities are being monitored [20]. Choi highlighted various IoT components, their associated vulnerabilities, and the types of threats or attacks they face, according to Table 11.1 summary. The table shows that the main dangers or vulnerabilities of IoT components include physical attacks, RFID integration, wireless sensor networks (WSNs) integration, DoS/DDoS attacks, and unauthorized data access.

Research acknowledges that AI-driven IoT is inherently pervasive, with various devices collecting data on users and their surroundings to provide specific services. The collected data are then processed by healthcare service providers, often under the users' control. However, even though data anonymization techniques, such as swapping out personal data for randomly created, unique IDs, are employed, they often fall short of ensuring true anonymity. It has been shown that individuals' identities can still be deduced from anonymized datasets. A notable example of this type of privacy breach involves a case in Massachusetts where a group of insurance

TABLE 11.1
Health Care Applications

Elements of Internet of Things (IoT) Systems	Security Flaws	Types of Threats and Attack Methods
Tangible items	Devices within this layer possess constrained computational, communication, and storage capabilities.	Attacks targeting physical components or infrastructure
	Since nodes are spread across remote locations, adversaries can readily access these devices and carry out malicious activities, include taking out security keys and credentials or resetting the devices.	Integration of radio frequency ID technology. Integration of wireless sensor networks (WSNs). Unapproved access to data and issues related to access control.
Technologies for data transmission and exchange	The network infrastructure is highly dynamic.	Wireless communications within personal area networks and local area networks.
	Devices operate with low power. The network experiences high rates of data loss.	Wireless communications across wide area networks (WAN).
	Choosing appropriate security techniques for each network element presents challenges.	Secure communication protocols for IoT in environments with limited resources.
	The network's defense capabilities differ across various networks.	Securing data during transmission.

commissions purchased health insurance for state workers and created hospital record visits that were made available to scholars for research purposes. To protect patient privacy, certain fields like addresses, names, and social security numbers were removed from the data, but information such as ZIP codes, gender, and birthdates remained, which still posed a risk to privacy [21].

11.6 CONCLUSION

In summary, in the world of AI-driven finance, striking a balance between data privacy, security, and ethics is crucial. Financial institutions must focus on safeguarding personal and financial information while adhering to ethical standards and societal values. By embedding strong data security mechanisms, adhering to privacy by design principles, and following ethical guidelines, these institutions can promote the responsible and trustworthy use of AI. Transparency, empowering users, and securing informed consent are key to fostering a privacy-aware approach. AI systems must be routinely audited and monitored in order to detect and reduce any dangers, biases, and unexpected consequences. Developing best practices and regulatory frameworks requires close cooperation with regulators and industry stakeholders. Employees with ongoing education and training are better able to foster a culture of ethical awareness, security, and privacy, ensuring they make informed decisions and follow best practices. Engaging in open communication with the public builds trust and addresses concerns. Maintaining a balance between data privacy, security, and ethics requires continuous effort and adaptability as technology advances and new challenges arise. By focusing on these priorities, financial institutions can build trust, protect individual rights, and ensure that AI-driven finance serves the greater good.

REFERENCES

1. Kim, H., Ben-Othman, J., Mokdad, L., Son, J., & Li, C. 2020, Research challenges and security threats to AI-driven 5G virtual emotion applications using autonomous vehicles, drones, and smart devices, IEEE Network 34, 288–294.
2. Butpheng, C., Yeh, K.-H., & Xiong, H. 2020, Security and privacy in IoT-cloud-based e-health systems: A comprehensive review, Symmetry 12, no. 7: 1191.
3. Si, S. L., You, X. Y., Liu, H. C., & Zhang, P. 2018, DEMATEL technique: A systematic review of the state-of-the-art literature on methodologies and applications, Math Probl Eng 2018: 1–33.
4. Gonaygunta, H. 2023, Machine learning algorithms for detection of cyber threats using logistic regression. *Department of Information Technology, University of the Cumberlands.*
5. Gonaygunta, H., & Sharma, P. 2021, Role of AI in product management automation and effectiveness.
6. Oluwaseyi, J. 2024, Machine Learning for Predictive Maintenance in Industrial Settings: Case Studies and Best Practices.
7. Zhang, Q. 2022, Advancements in privacy-preserving AI models for cyber security, IEEE Transactions on Information Forensics and Security 17, 2666–2679.
8. Rajeswari, K., Vivek, P., & Nandhagopal, J. 2020, Swing up and Stabilization of Rotational Inverted Pendulum by Fuzzy Sliding Mode Controller. In D. Hemanth, V. Kumar, S. Malathi, O. Castillo, B. Patrut (Eds.) Emerging Trends in Computing and

Expert Technology. COMET 2019. Lecture Notes on Data Engineering and Communications Technologies (Vol 35). Springer, https://doi.org/10.1007/978-3-030-32150-5_40.

9. Munn, Z., Peters, M. D., Stern, C., Tufanaru, C., McArthur, A., & Aromataris, E. 2018, Systematic review or scoping review? Guidance for authors when choosing between a systematic or scoping review approach, BMC Medical Research Methodology 18, no. 1: 1–7.
10. Wang, L. 2021, Hybrid privacy-preserving methodologies for AI-powered cybersecurity, ACM Transactions on Privacy and Security 24, no. 4: 1–28.
11. Nandha Gopal, J., & Muthuselvan, N. B. 2020, Current mode fractional order PID control of wind-based quadratic boost converter inverter system with enhanced time response, Circuit World 47, no. 4: 368–381.
12. Firouzi, F., Farahani, B., Barzegari, M., & Daneshmand, M. 2020, AI-driven data monetization: The other face of data in IoT-based smart and connected health, IEEE Internet of Things Journal 9, no. 8: 5581–5599.
13. Nadia-Ghomsheh, A., Farahani, B., & Kavian, M. 2020, A hierarchical privacy-preserving IoT architecture for vision-based hand rehabilitation assessment, Multimedia Tools and Applications 80, no. 20: 31357–31380.
14. Puaschunder, J. M. 2019, The legal and international situation of AI, robotics and big data with attention to healthcare. In: Report on behalf of the European Parliament European liberal Forum.
15. Nandha Gopal, J., BalasubramanianMuthuselvan, N., & Muthukaruppasamy, S., 2021 Model predictive controller–based quadratic boost converter for WECS applications. International Transactions on Electrical Energy Systems 31, no. 12: e13133.
16. Yaseen, M., Durai, P., Gokul, P., Justin, S., & Anand, A. J., "Artificial Intelligence Based Automated Appliances in Smart Home," 2023 Seventh International Conference on Image Information Processing (ICIIP), Solan, India, 2023, pp. 442–445.
17. Wang, J. 2020, Overcoming computational overhead in privacy-preserving techniques for AI-powered cybersecurity. Proceedings of the IEEE International Conference on Cybersecurity and Privacy (ICCP) (pp. 221–235).
18. Zhang, Q. 2022, Advancements in privacy-preserving AI models for cybersecurity, IEEE Transactions on Information Forensics and Security 17, 2666–2679.
19. Anuradha, T., Jasphin Jeni Sharmila, P., Kanimozhiraman, K., Kalaiselvi, C. K., Shruthi, & Jose, A. A., 2024, "Automatic Categorization of Emails into Folders Based on the Content of the Messages," Third International Conference on Intelligent Techniques in Control, Optimization and Signal Processing (INCOS), Krishnankoil, Virudhunagar district, Tamil Nadu, India, pp. 1–6.
20. Gopal J, N., & Muthuselvan, N. B. 2021, Educational tool for analysis of PI and fractional order PI controlled quadratic boost converter system using MATLAB/Simulink, The International Journal of Electrical Engineering and Education. https://doi.org/10.1177/00207209211013435.
21. Hemalatha, S., Mahalakshmi, M., Vignesh, V., Geethalakshmi, M., Balasubramanian, D., & Jose, A. A., "Deep Learning Approaches for Intrusion Detection with Emerging Cybersecurity Challenges," 2023 International Conference on Sustainable Communication Networks and Application (ICSCNA), Theni, India, 2023, pp. 1522–1529.

12 Artificial Intelligence in Financial Fraud Detection

M. Narender and A. Jose Anand

12.1 INTRODUCTION

Financial services fraud detection has always been problematic, requiring continuous alteration to the continuously altering monetary crime scene [1]. The seriousness of financial fraud puts the truthfulness and constancy of the whole monetary system in jeopardy in addition to posing significant dangers to specific customers [2]. Traditional fraud recognition practices have been proven to be insufficient over time in the face of more complexes and scientifically advanced fraudulent actions [3]. Fraud detection is the process of identifying and stopping dishonest behaviors intended to manipulate financial transactions for illegal benefit. The capacity to differentiate between genuine and fraudulent transactions is crucial in the complex world of finance, where enormous amounts of money are transferred on everyday basis [4]. The defense of people's and companies' monetary interests as well as the general constancy of the monetary sector climaxes the standing of fraud uncovering.

The antiquity of fraud recognition is categorized by an ongoing battle to keep up with shrewd con artists who adapt their approaches to take benefit of holes in arrangements that are already in place. Conventional techniques, which mostly relied on human inspection and rule-based procedures, were somewhat successful but eventually fell short when faced with more complex schemes. The evolution of fraud detection methods is comparable to a game of cat and mouse, in which adversaries exploit weaknesses in defenses and creatively counter new threats [5]. The basis for the rationale behind incorporating artificial intelligence (AI) into fraud protection is the requirement for complex, adaptable, and immediate detection abilities [6]. With its machine learning algorithms and data processing power, AI transforms the fraud prevention landscape. Since AI systems can analyze large datasets, identify intricate patterns, and adapt to evolving fraud schemes, they are an effective tool in the fight against financial crimes. The need to improve fraud detection efficacy and efficiency as well as to keep up with more complex fraudulent [7] activities that are difficult for conventional techniques to handle is what motivates the incorporation of AI. We explore the historical background, the changing field of AI-driven fraud detection, ethical issues, practical applications, and emerging trends that will shape AI's role in defending the financial services sector against fraudulent activity in this in-depth analysis. AI applications have garnered significant interest in the scientific community in recent years and have affected nearly every part of our lives, encouraging automation and innovation across a broad spectrum of sectors.

DOI: 10.1201/9781003521020-12

12.2 EVOLUTION OF AI IN FRAUD DETECTION

The rise of AI in fraud detection illustrates the shift from manual processes to complex automated systems. Due to the complexity of financial crimes, quick decisions and adaptability are now essential. Because of this, integrating AI is a crucial development. The transition from manual to automatic detection was one of the most significant phases in the evolution of fraud detection. Even though they were somewhat effective, manual processes were labor-intensive, time-consuming, and prone to human error. As the volume and complexity of financial transactions increased, there was a pressing need for tools that could assess massive datasets rapidly and identify fraudulent activity in real time [8]. Automated technology improved the efficacy, speed, and scalability of the fraud detection system. AI played a crucial role in automating the analysis of transactional data, enabling financial institutions to detect patterns and irregularities that could potentially indicate fraudulent activity with unprecedented accuracy. An AI model is trained for supervised learning using labeled datasets, enabling it to recognize patterns associated with both legitimate and fraudulent transactions [9]. This model might then predict new, unseen data. On the other hand, in the absence of labeled datasets, unsupervised learning uses natural structures in the data to identify patterns or abnormalities. Both methodologies help identify fraud by distinguishing between normal and suspicious behavior [10].

The discipline of fraud detection has been revolutionized by deep learning (DL) techniques, especially neural networks, which enable computers to automatically create hierarchical representations of data. Neural networks are particularly adept at handling complicated and nonlinear interactions, which makes them suitable for identifying intricate patterns that could be indicators of fraud [11]. Their ability to independently extract features from data has significantly increased the accuracy of fraud detection tools. Natural language processing (NLP) is another area of AI that has been used to detect fraud. NLP algorithms can be used to examine textual data, such as conversation logs or transaction descriptions, in order to identify linguistic patterns connected to fraudulent conduct [12]. NLP's ability to interpret linguistic nuances contributes to a more comprehensive and advanced fraud detection approach. The hallmark of AI's advancement in fraud detection is the replacement of static rule-based systems with dynamic, learning-driven models. These algorithms make use of enormous datasets, adapt dynamically to novel hazards, and give financial institutions a competitive advantage over crafty con artists. A powerful arsenal for the ongoing battle against financial crimes is the combination of DL techniques, NLP, and supervised and unsupervised learning [13].

12.3 FRAUD DETECTION WITH AI: CONCEPTS AND TECHNIQUES

AI-driven systems detect fraud and employ complex algorithms to examine trends and irregularities present in transactional data. AI continually modifies conventional rule-based systems, modifies its models based on past data to detect fraudulent activities that can elude detection by traditional means. AI comprises many methods to detect fraud [14].

12.3.1 Detection of Anomalies Automatically by AI

One important technique in AI-driven fraud detection is anomaly detection. AI systems closely scrutinize transactional data to find abnormal patterns, such as unexpected transaction quantities or geographic variances in transaction locations over a short period of time. These anomalies typically serve as red flags for potential fraud. By using machine learning, these systems may identify irregularities that would be difficult for hackers, crackers, and brute force analysts to locate manually. This increases the overall efficacy and accuracy of fraud detection operations [15]. AI has raised the bar for financial industry fraud detection and prevention by demonstrating remarkable performance in identifying irregularities. AI is considered effective because of its ability to analyze large amounts of data, recognize intricate patterns, and adapt to ever-changing dangers. Using complex algorithms and machine learning approaches, AI has proven to be able to detect irregularities with unparalleled efficiency and precision [16]. One of the key benefits of AI is its ability to identify intricate patterns that may elude the detection of fraud using traditional methods. Machine learning algorithms, particularly those that use unsupervised learning techniques, are capable of identifying minor deviations from the norm in large datasets [17].

These anomalies might include anomalous user behaviors, strange spending locations, or inconsistent transaction patterns. Because AI can learn new patterns automatically, it is very helpful in identifying fraud that evolves over time. AI systems can also consider multiple variables at once, including transaction history, user behavior, and contextual information. This thorough approach enables AI to analyze complex relationships and spot anomalies that might be indicators of fraud. The continuous learning component ensures a dynamic defense against evolving fraud schemes by making sure the system adjusts to new threats [18]. Numerous case studies demonstrate the value of AI in identifying anomalies and preventing fraud in the financial services sector. For instance, a machine learning model was employed by a big bank to analyze transaction data and identify unusual trends that might indicate fraud. The system was able to detect and thwart a sophisticated fraud scheme that was going to use hacked account information, saving the bank and its clients a great deal of money [19].

In a different case, a credit card company used AI algorithms to look at client behavior and transaction patterns. The AI system's real-time anomaly identification led to the immediate suspension of compromised accounts and the halting of illicit transactions [20]. These illustrations show how AI's quick processing and analysis of large datasets helps prevent fraud before it happens. When it comes to fraud detection, AI is unquestionably better than conventional techniques. It could be challenging for conventional methods, which usually rely on rules and preset criteria, to adapt to novel fraud tactics [21].

Rule-based systems are less successful in recognizing complex and quickly changing patterns since they usually define static thresholds for certain parameters. Conversely, AI uses dynamic algorithms that change in response to real-time inputs. Because of their flexibility, AI systems are able to anticipate new fraud tendencies and adapt to them without the need for human involvement [22].

AI-driven techniques are better at spotting irregularities and stopping fraud because they can assess several factors at once, learn from past data, and recognize complicated linkages. In summary, AI's capacity to revolutionize fraud detection in financial services is demonstrated by its efficacy in recognizing abnormalities.

AI improves the capacity of the industry to counteract increasingly complex fraud schemes by utilizing sophisticated algorithms and machine learning approaches. Case studies underscore the observable benefits of AI-driven fraud prevention, highlighting the technology's edge over conventional approaches and reaffirming its status as a potent instrument for safeguarding financial systems [23].

12.3.2 Behavioral Analysis

Another essential method used by AI-driven systems to detect fraud is also behavioral analysis. AI systems can identify anomalies that can point to fraudulent activity by tracking the patterns of client behavior over time. Alerts may be triggered, for example, by abrupt changes in spending patterns, such as an increase in high-value transactions or purchases made in strange places. This method, which doesn't only rely on strict rules, takes into account the context and unique client behavior patterns to provide a more sophisticated analysis of possible fraud [24].

12.3.3 Natural Language Processing

E-mails, social media postings, and customer service exchanges are examples of unstructured data sources that AI with NLP skills may interpret and identify signs of fraud from. NLP, for instance, may recognize suspect account activity by looking at social media postings or identify phishing efforts by examining the language used in emails. Financial organizations may now discover fraudulent activity that could go undetected using more conventional data analysis techniques, thanks to this capacity to handle and analyze unstructured data [25].

12.3.4 Continuous Learning

One characteristic that sets AI systems apart in fraud detection is continuous learning. By continuously learning from fresh data, these systems increase the accuracy of their fraud detection and remain abreast of changing fraud patterns and strategies. AI systems may improve their predicting powers and respond to new threats by continuously improving their models and absorbing new data. In the face of ever-evolving fraud strategies, this iterative learning process is essential to preserving the efficacy of fraud detection systems [26].

AI's place in banking and the prevention of financial fraud: AI is becoming a vital instrument in the banking and financial sectors for preventing fraud because of its ability to process large volumes of data rapidly. Technologies like DL, a type of machine learning that makes use of neural networks, have significantly enhanced fraud risk management using predictive analytics [27].

12.3.5 Enhanced Detection Accuracy

By using previous data trends to identify high-risk transactions or consumers, AI systems can improve proactive fraud detection procedures. For example, AI systems can flag suspect transactions for additional inquiry by examining transaction histories and recognizing trends linked to fraudulent operations. Financial organizations are able to identify and prevent fraud before it causes large financial losses because to this proactive strategy [28].

12.3.6 Real-time Monitoring

Another important benefit of AI-powered fraud detection systems is real-time monitoring. Fast detection and mitigation of fraudulent activities-including intricate schemes like account takeovers and card-not-present fraud, which offer serious obstacles to conventional fraud detection systems-are made possible by AI [29]. AI solutions have the capability to identify and address fraudulent behaviors immediately by continually monitoring transactions in real time. This minimizes the window of opportunity for fraudsters and reduces possible losses. AI-powered fraud detection systems are not without difficulties, despite these benefits [29]. They could produce false positives or false negatives; thus continual improvement and verification are required to maximize their dependability and effectiveness. It takes constant upgrades and advancements to the underlying algorithms to guarantee that AI systems can correctly discriminate between authentic and fraudulent activity [30]. Prospects for AI-assisted financial fraud detection in the future developments in the field of financial fraud detection using AI appear promising and will be fueled by Enhanced Capabilities for ML as fraudsters use more complex strategies, advances in machine learning algorithms will make it possible to detect fraudulent activity in real time. AI systems will get better at spotting minute trends and abnormalities linked to fraud as machine learning techniques advance. Improved machine learning skills will also make it easier to create fraud detection models that are more resilient and adaptable so they can keep up with new threats [31].

12.3.7 Advanced Natural Language Processing

NLP technology will be very helpful in analyzing different data sources to discover new fraud trends, which will enhance the proactive aspects of fraud detection systems. NLP will enable more accurate and comprehensive fraud detection by improving the understanding and analysis of unstructured data, such as emails and posts on social media. By expanding the pool of data available for fraud detection, this advancement will make it possible to identify new and innovative fraud techniques [32].

12.3.8 Blockchain Integration

By enhancing data transparency and integrity, blockchain technology has the potential to boost fraud detection and prevention efforts. Due to its decentralized nature and irreversibility, which makes it impossible for fraudsters to alter transaction data, blockchain provides a transparent and safe platform for financial transactions. Combining blockchain technology with AI-powered fraud detection systems may provide financial organizations with increased security and traceability, facilitating the discovery and advertences of fraudulent conduct [33].

12.3.9 Human Oversight and Ethical Considerations

In order to ensure fairness, openness, and regulatory compliance, human monitoring is still necessary. The ethical issues of utilizing AI for fraud detection are still quite significant. Even though AI can detect and prevent fraud quite effectively, ethical concerns like algorithmic bias and privacy still need to be taken care of. Human oversight is crucial to maintaining stakeholders' and customers' trust as well as ensuring that AI technologies are applied morally and responsibly [34].

12.4 THE RISE OF AI IN FINANCIAL FRAUD DETECTION

The increasing number and complexity of digital transactions has proven to be too much for traditional rule-based systems to handle. On the other hand, AI-powered systems use machine learning to instantly examine large datasets, improving detection skills in areas where conventional techniques are inadequate. By iteratively learning from past data, these systems adapt and improve continually, reducing risks and defending the interests of its users [35].

12.5 THE ROLE OF BIG DATA IN AI-POWERED FINANCIAL FRAUD DETECTION

As big data analytics (BDA) makes it possible to thoroughly analyze vast amounts of transactional data, it is essential for improving the efficacy of AI-powered financial fraud detection. The main benefits of BDA for fraud detection are discussed in next subsections.

12.5.1 Identification of Intricate Fraud Schemes

BDA may be used to identify intricate fraud schemes involving several accounts, financial institutions, or types of transactions. Massive amounts of data from several sources may include complex patterns and connections that may be used by BDA to identify fraudulent activities. This ability is particularly helpful in spotting coordinated fraud schemes involving several parties and transactions [36].

12.5.2 Behavioral Analysis

BDA may identify abnormalities, such as abrupt increases in transaction volumes or strange transaction locations that are suggestive of fraudulent activity by tracking and examining extensive consumer behavior patterns. A better comprehension of consumer behavior is made possible by behavioral analysis, which improves the accuracy of fraud detection. Financial organizations may increase the overall efficacy of fraud detection operations by using big data to spot minute behavioral changes that can indicate fraudulent activity [37].

12.5.3 Real-time Fraud Detection

The real-time analysis of big data improves the responsiveness of AI-driven fraud detection systems, making it possible to quickly detect and stop illegal transactions. Financial institutions may minimize potential losses and lessen the effect of fraud by detecting and responding to fraudulent activity immediately using real-time analysis. One of the main benefits of BDA for fraud detection is its capacity to handle and analyze massive amounts of data quickly [38]. Nevertheless, there are drawbacks to using big data, such as privacy issues and issues with data velocity, diversity, and volume. Strong data security protocols and regulatory compliance are necessary to preserve private client data and guarantee the ethical application of AI. To meet these issues and guarantee that big data is used responsibly for fraud detection, financial institutions need to put in place robust data governance procedures [39].

12.6 ETHICAL CONSIDERATIONS IN AI-DRIVEN FRAUD PREVENTION

Although AI has made tremendous progress in financial services fraud detection, ethical issues must be addressed to guarantee regulatory compliance, fairness, and transparency [40]. The integration of AI-powered systems into fraud detection necessitates resolving biases, guaranteeing openness in model operations, and complying with legal frameworks. An important ethical factor in AI-driven fraud prevention is the possibility of algorithmic biases. When AI models are trained on historical data that contains biases, the model may reinforce and even magnify such biases during the decision-making process [41]. For example, the AI model may unintentionally absorb and reinforce prejudices against specific demographics, such age, gender, or race, if previous data indicates such biases, this might result in unjust treatment.

Organizations must have policies in place to identify and lessen biases when developing and deploying AI models. To do this, models must be routinely audited for bias, training data must be representative and varied, and fairness measures must be included to evaluate the model's effects on various demographic groups. Furthermore, biases that may develop over time must be minimized and corrected by continuous monitoring and improvement [42]. One of the most important ethical factors in AI-driven fraud prevention is transparency. A lot of AI models, particularly intricate ones like neural networks, can function as "black boxes," making it difficult to decipher the logic underlying their judgments [43].

Concerns concerning accountability and the capacity to explain model results are raised by this lack of transparency, particularly when making important financial decisions. Transparency must be prioritized by organizations through the use of explainable AI (XAI) methodologies. Explainable models shed light on decision-making processes, allowing regulators, customers, and other stakeholders to comprehend the variables affecting fraud detection results [44]. Transparent AI promotes the moral application of AI in preventing fraud by enhancing accountability and fostering stakeholder and user confidence. The need of regulatory compliance increases as AI plays a bigger role in preventing fraud. Financial services are subject to several laws including fair lending practices, consumer protection, and data privacy. AI-based fraud detection systems must adhere to specific guidelines for moral and legal use [45].

Companies must understand how the financial services industry's AI regulations are evolving. Compliance initiatives should incorporate data protection laws such as the General Data Protection Regulation (GDPR) and the California Consumer Privacy Act (CCPA) [46]. Furthermore, compliance with anti-discrimination laws, fair lending practices, and other relevant financial regulations is crucial to ensuring that AI-driven fraud prevention conforms to moral and legal criteria.

12.7 AI IMPLEMENTATIONS IN VARIOUS SECTORS

Financial fraud detection has undergone a paradigm change thanks to AI and BDA, which make it possible to identify and stop fraudulent activity in a variety of industries in real time and with preventive measures. These technologies use advanced data analytics and state-of-the-art machine learning algorithms to swiftly evaluate massive datasets, spot complex fraud patterns, and effectively lower financial risks. Future advancements in machine learning, blockchain technology, and NLP ought to boost the complexity and effectiveness of AI-powered fraud detection systems. Transparency, ethical considerations, and regulatory compliance are still crucial for fostering confidence in AI-driven fraud detection systems. A ground-breaking advancement in the realm of reliable and flexible fraud prevention strategies is the application of AI and big data to financial fraud detection. By using these technologies, financial institutions can protect customer assets, uphold trust in the digital economy, and safeguard their operations against the increasing threat of fraud.

The further advancement of AI and BDA will have a significant impact on financial fraud detection in the future as they will enable more effective and prompt response against fraudulent conduct in an increasingly digital world [47, 48].

12.8 REAL-WORLD EXAMPLES OF AI IN FRAUD PREVENTION

The way corporations fight fraud is now revolutionized by AI. The real-world instances that demonstrate how AI is crucial to the identification and prevention of fraud are discussed in subsequent text.

12.8.1 AI-Powered Fraud Detection in Banking and Finance

AI-powered fraud detection has proven to be extremely beneficial for banks and other financial organizations. Unauthorized transactions, fraudulent loan applications, and account takeovers may all be avoided with the use of AI technologies. Banks may use AI to examine massive volumes of contract data in real time and spot fraudulent activity designs that might have gone undetected in the past. This allows banks to respond quickly to stop losses from happening. AI is another tool that Fintech companies may use to protect themselves against fraud. For instance,

MasterCard's Decision Intelligence uses AI to assess the risk of fraud and analyze cardholder spending patterns in real time, allowing them to flag questionable contacts before they are approved. AI may assist banks and other financial organizations in identifying possible weaknesses in their systems in addition to detecting fraudulent activity. They are able to put policies in place to stop fraud before it happens because to their proactive strategy [49].

12.8.2 AI Solutions for Ecommerce Fraud Prevention

E-commerce businesses frequently struggle with preventing fraud because they run the risk of fraudulent transactions and erroneous chargeback's. AI successfully identifies questionable orders and stops fraud by evaluating customer behavior, device usage designs, contract data, and other data. AI can detect patterns of deceitful activity and thwart fraudulent transactions by evaluating data from previous transactions [50]. AI can also assist online retailers in locating legal transactions that were mistakenly reported as fraudulent. AI can assess the probability of a valid transaction by evaluating data from many sources. This process lowers the number of false positives and enhances the overall precision of fraud protection schemes.

12.8.3 AI in Insurance Fraud Detection and Prevention

Insurance firms frequently have to deal with false claims which can result in large losses. Insurance firms may save money and lessen the general influence of deception by identifying deceitful claims before they are paid out, thanks to AI-powered fraud detection. AI may detect patterns of fraudulent conduct and flag possibly fraudulent claims for additional inquiry by evaluating data from many sources, including policyholder evidence, claim details, social media activity, medical records, and behavioral data from previous entitlements. AI may assist insurance businesses in identifying any weaknesses in their systems in addition to identifying fraudulent claims. AI can detect regions where deception is most likely to happen and put policies in place to stop fraud before it starts by examining data from previous claims.

12.8.4 AI in the Ride-hailing Industry

By examining information like location, booking trends, and payment methods, AI is able to recognize fraudulent drivers and riders. This lessens the likelihood of problems like phony reviews, ghost drivers, or fraudulent rides. In order to identify fraud efforts like phony payment methods, the unlawful use of credit cards that have been stolen, or account takeover attempts, it can also evaluate real-time data. Finally, AI is capable of behavioral analysis and can spot odd trends that might point to fraud. For instance, if a rider starts scheduling rides at odd hours or if a driver twitches pouring in a dissimilar location out of the blue.

Safeguard your business with AI-powered solutions from Dojah; harness the power of AI to combat fraud and stay ahead of fraudsters. We may utilize a variety of AI- and machine learning-powered technologies from Dojah to onboard and verify consumers at a reasonable cost.

12.9 FUTURE SCOPE

As the financial services industry battles increasingly sophisticated fraud threats, it is anticipated that the application of AI in fraud detection and prevention will continue to progress. Future trends and developments in this subject can be predicted by looking at emerging technologies, XAI innovations, federated learning, and collaborative efforts between regulatory authorities and financial firms. By utilizing cutting-edge technologies, fraud detection systems powered by AI are integrating complex biometric authentication methods. Biometrics, which include face recognition, fingerprint scanning, and behavioral biometrics, provide an additional layer of security by uniquely identifying individuals based on their physical attributes and behavioral tendencies.

Real-time biometric pattern analysis is done by AI algorithms to identify and stop fraudulent or illegal activity. Graph analytics and network analysis are becoming useful for identifying intricate fraud schemes involving several linked businesses. AI systems are able to recognize suspicious patterns that suggest organized fraud networks by displaying linkages and connections within large datasets. This technology improves the capacity to identify and stop fraud that might otherwise go undetected when utilizing conventional techniques.

One important trend in improving the interpretability and transparency of AI models is the development of XAI. Financial institutions are looking to XAI to give transparent justifications for the choices made by AI algorithms as regulatory oversight grows. It is essential to comprehend how AI comes to certain conclusions in order to establish compliance, foster productive relationships with stakeholders, and enable efficient coordination between automated systems and human specialists. It is anticipated that future advancements in XAI would concentrate on improving the interpretability and usability of AI models. This involves creating interactive dashboards, visualization tools, and clearer explanations of intricate AI judgments.

Financial institutions will give XAI more priority in order to comply with regulations, resolve moral dilemmas, and increase end-user confidence. Decentralized machine learning techniques like federated learning have the potential to improve financial institution cooperation without sacrificing data privacy. This technique allows AI models to be trained locally using data from specific universities; only model updates are shared. This preserves the localization of sensitive data while enabling collaborative learning across an institutional network. Federated learning fosters a cooperative environment for thwarting fraud by addressing privacy and data security issues.

Future developments in AI-driven fraud detection will likely see financial institutions working together more closely to exchange threat intelligence and best practices. Working together makes it possible to respond proactively to new fraud tendencies and guarantees that knowledge gained from one institution may benefit the whole financial system. Initiatives like cross-institutional alliances and information-sharing consortiums will be crucial in protecting the sector from changing dangers. It is anticipated that regulatory organizations would become increasingly involved in influencing the development of AI-driven fraud detection. The creation of precise policies, norms, and frameworks for the moral and responsible application of AI in the financial services industry is one area of anticipated progress. Regulatory agencies will work with industry participants to develop a coordinated strategy that strikes a balance between the requirement for strong protections and innovation.

Regulatory agencies will have to constantly adjust to the changing landscape of fraud and AI. This entails keeping up with technical developments, evaluating the moral implications of AI models, and making sure that laws continue to effectively protect the financial system and customers. Regulatory measures that are adaptable will be facilitated by continuous communication between industry participants and regulators through collaborative forums. In conclusion, future trends and advancements in AI-driven fraud detection and prevention include the integration of emerging technologies, advancements in XAI, federated learning for privacy-preserving cooperation, and collaborative efforts between financial institutions and regulatory bodies. As the financial landscape shifts, these improvements will contribute to the creation of stronger, more transparent and cooperative frameworks for preventing fraud in the digital age.

12.10 CONCLUSION

It is clear from the application of AI in financial services fraud detection and prevention that the battle against financial crimes has entered a revolutionary new phase. This conclusion summarizes the main findings, acknowledges the revolutionary influence of AI, and looks at the consequences for financial services fraud prevention going forward. Important revelations have been made throughout the extensive chapter, emphasizing the progression of fraud detection from manual techniques to AI-powered solutions. The adoption of AI models that make use of machine learning algorithms, DL methods, and NLP was prompted by the evolution context's revelation of the shortcomings of rule-based approaches. Examined the ethical implications, practical applications, and efficacy of AI in detecting anomalies. The technology demonstrated complex pattern recognition, decreased false positives and negatives, and addressed biases. The significance of resolving prejudices, maintaining openness, and adhering to legal frameworks was highlighted by ethical considerations. Emerging technologies, developments in XAI and federated learning, and cooperative initiatives between financial institutions and regulatory agencies were among the future topics covered. It is impossible to overestimate the revolutionary influence of AI on fraud detection.

AI has vastly improved the efficiency and accuracy of fraud prevention operations due to its capacity to scan large datasets in real time, recognize complex patterns, and adjust to developing fraud schemes. Predictive analytics, DL methods, and machine learning algorithms have evolved into essential weapons in the financial services sector's armory against complex financial crimes. AI has improved the industry's capacity to proactively identify new risks in addition to automating and streamlining the fraud detection process. The combination of graph analytics, XAI, and biometric identification has strengthened the fraud prevention ecosystem.

The sector has strengthened its defenses against a constantly changing threat landscape through cooperative sharing of threat intelligence and cross-institutional alliances. There are significant ramifications for how financial institutions will avoid fraud in the future. The sector is expected to see more developments in AI-driven technologies, such as the integration of federated learning for privacy-preserving cooperation, the adoption of developing technologies, and the improvement of XAI. In order to provide a coordinated strategy that strikes a balance between innovation and consumer protection, regulatory organizations are anticipated to play a critical role in establishing moral and responsible AI practices. Financial institutions would probably become more resistant to new fraud risks as long as they support cooperative efforts and information-sharing initiatives.

Ongoing developments involving proactive industry actions, and continuous modification of regulatory norms and guidelines will help ensure that AI-driven fraud prevention is not only efficient but also compliant with ethical standards and legal requirements. In conclusion, a paradigm change in the financial services industry is indicated by the revolutionary influence of AI on fraud detection and prevention. Future frameworks that successfully combat fraud in an increasingly digital and linked world are expected to be more secure, transparent, and collaborative as long as the industry continues to leverage the potential of AI.

REFERENCES

1. Akindote, O. J., Adegbite, A. O., Dawodu, S. O., Omotosho, A., Anyanwu, A. and Maduka, C. P. 2023. Comparative review of big data analytics and GIS in healthcare decision-making. World Journal of Advanced Research and Reviews, 20(3), pp. 1293–1302. https://doi.org/10.30574/wjarr.2023.20.3.2589.
2. Al-Anqoudi, Y., Al-Hamdani, A., Al-Badawi, M. and Hedjam, R. 2021. Using machine learning in business process re-engineering. Big Data and Cognitive Computing, 5(4), pp. 61. https://doi.org/10.3390/bdcc5040061.
3. Ali, S., Abuhmed, T., El-Sappagh, S., Muhammad, K., Alonso-Moral, J. M., Confalonieri, R., Guidotti, R., Ser, J. D., Díaz-Rodríguez, N. and Herrera, F. 2023. Explainable artificial intelligence (XAI): What we know and what is left to attain trustworthy artificial intelligence. Information Fusion, 99(101805), pp. 101805. https://doi.org/10.1016/j.inffus.2023.101805.
4. Antwarg, L., Miller, R. M., Shapira, B. and Rokach, L. 2021. Explaining anomalies detected by autoencoders using Shapley additive explanations. Expert Systems with Applications, 186, pp. 115736. https://doi.org/10.1016/j.eswa.2021.115736.
5. Arrieta, B. A., Díaz-Rodríguez, N., Del Ser, J., Bennetot, A., Tabik, S., Barbado, A., Garcia, S., Gil-Lopez, S., Molina, D., Benjamins, R., Chatila, R. and Herrera, F. 2020. Explainable artificial intelligence (XAI): Concepts, taxonomies, opportunities and challenges toward responsible AI. Information Fusion, 58(1), pp. 82–115. https://arxiv.org/pdf/1910.10045.pdf.
6. Bracke, P., Datta, A., Jung, C. and Sen, S. 2019. Machine learning explainability in finance: An application to default risk analysis. SSRN Electronic Journal. https://doi.org/10.2139/ssrn.3435104.
7. Buhrmester, V., Münch, D. and Arens, M. 2021. Analysis of explainers of black box deep neural networks for computer vision: A survey. Machine Learning and Knowledge Extraction, 3(4), pp. 966–989. https://doi.org/10.3390/make3040048.
8. Bussmann, N., Giudici, P., Marinelli, D. and Papenbrock, J. 2020. Explainable AI in fintech risk management. Frontiers in Artificial Intelligence, 3. https://doi.org/10.3389/frai.2020.00026.

9. Vaishnavi, R., Anand, J. and Janarthanan, R. 2009, Efficient Security for Desktop Data Grid using Cryptographic Protocol, IEEE International Conference on Control, Automation, Communication and Energy Conservation, Kongu Engineering College, Erode, Vol. 1, pp. 305–311, 4-6 June 2009.
10. Dargan, S. and Kumar, M. 2020. A comprehensive survey on the biometric recognition systems based on physiological and behavioral modalities. Expert Systems with Applications, 143, pp. 113114. https://doi.org/10.1016/j.eswa.2019.113114.
11. Dhanorkar, S., Wolf, C. T., Qian, K., Xu, A., Popa, L. and Li, Y. 2021. Who needs to know what, when? Broadening the Explainable AI (XAI) Design Space by Looking at Explanations Across the AI Lifecycle. Designing Interactive Systems Conference 2021. https://doi.org/10.1145/3461778.3462131.
12. Díaz-Rodríguez, N., Del Ser, J., Coeckelbergh, M., López de Prado, M., Herrera-Viedma, E. and Herrera, F. 2023. Connecting the dots in trustworthy artificial intelligence: From AI principles, ethics, and key requirements to responsible AI systems and regulation. Information Fusion, 99, pp. 101896. https://doi.org/10.1016/j.inffus.2023.101896.
13. Enholm, I. M., Papagiannidis, E., Mikalef, P. and Krogstie, J. 2021. Artificial intelligence and business value: A literature review. Information Systems Frontiers, 24(5), pp. 1709–1734. https://doi.org/10.1007/s10796-021-10186-w.
14. Fritz-Morgenthal, S., Hein, B. and Papenbrock, J. 2022. Financial risk management and explainable, trustworthy, responsible AI. Frontiers in Artificial Intelligence, 5(1). https://doi.org/10.3389/frai.2022.779799.
15. Gichoya, J. W., Thomas, K. J., Celi, L. A., Safdar, N. M., Banerjee, I., Banja, J. D., Seyyed-Kalantari, L., Trivedi, H. and Purkayastha, S. 2023. AI pitfalls and what not to do: Mitigating bias in AI. British Journal of Radiology, 96(1150). https://doi.org/10.1259/bjr.20230023.
16. Anuradha, T., Jasphin Jeni Sharmila, P., Kanimozhiraman, Kalaiselvi, K., Shruthi, C. K. and Jose A. A., 2024. Automatic Categorization of Emails into Folders Based on the Content of the Messages, 2024 Third International Conference on Intelligent Techniques in Control, Optimization and Signal Processing (INCOS), Krishnankoil, Tamil Nadu, India, 2024, pp. 1–6.
17. Roseline, J. F., Naidu, G., Pandi, V. S., alias Rajasree, S. A. and Mageswari, N. 2022. Autonomous credit card fraud detection using machine learning approach. Computers and Electrical Engineering, 102, p. 108132.
18. Ryman-Tubb, N. F., Krause, P. and Garn, W. 2018. How artificial intelligence and machine learning research impacts payment card fraud detection: A survey and industry benchmark. Engineering Applications of Artificial Intelligence, 76, pp. 130–157.
19. Sina, A. 2023. Open AI and its impact on fraud detection in financial industry. Journal of Knowledge Learning and Science Technology, 2(3), pp. 263–281.
20. Soviany, C. 2018. The benefits of using artificial intelligence in payment fraud detection: A case study. Journal of Payments Strategy and Systems, 12(2), pp. 102–110.
21. Stojanović, B., Božić, J., Hofer-Schmitz, K., Nahrgang, K., Weber, A., Badii, A., Sundaram, M., Jordan, E. and Runevic, J. 2021. Follow the trail: Machine learning for fraud detection in Fintech applications. Sensors, 21(5), p. 1594.
22. Tiwari, P., Mehta, S., Sakhuja, N., Kumar, J. and Singh, A. K. 2021. Credit card fraud detection using machine learning: a study. arXiv preprintarXiv:2108.10005.
23. Ashok, G. and Jose, A. A., 2023. Modified Image Encryption Algorithm Based on Chaotic Cryptography, 2023 7th International Conference on Electronics, Communication and Aerospace Technology (ICECA), Coimbatore, India, 2023, pp. 1506–1512.
24. Vesna, B. A. 2021. Challenges of financial risk management: AI applications. Management: Journal of Sustainable Business and Management Solutions in Emerging Economies, 26(3), pp. 27–34.

25. Yee, O. S., Sagadevan, S. and Malim, N. H. A. H. 2018. Credit card fraud detection using machine learning as data mining technique. Journal of Telecommunication, Electronic and Computer Engineering (JTEC), 10(1-4), pp. 23–27.
26. Adaga, E. M., Egieya, Z. E., Ewuga, S. K., Abdul, A. A. and Abrahams, T. O., 2024. Philosophy in business analytics: A review of sustainable and ethical approaches. International Journal of Management & Entrepreneurship Research, 6(1), pp. 69–86.
27. AL-Dosari, K., Fetais, N. and Kucukvar, M., 2022. Artificial intelligence and cyber defense system for Banking industry: A qualitative study of AI applications and challenges. Cybernetics and Systems, 2, pp. 302–330.
28. Ali, A., Abd Razak, S., Othman, S. H., Eisa, T. A. E., Al-Dhaqm, A., Nasser, M., Elhassan, T., Elshafie, H. and Saif, A., 2022. Financial fraud detection based on machine learning: A systematic literature review. Applied Sciences, 12(19), p. 9637.
29. Bao, Y., Hilary, G. and Ke, B., 2022. Artificial intelligence and fraud detection. Innovative Technology at the Interface of Finance and Operations: Volume I, pp. 223–247.
30. Carcillo, F., Le Borgne, Y. A., Caelen, O., Kessaci, Y., Oblé, F. and Bontempi, G., 2021. Combining unsupervised and supervised learning in credit card fraud detection. Information Sciences, 557, pp. 317–331.
31. Chakraborty, G., 2020. Evolving profiles of financial risk management in the era of digitization: The tomorrow that began in the past. Journal of Public Affairs, 20(2), p. e2034.
32. Chhabra Roy, N. and Prabhakaran, S., 2023. Internal-led cyber frauds in Indian banks: An effective machine learning–based defense system to fraud detection, prioritization and prevention. Aslib Journal of Information Management, 75(2), pp. 246–296.
33. de Bruijn, H., Warnier, M. and Janssen, M., 2022. The perils and pitfalls of explainable AI: Strategies for explaining algorithmic decision-making. Government Information Quarterly, 39(2), p. 101666.
34. Dugauquier, D., Bochove, G. V., Raes, A. and Ilunga, J. J., 2023. Digital payments: Navigating the landscape, addressing fraud, and charting the future with confirmation of payee solutions. Journal of Payments Strategy & Systems, 17(4), pp. 359–371.
35. Felzmann, H., Fosch-Villaronga, E., Lutz, C. and Tamò-Larrieux, A., 2020. Towards transparency by design for artificial intelligence. Science and Engineering Ethics, 26(6), pp. 3333–3361.
36. Fritz-Morgenthal, S., Hein, B. and Papenbrock, J., 2022. Financial risk management and explainable, trustworthy, responsible AI. Frontiers in Artificial Intelligence, 5, p. 779799.
37. Gautam, A., 2023. The evaluating the impact of artificial intelligence on risk management and fraud detection in the banking sector. AI, IoT and the Fourth Industrial Revolution Review, 13(11), pp. 9–18.
38. Gichoya, J. W., Thomas, K., Celi, L. A., Safdar, N., Banerjee, I., Banja, J. D., Seyyed-Kalantari, L., Trivedi, H. and Purkayastha, S., 2023. AI pitfalls and what not to do: Mitigating bias in AI. The British Journal of Radiology, 96(1150), p. 20230023.
39. Habbal, A., Ali, M. K. and Abuzaraida, M. A., 2024. Artificial intelligence trust, risk and security management (AI TRiSM): Frameworks, applications, challenges and future research directions. Expert Systems with Applications, 240, p. 122442.
40. Hassan, M., Aziz, L. A. R. and Andriansyah, Y., 2023. The role artificial intelligence in modern Banking: An exploration of AI-driven approaches for enhanced fraud prevention, risk management, and regulatory compliance. Reviews of Contemporary Business Analytics, 6(1), pp. 110–132.
41. Hemalatha, S., Mahalakshmi, M., Vignesh, V., Geethalakshmi, M., Balasubramanian, D. and A, J. A., "Deep Learning Approaches for Intrusion Detection with Emerging Cybersecurity Challenges," 2023 International Conference on Sustainable Communication Networks and Application (ICSCNA), Theni, India, 2023, pp. 1522–1529.

42. Javaid, M., Haleem, A., Singh, R. P. and Suman, R., 2022. Artificial intelligence applications for industry 4.0: A literature-based study. Journal of Industrial Integration and Management, 7(01), pp. 83–111.
43. Pandey, M. and Sergeeva, I., 2022. Artificial intelligence impact evaluation: Transforming paradigms in financial Institutions., 22(1), pp. 147–164.
44. Max, R., Kriebitz, A. and Von Websky, C., 2021. Ethical Considerations About the Implications of Artificial Intelligence in Finance. Handbook on Ethics in Finance, pp. 577–592, Springer, Switzerland.
45. Mohanty, B. and Mishra, S., 2023. Role of artificial intelligence in financial fraud detection. Academy of Marketing Studies Journal, 27(S4).
46. Anuradha, T., Jasphin Jeni Sharmila, P., Kalaiselvi, K., Shruthi, C. K. and A, J. A., "Automatic Categorization of Emails into Folders Based on the Content of the Messages," 2024 Third International Conference on Intelligent Techniques in Control, Optimization and Signal Processing (INCOS), Krishnan Koil, Virudhunagar district, Tamil Nadu, India, 2024, pp. 1–6.
47. Rangaraju, S., 2023. Secure by intelligence: Enhancing products with AI-driven security measures. EPH International Journal of Science and Engineering, 9(3), pp. 36–41.
48. Vyas, B., 2023. Java in action: AI for fraud detection and prevention. International Journal of Scientific Research in Computer Science, Engineering and Information Technology, pp. 58–69.
49. Wei, S. and Lee, S., 2024. Financial anti-fraud based on dual-channel graph attention network. Journal of Theoretical and Applied Electronic Commerce Research, 19(1), pp. 297–314.
50. Williamson, S. M. and Prybutok, V., 2024. Balancing privacy and progress: A review of privacy challenges, systemic oversight, and patient perceptions in AI-driven healthcare. Applied Sciences, 14(2), p. 675.

13 Graph-Based Intelligent Cyber Threat Detection System

Julien Michel and Pierre Parrend

13.1 INTRODUCTION

As the atomic input element to machine learning (ML) algorithms, features are the foremost parameters when it comes to detection problems, both for supervised – like classification of known attacks – and unsupervised – like statistical anomaly detection of suspicious behaviours – models. Therefore, feature engineering is a crucial part of a detection system. This is especially true for threat detection purposes, considering big and critical internet networks where data collection throughput can rise up to terabytes per minute. Operators in security centres monitoring such networks must ensure they take the right decision [1]. Consequently, the information and model predictions must reach them as fast and as clearly as possible. Therefore, feature engineering techniques should be scalable. Features must be explainable and time robust.

The challenge is to find what make a feature engineering scalable while producing explainable feature for time robust classification of threat. This raises two main research questions: (1) Which are the characteristics to make features explainable and time-robust in the context of large internet networks? As graph structures are very representative of the actual fine-grained network behaviour, (2) how are graph-based approaches efficient as a support for feature extraction and which type of graph approaches are more relevant and performant for threat detection purposes?

13.1.1 Threat Detection

Threat detection is the detection of any element that could compromise and cause damage to an information system [2, 3]. In our context, we consider the system to be a large network and consider as threat any elements in the data space that would hinder the operation process of machine and service in the network it does not have an authorisation to access to (attacks against availability and integrity of the systems), or that would unlawfully disclose information (attacks against confidentiality) [4]. The main objective of the threat detection system is therefore to identify, then stop, any behaviour that would hinder the "normal" operation process in the network while not hindering itself [5].

However, threat behaviours are becoming more complex and continuously adapt to defender models [6]. In addition, they do a better job at hiding themselves and it

DOI: 10.1201/9781003521020-13

becomes increasingly time costly for defence operators to manually detect attacks. As such there is an intense pressure for scalable automatisation of detection and classification of threats, which also implies a strong limitation of false positives alerts to limit inefficient manual verifications [7]. These classification algorithms must be efficient in their discrimination of threats while scalable, explainable and time-robust.

13.1.2 Feature Engineering

Feature engineering is the full process starting with data collection and ending in the execution of detection algorithms [8]. It includes all transformations in the original feature space such as normalisation, cleaning, categorisation or encoding, as well as the derivation of new features from original features, like relative values, distances. For example, typical derived features in network datasets would categorisation of port or IP addresses, thresholding on size of packets, or ratio between number of packet and total message size [9]. The last step of feature selection is the ablation of feature from the feature space to reduce the search space, optimise analysis time and remove noisy features that lower detection capability. The feature space for detection is without doubt the most crucial factor in any detection system as any properties or constraints that are not respected by the feature engineering process will not be respected by the detection system as a whole [10]. It is even more important considering that having an efficient feature engineering process will lead to the possibility of having more diverse choice in classification algorithms while retaining high detection performances [11]. In addition, with regards to threat detection and more particularly in the presence adversarial actors, each feature is a potential vector of vulnerability and as such having an optimal feature space is crucial.

Graph-based representation are closely related to network behaviour. As such it is expected that they could lead to explainable approaches with regards to threat detection in internet networks [12]. Graph-based representations, especially unattributed connectivity graph that only rely on the topological aspects of the network, are particularly relevant to the last raised point as they require and depend on a minimal amount of feature in the original feature space [13].

13.2 STATE OF THE ART

13.2.1 Methodology

In this study, we opted for a methodology that would allow us to focus on the property for feature engineering in the context of threat detection concrete issues. As a starting point, we identify the key properties a detection system should strives for to be operable in a trustworthy and sustainable manner. From these identified properties, we analyse the current trend in recent research works, what challenges have been identified in the literature and what are the feature engineering approaches based on graph that are trying to answer them. From those properties and challenges, we identify their research goals about specific detection problems and formulate criteria for feature engineering in their realisation. The scope of this study on threat detection includes works relative to the definition of the relation between features and the identified properties: scalability, explainability, quality, stability, time robustness. We than analyse graph-based feature engineering techniques and how they take, or do not take, those properties in consideration. For each of those properties, we provide definitions from the literature and determine how the property is considered important to threat detection by the literature. We then make statistics on the number of studies addressing the issues of interest in the literature to the subjects of this paper, namely threat detection, which is the main objective, feature engineering, which is the mean we use to attain the objective and graph representation, as a support of the feature engineering. We take interest in the papers that intersects those subjects between 2019 and June 2024. To better understand the place of feature engineering and graph representation in threat detection, we used Google Scholar to search for papers including our keywords: explainability, scalability, feature stability, feature quality, concept drift, threat detection, feature engineering and graph representation. For each of the keywords we obtain the number of research paper including them. We repeat this process with intersection of the different properties with research paper on the threat detection topic and then compare general trends in the presence of the keywords in paper topics and how often they are discussed in a single research paper. In addition, we consider the proportion on feature engineering and graph representation research paper in the topic of threat detection.

13.2.2 Prevalence of Feature Engineering and Consecutive Learning Properties in the Literature

We enounce in this sub-section a brief overview on the state of the art for the considered key property in the context of threat detection. We rely on Google Scholar search for an estimation of the number of papers on those topics and think the tendencies we can observe to be informative to have a general idea on the context on threat detection and its relation to feature engineering and graph representation.

Figure 13.1 shows the proportion of papers in the threat detection domain that include graph representation or feature engineering as one of their topics. Both have seen a growth of more than 300% between 2019 and 2024, with a spike in 2023

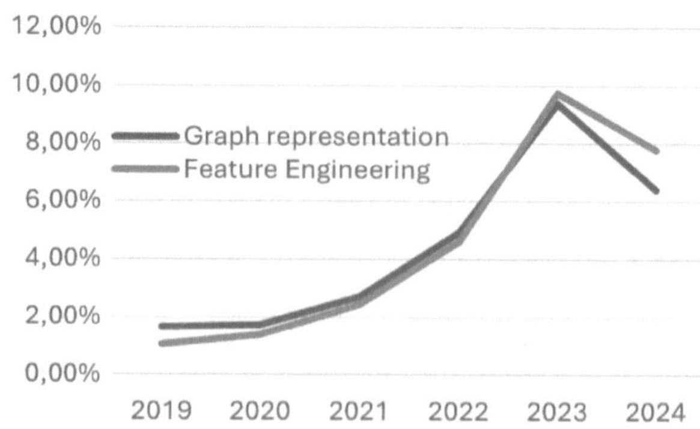

FIGURE 13.1 Proportion of paper on threat detection including graph representation or feature engineering between 2019 and June 2024.

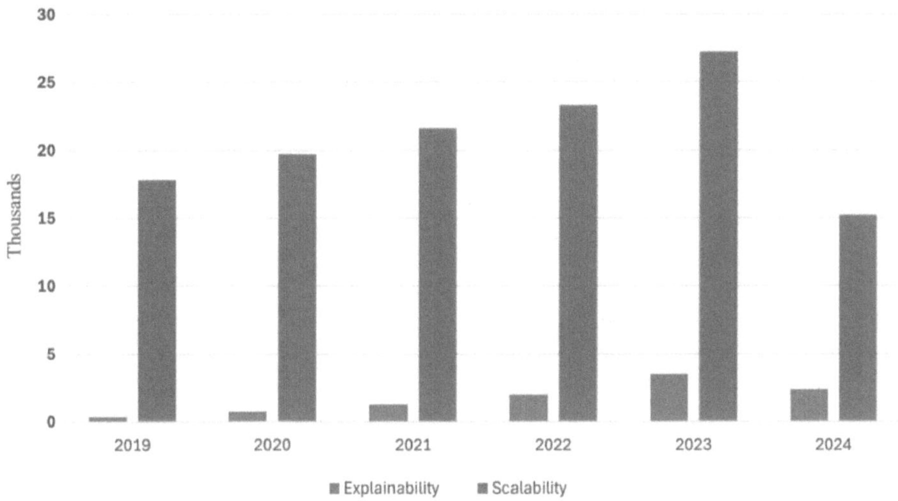

FIGURE 13.2 Research papers on threat detection including explainability and scalability between January 2019 and June 2024.

where research paper on threat detection including feature engineering or graph representation represented respectively 9.74% and 9.37% of paper on threat detection. We can observe that papers on threat detection have a similar evolution in their intersection with the topics of graph representation and feature engineering. The similarity in their growth could be related to the fact that they give tools for similar objectives of current landscape of the threat detection system.

As can be seen in Figures 13.2 and 13.3, all the properties considered have seen a growth in the number of papers between 2019 and June 2024 in the domain of threat detection. However, those properties are not equally spread, while in threat detection, scalability paper is considered in more than 27,000 papers in 2023, explainability is considered in 3,500 papers. Feature stability, feature quality and time robustness are respectively considered in 27, 99 and 768 papers in 2023. Over the five years all the properties have seen a growth in the number of paper and if the number of papers on those properties over 2024 remain constant we expect respectively 30,400, 4720, 898, 100 and 46 papers for scalability, explainability, time robustness, feature quality and feature stability in the topic of threat detection. Such difference is not surprising as those properties have not the same scope, nor relation to end point objectives. Scalability is often a requirement for a sustainable solution, while explainability is a desired properties that can be observed in a system by the users. Time robustness is property of system to its sustainability over time, while feature quality and stability are property to attain time robustness and explainability while ensuring scalability.

Graph representation and feature engineering are both topics that are being more considered for the detection of threat in the recent years, both seems to be important for more robust threat detection: feature engineering as the mean to ensure scalability, explainability and time robustness and graph representation as the support of feature engineering.

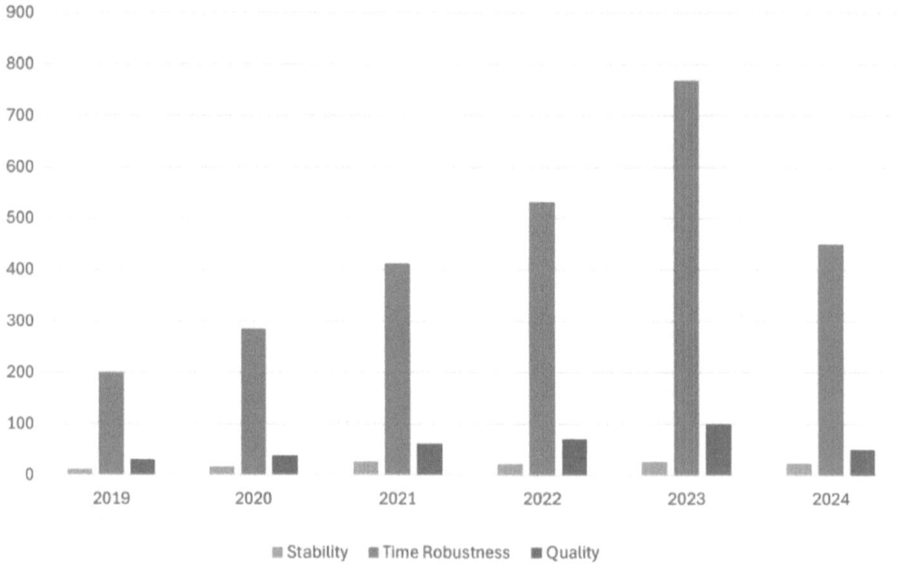

FIGURE 13.3 Research papers on threat detection including feature stability/quality and time robustness between January 2019 and June 2024.

13.3 KEY PROPERTIES FOR FEATURES IN THREAT DETECTION

In this section we introduce what we defined as five key properties for feature engineering in threat detection. We divide these in two categories: in the first category are the properties that are not tied to features but should be objectives for a threat detection approach to thrive, scalability, explainability and time robustness. We think those properties to be especially important consideration for any trustworthy and sustainable threat detection system. In the second category are the properties which are only related to the feature in the detection system, the feature quality and stability. While not totally disjointed, those properties are still different from one to another, and are a factor of utmost importance in ensuring the three previous properties.

13.3.1 Scalability

Scalability of a system is its capacity to function properly with an expected computational workload and within expected margins with future workloads. In the context of detection, scalability is the capacity to produce a prediction or decision under a time constraint considering a potentially higher volume of data. Thus, the importance of scalability of a system is directly related to its task. When working with increasing numbers of objects, it is required for a system to be scalable. Sustainability of such systems requires their scalability to maintain the quality of service. Eventually, if scalability is not insured, sustainability can be compromised, and the system must be replaced, leading to miscontent from users, new production costs or security issues [14]. Scalability is mostly constrained by the time of computation and the memory space, but in some cases can depends on structural design, for example with the limited number of IPv4 addresses. A system scalability is tied to its worse sub-process scalability. For a detection pipeline, it will be the step in the pipeline with the worst scalability.

As such in a threat detection system, feature engineering must be scalable. Threat detection requires handling of volumes of communication data which are ever growing with passing time [15]. Moreover, data are becoming increasingly diverse with several types of structure and structural constraints. Thus, feature engineering techniques should consider time, space and structural complexity to ensure operability of the threat detection system [16]. Threat detection environments data are often subjected to a high collection rate up to terabytes of data per minutes for big network [17]. Therefore, the time constraint is especially strong when considering those type of networks and security operating centre scenario which require prediction in less than a minute. Therefore, scalability is a main concern to threat detection and by extension to feature engineering in this context. With regards to scalability, the main challenge identified by the literature for threat detection are scalable model that retain high detection performance, scalable feature engineering for high detection performance and scalable data structure for scalable feature engineering.

13.3.2 Explainability

Explainability for an AI-based system is defined as the capacity for each part of this system to provide an explanation for its prediction, all the parameters used in the detection system, and the actions it has taken. It is a main concern in AI-driven solution as it is difficult to have a complete comprehension of machine learning model decision [18]. A system having a higher explainability level should be more trustworthy as the users would be reach an understanding of decision made by the system, and as such would have a better useability [19]. The main factor for this better useability is the capacity of the user to determine if using the AI-model results and his understanding will yield a better decision than his own. Thus, performances are a critical issue when considering explainability in AI-driven systems [20]. However, there is currently a lack in our capability of evaluation or quantification of explainability, therefore quantifying the relation between performances and the different layers of explainability an open issue [21]. Nevertheless, explainability has a major place in the current AI-driven detection landscape as it can be used to prove your detection system respect ethical concerns [22].

In the threat detection domain, explainability is especially important as trust is a main issue in the detection of attack as any attack detected by a system with low trustworthiness will be as good as not [16]. Attacks must be ensured to be detected efficiently. False positives have a remarkably high impact because they will lose time to operator in security centre to analyse the alert or interfere with the quality of services for certain users if an action is mistakenly taken for the false alert [23]. However, there is an interesting concern about explainability for AI detection models and their interaction with adversarial models. While it is quite known that adversarial example can produced for black box model, it can be thought that having more explainability in a model or its features could yield more adversarial example. In fact, recent works have shown that using feature explainability it is possible to detect which feature could be a liability in consideration to adversarial models [24]. Main challenges considered by the literature for the explainability of model are on the evaluation of the explainability of a model and about the eventual trade-off between explainability and detection performances.

13.3.3 Time Robustness

A detection system is time-robust if its detection performances are not compromised over time, therefore it is robust to concept drift. Concept drift is a problem by its unpredictable aspect and the diverse aspect of changes it encompasses. In addition, changes in detection target and in the detection environment, i.e. the data that are not supposed to be a detection target, are both important concepts. A concept is defined as an event with a certain probability of appearance in an environment [25]. In this context, concept drift is not only the change in the behaviour of a concept, but additionally how the distribution of concepts evolves in the environment and how some concepts disappear, or new concepts appear. This statistical definition of concept drift leads to the detection of concept drift properties as a mean to characterise it: its time of occurrences, severity and distribution at a given time [26]. Concept drift as such is also a problem to consider for unsupervised detection, but it also a mean to detect unexpected behaviours such as new potential threat [27].

In threat detection, concept drift is especially important as change in environment of detection can lead to an increase in the number of false alarms and leads to an unsustainable detection system for the detection of attacks in data streams [28]. In addition, changes in detection target will lead to a decrease in the detection of previously seen threats, meaning the detection system will lose gradually its efficiency if nothing is done [29]. Features can have a varying robustness to concept drift for the detection of threats, thus feature engineering is vital in producing AI-based model time robust [30]. Feature engineering can even go a step further in its consideration of concept drift, with evolving feature set which react to concept drift in the data stream analysis [31]. A shift in the data profile is detected, feature previously selected that are submitted to shift are reevaluated for selection with the objective for the selection to better correspond to current data profile. The research challenges identified in the literature with regards to time robustness are the detection of shift in a data profile, the extraction of feature for a time robust feature space and the automatisation of update for feature space in consideration to concept drift.

13.3.4 Feature Quality

Feature quality is a characteristic which is inherent to a specific detection purpose. Depending on the detection objectives such as false positive rate optimisation or overall performance, optimisation a single feature quality can vary greatly. To determine how qualitative a feature is, there is a strong need to understand all useful information it bears for the chosen purpose [32]. In addition, a feature quality regarding a detection of a specific target on a can differ depending on the considered feature set, since information from distinctive features in the same set can overlap. Feature engineering is relevant for maximisation of the quality of a feature when considering feature quality because the quality of an engineered feature can be higher than cumulated quality of the source features [33]. For example, you can have distinctive features that have a range of value which are all evenly distributed when looking at them and the different classes to detect. As such the quality of those single feature would be low. But then you could notice that by crossing them the distribution is not even with the classes to detect, then resulting in a feature of higher quality. Having features that do not contribute to the detection can be very detrimental to AI-based detection models, adding them to the feature set would make a drop in the detection performances. As such feature quality can be crucial in assembling a purposeful feature set [34], and to select the right features depending on specific objectives such as the detection of a particular target classes or lowering the false positive rate.

The relation of feature quality to threat detection is dual. Firstly, in terms of detection performances, having a better quality of feature leads to better results as we would keep only features that are beneficial to the detection performances. This is supposedly due to having feature more closely related to physical or digital reality [35]. Secondly, having a better feature quality for dataset could lead to an overall better dataset quality [36]. While the quality of a dataset is not only tied to its features, and those features do not have a direct influence on the general behaviour of the data in the dataset, they are the main interface between the data and the threat detection tools [37]. The main challenges with regards to feature quality according to the literature are the evaluation of feature quality and the evaluation of the impact of feature quality to detection performances.

13.3.5 Feature Stability

Feature stability is a property of features which suffers from a reduced consideration in the literature being twice as less present in research paper in the past five years than feature quality. However, it is tightly linked to explainability, as having a feature not stable would mean the information it brings is not stable, feature quality, as it is constant if the feature is stable, and vary if it is not, and time robustness, as if the whole feature set is stable than you are not subjected to concept drift anymore [38]. It is defined as a measure of the robustness of the feature, i.e. considering the whole dataset, how relevant the feature is to the detection objectives, such as the optimisation of true positive rate for binary detection [39]. Depending on set conditions, it is possible to determines different values of stability, using different feature sets or aggregates which can be relevant for example to detect cyclo-stationarity. Empirically a more stable feature should be more qualitative, more explainable, and eventually more time robust. In addition, when considering feature stability for feature selection, it should result in a more stable feature selection process as you would not need to reconsider stable features [40].

This last observation holds true for some threat detection issues like phishing detection [41]. Moreover, there is another form of information that can be extracted from feature stability and is especially important in threat detection which is how will the feature behave when there is a shift in the trend of the data. Threat detection environments are very subjected to change in the values of the feature space over time in both the general environment and the targets to detect in this environment, i.e. the concept drift [42–46]. In this context, what is of interest is not the general stability of the feature, but how a robust or stable feature can maximise stability for the environment to differentiate threat from normal behaviours. Generating or extracting features which are stable in consideration to evolution of data with passing time is a key property for a time robust threat detection and the concept drift of the data [47]. The main challenges considering feature stability for threat detection identified in the literature are how correlated is a feature space stability to time robustness and how to ensure feature stability with passing time.

13.4 GRAPH-BASED APPROACHES

In this section, different kinds of graph-based approaches for feature engineering that we have considered are detailed. We give a general idea of their exploitation for threat detection purposes and point out their advantages and limitations.

13.4.1 Landscape of Graph-Based Feature Engineering

The main objective in feature engineering is to render new angles of information accessible and operable for a specific purpose. Graph-based feature engineering techniques have thus evolved to open the access to information that was not available beforehand. The main point of graph representation is the capacity of showing behaviour of interactions between distinctive objects in the data space [48]. In addition, it leads to a multitude of graph representations, themselves leading to distinct types of structures on different scales, temporality, information layers and means of accessing that information. These graphs representations type and different possibles processing operations are interchangeable in their association. We propose in our taxonomy of feature generation techniques (Figure 13.4) to make a distinction between automatic feature generation, i.e. feature generated by a learning model, and more classical feature extraction methods as they usually are both significantly different in the resultant features. They do however have a similitude in their purpose to benefit from data structure closely related to real-world structure like social networks or transportation networks for examples [49].

13.4.2 Types of Metrics

In this section the different kinds of graph metrics, i.e. the parameter from graph that can be extracted as new features, are detailed. There are various means to produce feature from graph structured data [50], which depend on a range of factors that are detailed in the next sub-sections, where we explain advantages and limitations for each of them.

13.4.2.1 Global and Local Metrics

The first important parameter in graph feature extraction for threat detection is the locality of the metrics. Indeed, depending on their locality, metrics can be tied

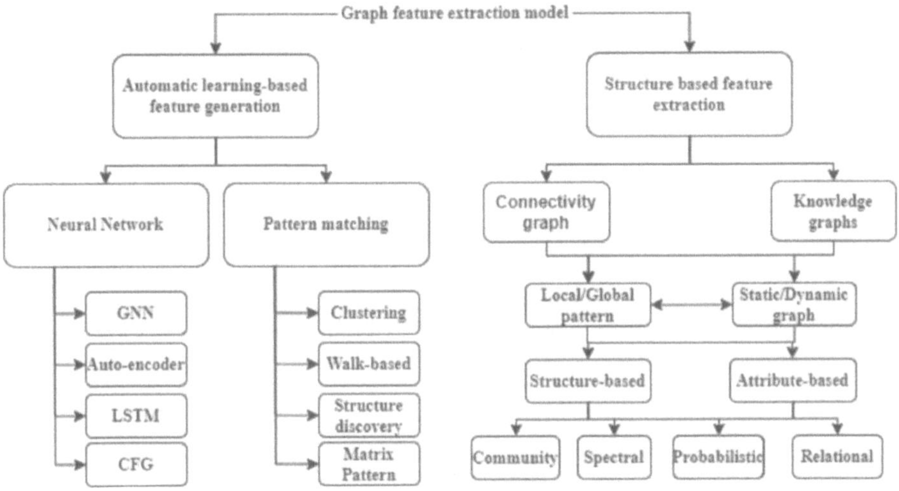

FIGURE 13.4 Graph-based feature generation taxonomy.

to vastly different threat behaviours [51]. The locality of a metrics depends on the objects needed to compute it. For example, a global metric is a metric that need to consider elements in the whole graph to be computed, at most it would refer to all the nodes and edges contained in the graph. Invertedly, the most local metrics would be information tied to a single node or edge. In general, either node or edge corresponds to a single data point in a tabular view, meaning such information are not related to graph topological behaviours. Thus, we in general a metric is local when it considers a single node or edge and its direct neighbours. There are multiple locality levels between local and global including connected component level and or any partition-based sub-graph level. Each locality has its importance when trying to detect specific threat behaviours as some threats could have a visible impact only considering a smaller part of the graph, while others could only be detected while looking at a bigger scale. Those graph topological behaviours are linked to spatial behaviours of the data, and actively link attack behaviours with specific elements in the graph. Thus, there is a need for any graph feature engineering for threat detection to determine which locality is relevant to its various threat detection purposes.

13.4.2.2 Dynamicity and Temporality

While locality is important for its relation to the spatial behaviour of threats, attack behaviours are temporal events as well. Dynamicity is the parameter representing how behaviours evolve in a data structure. Graph structure can be adapted to show the appearance, disappearance, or transformation of events inside the structure. There are dynamic graph representations and there exist multiple representations which can represent different behaviours. For example, a complete dynamic graph can be represented as a full spatial graph with additional temporal edges between nodes having the same identifier at different temporality. The temporality is the given time of an event for a given dynamic graph representation. In the case of a complete dynamic graph, it is adapted for representing the evolution of the behaviour of a single node. However, it would show severe drawbacks in term of scalability for the representation of the dynamicity of global metrics. For the same data we can represent dynamic graphs with different parameters for temporality. Absolute time can be the parameter for temporality, although usually data are divided into slices of time and the temporality is affected by time windows. Another mean of representing a dynamic graph is to make series of graphs divided by those time windows. Data are assigned to the graph which corresponds to their time window. In the dynamic graph, locality of the behaviour and evolution of those behaviour is considered to better represent and detect the spatiotemporal events. Different combinations of locality and dynamicity will lead to different metrics more representative of specifics threats.

13.4.2.3 Attributed Graph

Nodes and edges in a graph can contain properties apart from their identifier. Those properties are named attributes. If a graph is made of nodes and edges without any attribute, it is said to be unattributed, otherwise this is an attributed graph. In the case of an unattributed graph, we are only able to compute metrics based on the topological aspect of the graph. The purpose of an attributed graph is to be able to build relations between objects in the graph based on their attributes. To compute the graph metrics, we add another constraint based on graph elements attributes. This way, we can add a bias correlated to the attributes to the topological metrics. However, while biases are necessary for detection, it can also be detrimental and can lead to overfitting for example. Thus, attribution in graphs should be thought carefully, as one of the purposes of graph representation is to be free of some data bias. We want graph representation to give detection criteria related to mandatory behaviours of specific threats while avoiding criteria related to behaviours of a specific threat on a specific period, but which could be easily modified at later times. Relying on more attributes lead to more leeway in the compromission of the time robustness of a model.

13.4.3 Learning-based Approaches

As can be seen on Figure 13.4, learning-based approaches are detached from other types of approaches in our taxonomy. The principal reasons for this separation are the fact that learning based approaches are mostly automatised and that the feature generation is directly tied to the model performances. We define by automatised the fact that before the computation of such feature generation model, the user has no prior knowledge of the features that will be extracted. While in more classical feature generation models, features to be extracted are defined and purposefully extracted, in learning-based approaches features happen to be extracted. There is a radical paradigm shift, leading to the second difference: instead of choosing features to be extracted in expectation to optimize model performances, we optimize the model in expectation of meaningful features. Thus, we make a clear distinction between those approaches in the taxonomy. Neural networks (NNs) have proven to be particularly efficient for the generation of feature for optimisation problems [52] and have been applied to threat detection models [53].

13.4.3.1 Graph Neural Network

Graph neural networks (GNNs) are the NNs processing graph structured data. They are primarily used in the detection of elements or groups of elements in a graph, i.e. nodes, edges, sub-graphs or connected components. They are mostly applied to attributed graphs as a mean to extract information from the graph attributes. This capacity of association of graph attributes and topology has led to various works for the generation of features using graphs. The main advantage of the use of GNN is their capacity to aggregate the data from a graph structure automatically and efficiently [54]. They may however need more classical feature engineering in prior for specific use-cases. As most NN-based models, they have a high specificity leading to approaches of feature generation not extensible to any use beside the specific case they have been modelled for [55]. GNNs have recently been applied to feature engineering for threat detection purposes [56].

13.4.3.2 Other Neural Networks

While less common for the extraction of features based on graph, other NN techniques have also been applied to this purpose. For example, to manage dynamic graph, long short-term memory (LSTM) which is an autoencoder model based on recurrent NNs (RNN) has been applied to generate features on temporal information in the graph and to assign them to static nodes [57]. The choice of LSTM compared to GNN is justified in the literature by a characteristic of GNN to over-focus on the topological aspect of the graph [58]. For threat detection purposes as well, like in the detection of specific messages in social networks we have seen use of NN as a mean to produce highly qualitative features [59]. Other types of RNN have proven to be efficient in engineering of features, notably applied on control flow graphs for unsupervised detection [60].

13.4.4 Topological Based Approaches

While not inherently unattributed approaches, topological-based approaches primarily focus on the connectivity inside the graph structure. They are interesting for feature extraction purposes as they tend to be scalable for pre-defined tasks and shows prominent level of interpretability [61]. Since many threat detection problems involve structures that can be precisely represented by graph like social networks, end-to-end network or Internet of Things devices networks, their topological aspect can be highlighted in the graph metrics. In addition, some approaches produce high-quality feature representing global behaviours [61], while others can better represent local behaviours in the graph structure. Thus, a broad range of behaviours can be tailored for the detection of specific threats.

13.4.4.1 Probabilistic Models

Probabilistic models may be the topological approach most closely related to learning-based models as they produce feature spaces based on rules. They differ from learning-based models from the sources of the rules as they are human made mathematical rules [49]. Some models make use of hierarchical structures in networks to compute probabilities of existence of edges between nodes in the graph. This is the link prediction. Instead of using this information as for link prediction, the probabilities are mapped to data in the feature space and can then be used by various learning algorithms. The major drawback of most of these models it that they are often based on Bayes rules and suffer from a serious lack of scalability. Another branch of probabilistic models are the stochastic models. The main advantage of stochastic models is that while they are automatable, they are highly parametrisable as well [62]. By their stochastic nature, the computation time is malleable and can provide an adaptable framework for the dynamicity of the data as they can update the feature space automatically.

13.4.4.2 Community Models

Graph community structures are a mean to partition a graph into clusters using the graph topology itself as the only parameter. A node is part of community if it is more closely related to the node in its community compared to other nodes in the graph structure depending on the community partitioning criteria. The most common partition criterion is the maximisation of the modularity. The appearance of community-based partition as another form of clustering comes from the realisation that for large networks, the clustering techniques were failing in distinguishing communities compared to the ground truth of the network-based datasets. Large networks tend to have a lot of noise, i.e. behaviour sufficiently different for being outlier, but not significative while considering a threat detection purpose. As such classical clustering approaches tend to make small clusters out of all those small-scale outlier behaviours. On the other hand, those uninteresting behaviours are statistically over-present when compared to threat behaviours, rendering more the detection of threats more difficult in this context. However, specific threat behaviours have shown to be closely related in a graph structure. Hence, community structures have emerged to highlight those behaviours. Additionally, small-scaled attacks have a lessened impact while looking at whole graph metrics, whereas they are more impactful on a community structure. Community-based approaches also prove to be quite time efficient and more explainable as they can tie behaviour to areas in the graph structure [27].

13.4.4.3 Spectral Models

Spectral models use the Laplacian matrix representation of a graph to extract features. Matrix representation for graphs can be very costly both in space and time complexity. Thus, spectral models are mostly applied on dynamic graphs, where the number of node and edge for each time window tend to be lower as activity on short period are more concentrated in the graph structure. Spectral models specifically access topological information inside the graph through the eigenvalues of the Laplacian matrix. These values are the main interest of spectral models because they provide information on the graph structure in an instantaneous manner, such as the number of connected components corresponding to zero in the eigenvalues. Moreover, similar to the community models, spectral models are independent from the original feature set of the dataset, i.e. they work well with unattributed graph. Thus, they are not as affected by noise and bias in the original feature space [63].

13.4.5 Relation-based Approaches

Relational-based approaches, while still relying on topological aspects of the graph structure are not dissociable from the attributed aspect. Indeed, those approaches are tied to heterogeneous graph structures, where nodes can be objects of several types and edges are the relation between those nodes. They can represent a relational database where any row in the database is a node and edge are foreign keys. However, a relational database is not required as input data [64]. As such, it is possible for relational-based approaches to be quite scalable using relational database for data storage [65]. In addition, by the nature of the relation between the different objects, features and rules generated are inherently explainable [66, 67]. Relational graph structures can represent variety of types of data and prove to be efficient at modelling data with multitude of objects classes as videos or natural language texts [68]. This proves to be particularly interesting in the detection of threats in a social network environment as they can make an efficient use of posts content [69]. In addition, providing more explainable features give a more trustworthy base on approaches for threat detection [70].

13.4.6 Knowledge Graphs

Similar to approaches based on relational graphs, approaches based on knowledge graphs are indissociable from the knowledge graph structure. They present similarity to relational graphs, notably by the facts that they are heterogeneous graphs where nodes are from different classes of objects or concepts in this case, and the edges represent relations between the objects. The main difference between relational and knowledge graphs is that a knowledge graph can be refined. More precisely, generated features from the knowledge inside the graph will lead to further analysis which in return will feed the knowledge graph resulting in a new knowledge graph [71]. Not all approaches using knowledge graphs for feature engineering have a process to update the knowledge graph. However, this raises the critical point about knowledge graphs: elements in the knowledge graph do not need to exist in the original data to be part of the knowledge graph. While the nodes can represent existing objects, they can represent more abstract concepts. In addition, knowledge graph-based approaches intend to be highly explainable representation and to perform efficiently together with tree-based learning models like random forest or XgBoost [72]. Knowledge graphs being structurally like relational graphs, feature engineering approaches are scalable, and this hold true for tree-based models.

13.5 DISCUSSIONS

In this section, we detail our reflection on the observations while analysing the literature and shed light to the lack of consideration about certain areas of feature engineering for threat detection. We try to propose leads on points we think should be improved in the future.

13.5.1 Limitations of GNN and Other NN Approaches

Studies on GNN and other NN approaches tend to focus on the performance optimisation of specific models. As such, key properties for an efficient feature engineering, independent of the learning model itself are rarely considered in this domain. While explainable GNN models exist, they are still exceedingly rare [73]. And since NN-based models are inherently less explainable, as a feature generation tool they produce features that are hardly interpretable. While for a specific detection process, it may be acceptable, it is completely unreliable for threat detection purposes as it is impossible to gauge the trustworthiness of a system based on those approaches. This is specially the case for prolonged detection over networks that are subject to concept drift. NN models being extremely specific, they are particularly sensitive to concept drift and therefore are not time robust. If in addition they are not explainable, it is hard to detect the breaking point where the detection system will stop to function properly. NN models additionally suffer from problems linked to their exploitation. They are not scalable and as such cannot handle a big volume of data under time constraint. In addition, they are not adapted to handle heterogeneous graphs or dynamic graphs. Moreover, they suffer from imbalance in the classes, which is inevitable as threats are in most cases a minority in the data space.

13.5.2 Use of Attributes in Graph Feature Engineering

Different approaches based on graphs for feature engineering use attributed graphs. While attributed graphs add another layer of information compared to unattributed graphs, the graph attributes represent either features from the original feature space or are derived from them. As such they suffer from part of the original feature space biases. While in a static context, the impact of this matter of fact could be minimal. In a detection environment subject to concept drift, this is crucial. This is a main argument for using graph representation for feature engineering. We want the features we extract from the graph to be representative of threat behaviours throughout the system life cycle and not of the behaviour the threat had at a specific time point. Behaviours issued from the original feature space are frequently those that could be easily modified by an attacker, and thus having a model that can avoid relying on these features produce more time robust predictions relying on more stable features. In addition, relying on a lower number of features tends to make more time robust models.

13.5.3 Consideration of Key-Properties in Current Landscape

The number of research papers addressing properties of features are scarce, especially in the threat detection domain. While scalability, explainability and time robustness properties that are not directly tied to features are discussed, feature quality and stability are hardly considered, even in feature engineering focused works. These can be explained, as we could observe that in most works that claim to address feature engineering, feature engineering was in fact not the focus of the work. Feature engineering is a mean to obtain better detection performance. Current landscape of feature engineering for threat detection lacks means for the evaluation of the feature engineering and feature spaces. Decision systems, powered by AI or not, are feature-driven and ultimately the main parameter of those system are the features. They are the prime target for adversarial behaviours as well. Therefore, the quality of feature spaces should be ensured, and we should ensure their conformity to the threat detection purpose.

13.6 CONCLUSION AND FUTURE WORKS

In this chapter, prominent graph representations and approaches for feature engineering purposes have been detailed with regards to threat detection and classification of attacks in large network environment. We synthetised definition for what we identify as key properties for feature engineering and robust threat detection, and analysed how they are considered in the current landscape of threat detection. We elicited the limitations of the current approaches for graph-based feature engineering while highlighting the relevant behaviour they display for time-robust, scalable, and explainable detection, such as the minimisation of original feature space as parameter for derived features, the assignment of local behaviour from graph structures to the data and smoothing of statistical anomalies in large network data environments.

For future works, we would like to evaluate concept of drift robustness, while primarily looking for clues in identifying and evaluating criteria for having a time-robust feature space. To this end we expect graph representation to produce stable and qualitative features.

REFERENCES

1. M. Alhanahnah, C. Stevens and H. Bagheri. 2020. Scalable analysis of interaction threats in IoT systems. *ISSTA 2020 – Proceedings of the 29th ACM SIGSOFT International Symposium on Software Testing and Analysis*: 272–285.
2. S. Hemalatha, M. Mahalakshmi, V. Vignesh, M. Geethalakshmi, D. Balasubramanian and J. Anand A. 2023. Deep Learning Approaches for Intrusion Detection with Emerging Cybersecurity Challenges. *2023 International Conference on Sustainable Communication Networks and Application (ICSCNA)*: 1522–1529.
3. R. Reshma and A. Jose Anand. 2023. Predictive and Comparative Analysis of LENET, ALEXNET and VGG-16 Network Architecture in Smart Behavior Monitoring. *2023 Seventh International Conference on Image Information Processing (ICIIP)*: 450–453.
4. C. Regan, M. Nasajpour, R.M. Parizi, S. Pouriyeh, A. Dehghantanha and K.-K.R. Choo. 2022. Federated IoT attack detection using decentralized edge data. *Machine Learning With Applications*, *8*, 100263.
5. Z. Deng, K. Chen, G. Meng, X. Zhang, K. Xu and Y. Cheng. 2022. Understanding Real-world Threats to Deep Learning Models in Android Apps. *Proceedings of the ACM Conference on Computer and Communications Security*: 785–799.
6. F. Tramèr, R. Shokri, A. San Joaquin, H. Le, M. Jagielski, S. Hong and N. Carlini. 2022. Truth Serum: Poisoning Machine Learning Models to Reveal Their Secrets. *Proceedings of the ACM Conference on Computer and Communications Security*: 2779–2792.
7. T. Kim, N. Park, J. Hong and S.W. Kim. 2022. Phishing URL Detection: A Network-based Approach Robust to Evasion. *Proceedings of the ACM Conference on Computer and Communications Security*: 1769–1782.

8. D. Chicco, L. Oneto and E. Tavazzi. 2022. Eleven quick tips for data cleaning and feature engineering. *PLoS Computational Biology, 18*(12), 1–10.
9. X. Larriva-Novo, V.A. Villagrá, M. Vega-Barbas, D. Rivera and M. Sanz Rodrigo. 2021. An IoT-focused intrusion detection system approach based on preprocessing characterization for cybersecurity datasets. *Sensors (Switzerland), 21*(2), 1–15.
10. X.I.A. Xin and D. Lo. 2018. Feature Generation and Engineering for Software Analytics. In *Feature engineering for machine learning and data analytics* (pp. 335–358).
11. P.W. Khan and Y.C. Byun. 2020. Genetic algorithm based optimized feature engineering and hybrid machine learning for effective energy consumption prediction. *IEEE Access, 8*, 196274–196286.
12. T. Pourhabibi, K.L. Ong, B.H. Kam and Y.L. Boo. 2020. Fraud detection: A systematic literature review of graph-based anomaly detection approaches. *Decision Support Systems, 133.*
13. E. Navruzov and A. Kabulov. 2022. Detection and analysis types of DDoS attack. *2022 IEEE International IOT, Electronics and Mechatronics Conference, IEMTRONICS 2022.*
14. A. Gupta, R. Christie and R. Manjula. 2017. Scalability in Internet of Things: Features, techniques and research challenges. *International Journal of Computational Intelligence Research, 13*, 7.
15. F. Pierazzi, G. Apruzzese, M. Colajanni, A. Guido and M. Marchetti. 2017. Scalable architecture for online prioritisation of cyber threats. *International Conference on Cyber Conflict, CYCON, 2017* 1–18.
16. W. Abdelghani, C.A. Zayani, I. Amous and F. Sèdes. 2019. *Trust Evaluation Model for Attack Detection in Social Internet of Things* (pp. 48–64).
17. W. Yao, H. Shi and H. Zhao. 2023. Scalable anomaly-based intrusion detection for secure Internet of Things using generative adversarial networks in fog environment. *Journal of Network and Computer Applications, 214.*
18. A. Barredo Arrieta, N. Díaz-Rodríguez, J. Del Ser, A. Bennetot, S. Tabik, A. Barbado, S. Garcia, S. Gil-Lopez, D. Molina, R. Benjamins, R. Chatila and F. Herrera. 2020. Explainable artificial intelligence (XAI): Concepts, taxonomies, opportunities and challenges toward responsible AI. *Information Fusion, 58*, 82–115.
19. J. Qu, J. Arguello and Y. Wang. 2021. A Study of Explainability Features to Scrutinize Faceted Filtering Results. *International Conference on Information and Knowledge Management, Proceedings*: 1498–1507.
20. B. Crook, M. Schlüter and T. Speith. 2023. Revisiting the Performance-Explainability Trade-Off in Explainable Artificial Intelligence (XAI). *2023 IEEE 31st International Requirements Engineering Conference Workshops (REW)*: 316–324.
21. G. Vilone and L. Longo. 2021. Notions of explainability and evaluation approaches for explainable artificial intelligence. *Information Fusion, 76*, 89–106.
22. R. Islam, W. Eberle and K. Ghafoor. 2020. Towards Quantification of Explainability in Explainable Artificial Intelligence Methods. *The Thirty-Third International Flairs Conference.*
23. A.A. Rida, R. Amhaz and P. Parrend. 2023. Metrics for Evaluating Interface Explainability Models for Cyberattack Detection in IoT Data. *International Conference on Complex Computational Ecosystems*: 180–192.
24. A. Hartl, M. Bachl, J. Fabini and T. Zseby. 2020. Explainability and Adversarial Robustness for RNNs. *2020 IEEE Sixth International Conference on Big Data Computing Service and Applications (BigDataService)*: 148–156.
25. G.I. Webb, R. Hyde, H. Cao, H.L. Nguyen and F. Petitjean. 2016. Characterizing concept drift. *Data Mining and Knowledge Discovery, 30*(4), 964–994.
26. J. Lu, A. Liu, F. Dong, F. Gu, J. Gama and G. Zhang. 2018. Learning under concept drift: A review. *IEEE Transactions on Knowledge and Data Engineering, 31*(12), 2346–2363.

27. B. Friedrich, T. Sawabe and A. Hein. 2023. Unsupervised statistical concept drift detection for behaviour abnormality detection. *Applied Intelligence, 53*(3), 2527–2537.
28. A. Guerra-Manzanares, M. Luckner and H. Bahsi. 2022. Android malware concept drift using system calls: Detection, characterization and challenges. *Expert Systems with Applications, 206*, 1–10.
29. S. Kumar, A. Viinikainen and T. Hamalainen. 2018. Evaluation of ensemble machine learning methods in mobile threat detection. *2017 12th International Conference for Internet Technology and Secured Transactions, ICITST 2017*: 261–268.
30. D.W. Fernando and N. Komninos. 2022. FeSA: Feature selection architecture for ransomware detection under concept drift. *Computers and Security, 116*, 1–19.
31. Z. Chen, Z. Zhang, Z. Kan, L. Yang, J. Cortellazzi, F. Pendlebury, F. Pierazzi, L. Cavallaro and G. Wang. 2023. Is It Overkill? Analyzing Feature-Space Concept Drift in Malware Detectors. *IEEE Security and Privacy Workshops (SPW)*: 21–28.
32. J. Liu, Y. Lin, M. Lin, S. Wu and J. Zhang. 2017. Feature selection based on quality of information. *Neurocomputing, 225*, 11–22.
33. S.S. Naqvi, W.N. Browne and C. Hollitt. 2016. Feature quality-based dynamic feature selection for improving salient object detection. *IEEE Transactions on Image Processing, 25*(9), 4298–4313.
34. M. Lorbach, R. Poppe, E.A. van Dam, L.P.J.J. Noldus and R.C. Veltkamp. 2015. Automated recognition of social behavior in rats: The role of feature quality. In V. Murino & E. Puppo (Eds.), *Image Analysis and Processing* (Vol. 9280, pp. 565–574). Springer International Publishing.
35. A. Alhowaide, I. Alsmadi and J. Tang. 2019. Features Quality Impact on Cyber Physical Security Systems. *2019 IEEE 10th Annual Information Technology, Electronics and Mobile Communication Conference (IEMCON)*: 0332–0339.
36. S. Picard, C. Chapdelaine, C. Cappi, L. Gardes, E. Jenn, B. Lefevre and T. Soumarmon. 2020. Ensuring Dataset Quality for Machine Learning Certification. *2020 IEEE International Symposium on Software Reliability Engineering Workshops*: 275–282.
37. M. Torres, R. Alvarez and M. Cazorla. 2023. A malware detection approach based on feature engineering and behavior analysis. *IEEE Access, 11*, 105355–105367.
38. J.E. van Timmeren, R.T.H. Leijenaar, W. van Elmpt, J. Wang, Z. Zhang, A. Dekker and P. Lambin. 2016. Test–Retest data for radiomics feature stability analysis: Generalizable or study-specific? *Tomography*, 2(4), 361–365.
39. L. Al-Shalabi. 2022. New feature selection algorithm based on feature stability and correlation. *IEEE Access, 10*, 4699–4713.
40. C. Huang. 2021. Feature Selection and Feature Stability Measurement Method for High-Dimensional Small Sample Data Based on Big Data Technology. *Computational Intelligence and Neuroscience, 2021.*
41. L. Al-Shalabi and Y. Hasan Jazyah. 2024. Phishing detection using hybrid algorithm based on clustering and machine learning. *International Journal of Computing and Digital Systems, 15*(1): 1–13.
42. R. Zuech and T.M. Khoshgoftaar. 2015. A survey on feature selection for intrusion detection. *Proceedings of the 21st ISSAT International Conference on Reliability and Quality in Design*: 150–155.
43. M. Singh, P. Bhambri and K. Kaur. 2005. Network security. In Proceedings of the National Conference on Future Trends in Information Technology (pp. 155–161). SJPMLIET, Radaur, Yamuna Nagar.
44. P. Bhambri, V.K. Sinha and I.S. Dhanoa. 2020. Development of cost effective PMS with efficient utilization of resources. *Journal of Critical Reviews, 7*(19), 781–786. Retrieved from http://www.jcreview.com/index.php?fulltxt=102893&fulltxtj=197&fulltxtp=197-1594892754.pdf.

45. P. Bhambri and M. Singh. 2006. Artificial intelligence. In National Seminar on E-Governance: Pathway to Progress. (p. 14). SSIET, Derabassi.
46. J. Kaur, P. Bhambri and S. Kaur. 2019. SVM classifier based method for software defect prediction. *International Journal of Analytical and Experimental Model Analysis, 11*(10), 2772–2776. Retrieved from http://ijaema.com/Volume-11-Issue-10-October-2019-1/.
47. D. Angioni, L. Demetrio, M. Pintor and B. Biggio. 2022. *Robust Machine Learning for Malware Detection over Time.*
48. J. Cheng. 2023. Graph Feature Management: Impact, Challenges and Opportunities. *Proceedings of the 6th Joint Workshop on Graph Data Management Experiences & Systems (GRADES) and Network Data Analytics (NDA).*
49. E.C. Mutlu, T. Oghaz, A. Rajabi and I. Garibay. 2020. Review on learning and extracting graph features for link prediction. In *Machine Learning and Knowledge Extraction* (Vol. 2, Issue 4, pp. 672–704). MDPI.
50. T. Siameh. 2017. *Graph Analytics Methods In Feature Engineering.*
51. G. Giakkoupis, A.-M. Kermarrec, O. Ruas and F. Taiani. 2021. Cluster-and-Conquer: When Randomness Meets Graph Locality. *2021 IEEE 37th International Conference on Data Engineering (ICDE)*: 2027–2032.
52. X. Chen, B. Qiao, W. Zhang, W. Wu, M. Chintalapati, D. Zhang, Q. Lin, C. Luo, X. Li, H. Zhang, Y. Xu, Y. Dang, K. Sui and X. Zhang. 2019. Neural Feature Search: A Neural Architecture for Automated Feature Engineering. *2019 IEEE International Conference on Data Mining (ICDM)*: 71–80.
53. E. Esenogho, I.D. Mienye, T.G. Swart, K. Aruleba and G. Obaido. 2022. A neural network ensemble with feature engineering for improved credit card fraud detection. *IEEE Access, 10*, 16400–16407.
54. Z. Wang, F. Yang, Q. Xu, Y. Wang, H. Yan and M. Xie. 2023. Capacity estimation of lithium-ion batteries based on data aggregation and feature fusion via graph neural network. *Applied Energy, 336*, 1–11.
55. S.O. Kayode and K. Sheriffdeen. 2024. *Enhancing Graph Neural Network Performance through Advanced Node and Edge Feature Engineering Techniques.*
56. C. Xue, X. Wang, Y. Zhou, P. Palangappa, R. Brugarolas Brufau, A.D. Kakne, R. Motwani, K. Ding and J. Zhang. 2023. Graph Enhanced Feature Engineering for Privacy Preserving Recommendation Systems. *ACM International Conference Proceeding Series*: 44–51.
57. S. Khoshraftar, S. Mahdavi, A. An, Y. Hu and J. Liu. 2019. Dynamic Graph Embedding via LSTM History Tracking. *2019 IEEE International Conference on Data Science and Advanced Analytics (DSAA)*: 119–127.
58. W. Hong, J. Yin, M. You, H. Wang, J. Cao, J. Li, M. Liu and C. Man. 2023. A graph empowered insider threat detection framework based on daily activities. *ISA Transactions, 141*, 84–92.
59. D. Zimbra, M. Ghiassi and S. Lee. 2016. Brand-related twitter sentiment analysis using feature engineering and the dynamic architecture for artificial neural networks. *Proceedings of the Annual Hawaii International Conference on System Sciences, 2016-March*: 1930–1938.
60. L. Massarelli, G.A. Di, L. Cini, F. Petroni, L. Querzoni and R. Baldoni. 2019. Investigating Graph Embedding Neural Networks with Unsupervised Features Extraction for Binary Analysis. *Proceedings of the 2nd Workshop on Binary Analysis Research.*
61. A. Grover and J. Leskovec. 2016. Node2vec: Scalable feature learning for networks. *Proceedings of the ACM SIGKDD International Conference on Knowledge Discovery and Data Mining*, 13–17 August 2016, pp. 855–864.
62. Y. Huang, Y. Zhou, M. Hefenbrock, T. Riedel, L. Fang and M. Beigl. 2023. Automatic Feature Engineering Through Monte Carlo Tree Search. *Lecture Notes in Computer*

Science (Including Subseries Lecture Notes in Artificial Intelligence and Lecture Notes in Bioinformatics), 13715 LNAI: 581–598.

63. W. Zheng, X. Zhu, Y. Zhu, R. Hu and C. Lei. 2018. Dynamic graph learning for spectral feature selection. *Multimedia Tools and Applications*, *77*(22), 29739–29755.
64. H. Zhang, Q. Gan, D. Wipf and W. Zhang. 2023. *GFS: Graph-based Feature Synthesis for Prediction over Relational Databases*.
65. M.O. Çakıroğlu, H. Kurban, P. Sharma, O. Kulekci, E.K. Buxton, M. Raeeszadeh-Sarmazdeh and M. Dalkilic. 2024. An extended de Bruijn graph for feature engineering over biological sequential data. *Machine Learning: Science and Technology*, Vol. 5, No. 3. Pp. 1–15.
66. J. Clearman, R.R. Fayzrakhmanov, G. Gottlob, Y. Nenov, S. Reissfelder, E. Sallinger and E. Sherkhonov. 2019. Feature engineering and explainability with vadalog: A recommender systems application. *Datalog*, *2*, 39–43.
67. L. Xue, C.F. Lynch and M. Chi. 2016. Unnatural Feature Engineering: Evolving Augmented Graph Grammars for Argument Diagrams. *International Educational Data Mining Society, Paper Presented at the International Conference on Educational Data Mining (EDM)*.
68. U. Khurana, H. Samulowitz and D. Turaga. 2018. Feature engineering for predictive modeling using reinforcement learning. *Proceedings of the AAAI Conference on Artificial Intelligence*, *32*(1), 1–8.
69. W. Herzallah, H. Faris and O. Adwan. 2018. Feature engineering for detecting spammers on Twitter: Modelling and analysis. *Journal of Information Science*, *44*(2), 230–247.
70. J.S. Tharani, Z. Hou, E.Y.A. Charles, P. Rathore, M. Palaniswami and V. Muthukkumarasamy. 2024. Unified feature engineering for detection of malicious entities in blockchain networks. *IEEE Transactions on Information Forensics and Security*, 19, 8924–8938.
71. M. Atzmueller and E. Sternberg. 2017. Mixed-initiative feature engineering using knowledge graphs. *Proceedings of the Knowledge Capture Conference, K-CAP 2017*.
72. L. Li, H. Yang, Y. Jiao and K.Y. Lin. 2020. Feature generation based on knowledge graph. *IFAC-PapersOnLine*, *53*(5), 774–779.
73. H. Kim, B.S. Lee, W.Y. Shin and S. Lim. 2022. Graph anomaly detection with graph neural networks: Current status and challenges. *IEEE Access*, *10*, 111820–111829.

14 Future Trends in Artificial Intelligence Driven Security

Utpal Ghosh and Uttam Kr. Mondal

14.1 INTRODUCTION

Nowadays, wireless acoustic sensor networks (WASNs) are gaining traction as a promising technology with diverse applications such as environmental monitoring, surveillance, and healthcare. These networks comprise compact, cost-effective sensor nodes outfitted with microphones which can capture audio signals from the surrounding environment. However, ensuring the security of data transmission in WASNs poses significant challenges due to the inherent vulnerabilities of wireless communication channels. The secure transmission of audio data over a WASN is crucial to protect sensitive information from unauthorized access and tampering. Traditional cryptographic techniques such as encryption and decryption are commonly used to achieve data security. However, applying conventional encryption methods directly to audio data in WASNs can be inefficient and may introduce significant overhead because of the substantial amount of data and the limited computational resources of sensor nodes.

In response to these challenges, this chapter introduces an innovative technique for the secure transmission of audio over WASN using the recursive key rotation (RKR) algorithm. The RKR algorithm is a cryptographic method that involves rotating characters in a text by a certain number of positions, recursively applying the rotation until a specified depth is reached. By leveraging the unique characteristics of the RKR algorithm, author aims to achieve efficient and lightweight encryption of audio data in WASNs, thereby ensuring secure transmission while minimizing computational overhead. The proposed technique offers several advantages over traditional encryption methods. First, the RKR technique is ideal for sensor nodes with limited resources, as it requires minimal computational resources and memory overhead [1]. In addition, the recursive nature of the algorithm enhances the security of data transmission by continuously changing the encryption key, making it more resilient to cryptographic attacks. Moreover, the lightweight nature of the RKR algorithm enables real-time encryption and decryption of audio data, making it suitable for time-critical applications in WASNs. To ascertain the efficacy of the proposed method, extensive simulations and experiments will be conducted using a realistic WASN setup. The performance of suggested technique will be evaluated in terms of data security, computational overhead, and transmission efficiency. In addition, the implementation of convolutional neural network (CNN) model within this

 DOI: 10.1201/9781003521020-14

technique has improved the efficiency of security as well as compressed the audio signal. Furthermore, comparative analysis will be conducted to assess the superiority of the RKR-based approach over conventional encryption techniques in WASNs. The primary aim of this proposed study is to suggest and evaluate a novel technique that integrates the RKR algorithm into the secure transmission of audio over WASN. The proposed technique aims to enhance the confidentiality, integrity, and authenticity of audio data while considering the resource constraints inherent in sensor nodes. In summary, this chapter presents a novel technique for secure transmission of audio over WASN using the RKR algorithm. By leveraging the unique properties of the RKR algorithm and applying the artificial intelligence-based CNN model, this proposed method aims to address the challenges associated with securing audio data transmission in WASNs while minimizing computational overhead and ensuring real-time performance.

This research contributes a comprehensive investigation into the application of the RKR algorithm comprising CNN model for securing audio transmissions in WASNs. The subsequent sections of this chapter will delve into the literature survey, the proposed methodology, experimental setup, and results obtained from the proposed technique. This study concludes by discussing findings, potential directions, and implications for future research [2].

14.2 LITERATURE REVIEW

The secure transmission of audio data within these networks is paramount to ensure confidentiality, integrity, and authenticity. This literature review aims to scrutinize existing research in the field, focusing on cryptographic techniques and security mechanisms, with the objective of laying the groundwork for the development of a novel technique leveraging the RKR algorithm. The study by Ismail et al. [3] explores lightweight cryptography techniques for enhancing security in wireless sensor networks, providing insights into the feasibility of lightweight algorithms like RKR. Study by Faris et al. [4] offers a comprehensive overview of lightweight cryptographic algorithms, including their suitability for securing audio transmissions in wireless sensor networks. A detailed survey of security challenges and solutions in wireless sensor networks has been provided that offering valuable insights into encryption techniques applicable to audio data transmission [5]. Article by Harn et al. [6] proposes a lightweight symmetric key cryptography approach for securing data in wireless sensor networks, which could complement RKR-based encryption techniques. Researchers present a comprehensive survey of security mechanisms for wireless sensor networks, shedding light on the importance of secure data transmission protocols [7]. Article by Li et al. [8] presents a data transmission scheme for wireless sensor networks that prioritizes both security and efficiency that leverage chaotic compressive sensing, highlighting the importance of robust encryption techniques for protecting transmitted data. The researchers of reference [9] provide a comprehensive survey of data security issues in wireless sensor networks, discussing various cryptographic algorithms and their applicability to secure data transmission, which can provide context for evaluating the effectiveness of the RKR algorithm.

Article by Fang et al. [10] introduces a data aggregation scheme designed to be both efficient and secure in wireless sensor networks, emphasizing the need for robust security measures to protect aggregated data during transmission. The researchers of Salau et al. [11] conduct a survey of security threats and defenses in wireless sensor networks, discussing various cryptographic techniques and intrusion detection methods, which can provide insights into the security challenges. The authors of reference [12] propose a lightweight and secure scheme for transmitting data in wireless sensor networks, highlighting the importance of minimizing overhead while ensuring robust encryption to protect sensitive data, which aligns with the objectives of the RKR algorithm. Article by Sadkhan and Salman [13] presents a lightweight security mechanism tailored for wireless sensor networks, emphasizing the importance of minimizing resource consumption while maintaining robust security measures, which is pertinent to the objectives of the RKR algorithm. This research introduce a secure data transmission scheme based on chaotic compressive sensing specifically designed for WSNs, highlighting the utilization of chaotic dynamics to enhance data security [14]. The researchers of reference [15] provides a comprehensive survey on efficient and secure data aggregation techniques in WSNs, discussing various cryptographic algorithms and optimization strategies. This study provides an overview of security concerns in wireless sensor networks, encompassing common attack vectors and corresponding countermeasures [16]. A distributed clustering approach has been proposed for ad hoc sensor networks, blending energy efficiency and hybrid techniques., which can offer insights into energy-efficient mechanisms relevant to the RKR algorithm's energy-efficient transmission of audio data [17]. A novel CNN -based audio encoding model improves compression [18], while several CNN models for images have debuted [19, 20]. Article by Hemalatha et al. [21] demonstrates that deep learning approaches significantly enhance the accuracy and efficiency of intrusion detection systems amidst evolving cybersecurity threats. The study highlights the potential of these advanced techniques to address complex security challenges, ensuring more robust protection against intrusions in various network environments. The research article by Reshma and Anand [22] concludes that VGG-16 outperforms LENET and ALEXNET in terms of accuracy and reliability for smart behavior monitoring applications. Their comparative analysis underscores VGG-16's superior performance in handling complex image processing tasks, making it the preferred choice for such applications.

14.3 PROPOSED TECHNIQUE

The RKR algorithm for WASNs plays a crucial role in enhancing security by periodically updating encryption keys used for communication. This algorithm helps mitigate the risk of key compromise and unauthorized access to sensitive data transmitted over the network. By regularly rotating keys, the algorithm helps maintain the confidentiality and integrity of communications within the WASN, thereby enhancing overall network security. In addition, the recursive nature of the algorithm allows for efficient and scalable key management across large-scale WASN deployments. The RKR adds an additional layer of security by ensuring that even if a key is compromised, it becomes obsolete after a certain period. Proposed WASNs are often deployed in dynamic and unpredictable environments. RKR allows the network to adapt to changes in the security landscape by providing a mechanism to update keys based on predetermined intervals or triggered events. By frequently changing encryption keys in this present technique, the network becomes more resilient against various cryptographic attacks, including those attempting to exploit vulnerabilities associated with static keys. This proactive approach enhances the overall security posture of the WASN. RKR algorithms can be designed to efficiently distribute new keys across the network. This ensures that all nodes receive updated keys in a timely manner, minimizing the risk of communication disruptions due to outdated or compromised keys. RKR algorithms need to strike a balance between enhancing security and minimizing the computational overhead associated with key management [23]. These proposed efficient algorithms ensure that the benefits of key rotation outweigh the costs. The CNN encoder and decoder network enrich the degree of security as well also improve the encoding and decoding time of audio signal.

The proposed system is mainly divided into two phase. During the operation of first phase, an audio file has been taken in .wav file as input, then this audio file (.wav) is converted into a text (.txt) or document (.docx) files by applying SpeechRecognition function through Python program. As a result, after conversion a text or document file has been generated which consists of the corresponding transcription character string. After that implement the RKR symmetric encryption algorithm to this text or document file. As a result, an encrypted ciphertext has been generated which has to be now transmitted over WASN. The ciphertext produced by the RKR algorithm is subjected to training with a CNN model, allowing the network to iteratively optimize its representations of the input data. The CNN comprises an input layer, multiple hidden layers and an output layer. In the proposed model, three two-dimensional convolutional layers, pooling layers, normalization layers and fully connected layers are utilized to extract features from the input signals. CNNs operate by utilizing convolution and pooling layers on input audio signals. Convolution layers extract features by sliding small filters over the audio and calculating dot products with the input, while pooling layers subsequently downsample the convolution layer output to enhance computational efficiency by reducing data dimensionality [24]. The CNN encoder network within WASN encrypts data, which is then decoded by the CNN decoder network. Comprising activation layers, up-sampling layers and three deconvolution layers, the decoder network reconstructs the encrypted data. After transmission of the data over network securely, it has to decrypt into the original plaintext for understanding the input audio or message using RKR symmetric decryption algorithm and then convert this plaintext file into its corresponding audio file by applying SpeechRecognition technique through Python code. During the implementation of RKR symmetric encryption, each character of the character string of text or document file has been converted into a byte. Figure 14.1 depicts the overall functional diagrammatic representation of the proposed system.

Figure 14.2 showcases the working flow representation of the proposed technique step by step.

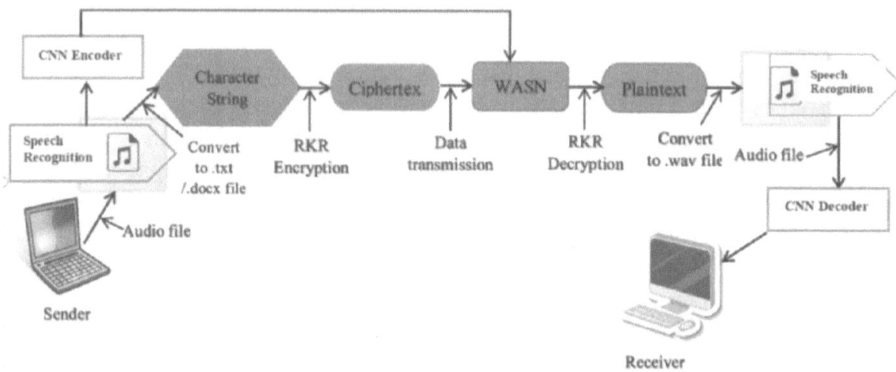

FIGURE 14.1 Schematic block diagram of the present system.

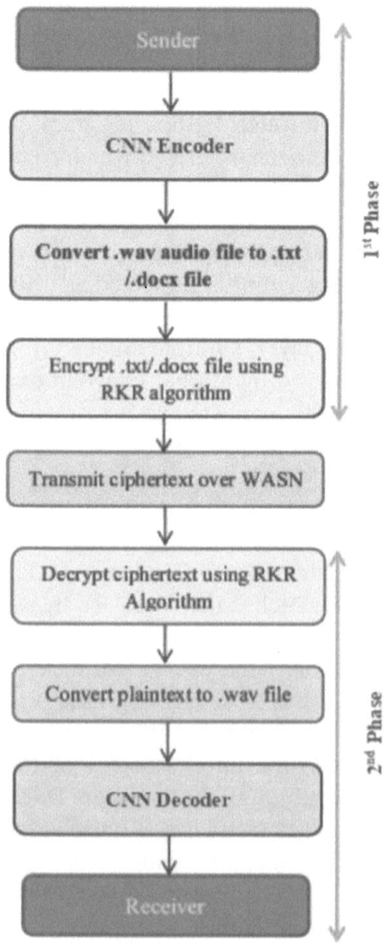

FIGURE 14.2 Working flow diagram of the proposed technique.

The implementation of the RKR symmetric encryption and decryption algorithm follows a specific scheme:

i. The plaintext, typically a text or document file, is considered and converted into blocks of bits of various sizes such as 4/8/16/32/64/128/256.
ii. The source stream is split into two equal sections.
iii. Deriving a key value from the source stream, typically selected as half the size of the source stream.
iv. Performing modulo-2 addition (XOR) between the first half of the source stream and the key value to generate the first intermediate block.
v. Applying XOR operation between the second half of the source stream and the reversed key value to produce the second intermediate block.
vi. This iterative process continues for various intermediate blocks until the entire source stream is reconstructed. Upon a finite number of successful iterations, the complete source stream is regenerated.
vii. The decryption procedure of the RKR algorithm closely mirrors the encryption process.

Concerning the encryption of the entire bit stream, different methodologies can be employed based on how the blocks are structured. These are discussed in subsequent subsections.

14.3.1 Blocks with Equal Size

If all blocks have equal lengths, and intermediate blocks after a fixed number of iterations are considered as the corresponding encrypted blocks, then the same number of iterations will be required for encrypting the entire bit stream. The encryption key will consist of the fixed block size, the fixed number of iterations for all blocks, and the key value used during encryption.

14.3.2 Blocks with Different Sizes

For blocks of different lengths, unique blocks will require varying numbers of iterations to complete corresponding cycles. The least common multiple (LCM) of these iteration counts will determine the total number of iterations needed to complete the cycle for the entire stream. If 'i' iterations are used to encrypt the entire stream, an additional (P – i) iterations will be needed to decrypt the encrypted stream.

Implementing diverse encryption policies, such as designating various intermediate blocks as encrypted blocks for different source blocks, increases the complexity of the encryption key, thereby enhancing security. Algorithm 14.1 furnishes a clear understanding of the RKR symmetric encryption and decryption algorithm.

Algorithm 14.1 Detailed implementation algorithm for RKR encryption and decryption process

1. Declare a function encrypt(text, key):
   a. If the length of text is less than or equal to 1, return text

   b. Calculate the mid index as the length of text divided by 2
   c. Encrypt the left half of the text using the key: left_text = encrypt(text[:mid], key)
   d. Encrypt the right half of the text using the key: right_text = encrypt(text[mid:], key)
   e. Concatenate the encrypted left and right halves: encrypted_text = left_text + right_text
   f. Rotate the key to the left by one position: new_key = rotate_key_left(key)
   g. Return the XOR of encrypted_text and new_key as the result
2. Define a function decrypt(encrypted_text, key):
   a. If the length of encrypted_text is less than or equal to 1, return encrypted_text
   b. Calculate the mid index as the length of encrypted_text divided by 2
   c. Rotate the key to the left by one position: new_key = rotate_key_left(key)
   d. Decrypt the left half of the encrypted text using the new key: left_text = decrypt(encrypted_text[:mid], new_key)
   e. Decrypt the right half of the encrypted text using the new key: right_text = decrypt(encrypted_text[mid:], new_key)
   f. Concatenate the decrypted left and right halves: decrypted_text = left_text + right_text
   g. Return the XOR of decrypted_text and key as the result
3. Define a function rotate_key_left(key):
   a. Extract the first character of the key: first_char = key[0]
   b. Rotate the remaining characters of the key to the left by one position: rotated_key = key[1:] + first_char
   c. Return the rotated_key
4. Define a function rotate_key_right(key):
   a. Extract the last character of the key: last_char = key[-1]
   b. Rotate the remaining characters of the key to the right by one position: rotated_key = last_char + key[:-1]
   c. Return the rotated_key
5. Specify a main function to demonstrate the usage of the encrypt and decrypt functions:
   a. Generate a random key of fixed length
   b. Prompt the user to input the text to be encrypted
   c. Encrypt the input text using the generated key: encrypted_text = encrypt(input_text, generated_key)
   d. Print the encrypted text
   e. Decrypt the encrypted text using the generated key: decrypted_text = decrypt(encrypted_text, generated_key)
   f. Print the decrypted text
6. Call the main function to start the encryption and decryption process

After successful transmission of encrypted data over WASN, in the second phase of proposed technique, decryption process of the ciphertext to the original plaintext has been performed and the decrypted data is stored into a text or document file.

After that, this file has been converted into the corresponding .wav audio file using SpeechRecognition function through Python program which is further transmitted to the receiver's end.

14.4 IMPLEMENTATION AND RESULTS ANALYSIS

The experimental results validate the efficacy of the proposed technique in achieving both security and energy efficiency in audio transmission across WASNs. The RKR-based encryption mechanism ensures robust protection of audio data against unauthorized access and eavesdropping attacks. The simulation work has been done through Python version 3.12.1. To perform the first phase and second phase of the proposed technique, a sample .wav file named 'speech.wav' has been taken which consists of the character string "good evening ladies and gentlemen we like to welcome you to play then you reduce broadcast". During first phase, the audio is converted to its corresponding transcription (character string) through implementing SpeechRecognition using Python code and storing the character string into a .txt file named as 'Output_Char.txt'. After that, apply the RKR algorithm to the .txt file and consider the .txt file as plaintext and it has been encrypted to its corresponding ciphertext named as 'cipher'. Figures 14.3 and 14.4 showcase the transcription character string of the audio file and the encrypted Unicode formation of the sample audio, respectively.

Figures 14.5 and 14.6 depict the snapshot of the encrypted data representation and binary representation of the ciphertext, respectively, which has been decrypted later.

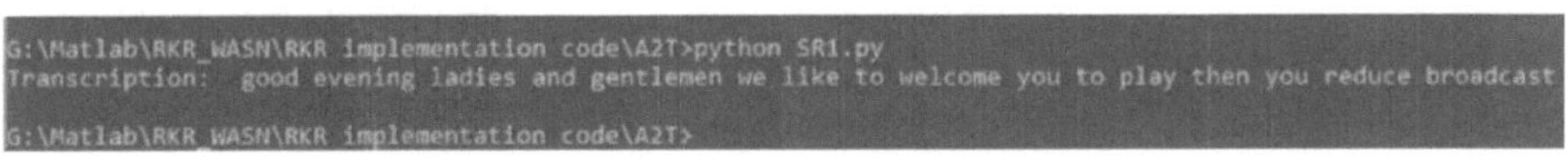

FIGURE 14.3 The transcription character string of the audio file.

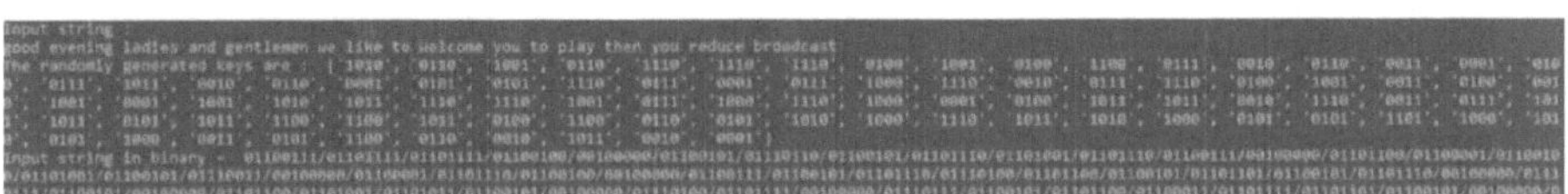

FIGURE 14.4 Encrypted Unicode formation of the sample audio.

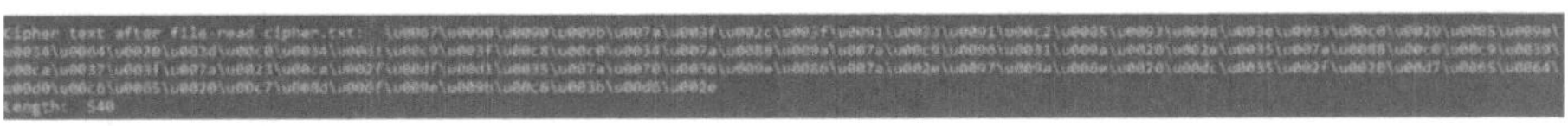

FIGURE 14.5 Encrypted data of the ciphertext.

FIGURE 14.6 Binary representation of the ciphertext.

```
Plain text :
 good evening ladies and gentlemen we like to welcome you to play then you reduce broadcast
Length of Plain text : 90
Decryption time 1.9450232982635498 seconds
Plain text written to plain.txt
```

FIGURE 14.7 Plaintext that decrypted from the ciphertext.

Figure 14.7 represents the corresponding plaintext into the text file named 'plaintext.txt' after successful decryption of the encrypted data. This simulation result also shows the encryption and decryption time for performing encryption and decryption algorithm process of the text file, which has been very useful during computing the performance analysis of this proposed technique.

14.5 PERFORMANCE ANALYSIS

The assessment of encryption/decryption time results is pivotal in gauging algorithmic efficiencies concerning execution. This study endeavors to establish a relationship between file size and the corresponding encryption/decryption time. To investigate the non-homogeneity between the original and encrypted files, a chi-square test has been employed. The "Pearsonian Chi-square test" [25] is utilized to determine whether the observations in encrypted files conform well to a hypothetical distribution. In this context, the chi-square distribution is applied with (256 – 1) = 255 degrees of freedom, where 256 denotes total count of classes of potential characters in both the source and encrypted files. If the observed statistic value exceeds the tabulated value at a given significance level, the null hypothesis is rejected. The "Pearsonian Chi-square" or the "Goodness-of-fit Chi-square" is expressed using equation (14.1):

$$X^2 = \sum\left\{\left(f_0 - f_e\right)^2 / f_e\right\} \tag{14.1}$$

The variables f_e and f_0 represent the character frequency in the source file and the corresponding encrypted file respectively. Based on this formula, the Chi-square values have been computed for sample pairs of source and encrypted files.

The time consumed in encrypting and decrypting files can be influenced by several factors, including the efficiency of the encryption algorithm, the size of the files being processed, and the hardware specifications of the machine where the code is executed. The efficiency of file encryption and decryption processes can be influenced by various factors. First, the encryption and decryption algorithms implemented in the code significantly impact processing time. Some algorithms are inherently faster or more computationally intensive than others. Second, the size and complexity of the files being encrypted or decrypted play a crucial role. Larger files generally require more processing time compared to smaller ones due to increased data volume and computation requirements. In addition, the hardware architecture of the machine executing the code can affect processing speed. Machines with faster processors, sufficient RAM, and optimized hardware configurations tend to execute encryption and decryption tasks more swiftly than those with lower specifications.

Therefore, optimizing encryption and decryption algorithms, considering file size management techniques, and utilizing hardware resources effectively are essential strategies to minimize processing time in file encryption and decryption operations.

14.5.1 Results for .txt File

Table 14.1 gives the results of implementing the technique on .txt file. Ten converted .txt files have been considered here which are converted from ten different .wav audio files. There sizes ranges from 78 kb to 1037 kb. The encryption time ranges from 3.690951 seconds to 46.931283 seconds. The decryption time ranges from 7.802668 seconds to 102.878593 seconds. The Chi square value is observed from 1779389 to 24803012 with degree of freedom being 255. The values of degree of freedom are expressed such that the source file from the sender which is encrypted as ciphertext for transmission over WASN and the final plaintext which is decrypted from the ciphertext remains same with respect to the size of file and the number of characters, respectively. This table also computes the consumption of energy for transmission of each file from sender to receiver and evaluates the data loss ratio. Data loss has been calculated based on the difference between the value of sample file size at sender's end and the output file size after successful transmission of data at receiver's end. These parameters prove that the present system is energy efficient, enhance data security and reduce the possibility of data loss. According to the experimental observation during encryption and decryption process, it has been examined that the length of the plaintext at sender's end (which is converted from sample .wav audio) is equal to the length of the final plaintext at the receiver's end (which is further converted to .wav audio file). As a result, it can be concluded that the data loss ratio is near zero.

A portion of the table is graphically depicted in Figure 14.8, illustrating a graphical correlation between the sample file size and encryption-decryption time for .txt files.

TABLE 14.1
Result for .txt Files for the Proposed Technique

File Name	File Size (in kb)	Encryption Time (Sec.)	Decryption Time (Sec.)	Output File Size (kb)	Chi Square Value	Degree of Freedom	Energy Consumption (kWh)	Data Loss (%)
t1.txt	78	3.690951	7.802668	78	1779389	255	0.13	0
t2.txt	116	5.738853	11.479705	116	2731815	255	0.17	0
t3.txt	231	10.715104	22.597628	231	5553544	255	.21	0
t4.txt	346	15.674385	34.612041	346	8258018	255	.27	0
t5.txt	461	21.023472	46.148698	461	10948916	255	.29	0
t6.txt	576	26.363543	56.580975	576	13520748	255	0.32	0
t7.txt	691	31.369780	75.399209	691	16615419	255	0.367	0
t8.txt	806	36.636649	82.603096	806	19480955	255	0.394	0
t9.txt	921	43.289955	91.847981	921	21859759	255	0.435	0
t10.txt	1037	46.931283	102.878593	1037	24803012	255	0.479	0

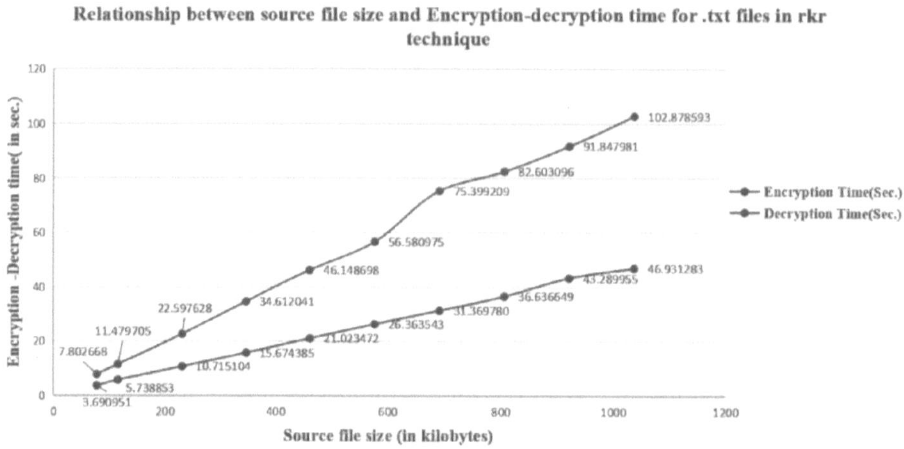

FIGURE 14.8 Relation between sample file size and encryption/decryption time for .txt files for the proposed technique.

This figure suggests a tendency for the encryption-decryption time to change in an almost linear fashion with the size of the sample file.

14.5.2 Results for .Docx File

Table 14.2 gives the result of implementing the technique on .docx file. Here ten .docx file have been considered which are converted from ten different .wav audio files. There sizes range from 71 kb to 580 kb. The encryption time ranges from 6.332366 seconds to 65.688301 seconds. The decryption time ranges from 14.032799 seconds to 145.917293 seconds. The Chi square value is observed from 3156613 to 31539854

TABLE 14.2
Result for .docx Files for the Proposed Technique

File Name	File Size (kb)	Encryption Time (sec.)	Decryption Time (sec.)	Output File Size (kb)	Chi Square Value	Degree of Freedom	Energy Consumption (kWh)	Data Loss (%)
e1.docx	71	6.332366	14.302799	47	3156613	255	0.19	0
e2.docx	127	14.211848	27.832035	55	6212360	255	0.256	0
e3.docx	184	19.497811	42.598572	63	9346064	255	0.291	0
e4.docx	240	26.403834	55.192321	71	12815357	255	0.32	0
e5.docx	297	31.783751	67.940027	79	15873544	255	0.368	0
e6.docx	353	38.901685	87.938537	87	18914712	255	0.389	0
e7.docx	410	46.249461	101.125988	95	22093670	255	0.42	0
e8.docx	467	52.360968	114.036586	103	25408507	255	0.457	0
e9.docx	523	59.184049	129.939484	111	28624185	255	0.517	0
e10.docx	580	65.688301	145.917293	118	31539854	255	0.546	0

with degree of freedom being 255. The values of degree of freedom are expressed such that the source file from the sender which is encrypted as ciphertext for transmission over WASN and the final plaintext which is decrypted from the ciphertext remain same with respect to the size of file and the number of characters, respectively. This table also computes the consumption of energy for transmission of each file from sender to receiver and evaluates the data loss ratio. Data loss has been calculated based on the difference between the value of sample file size at sender's end and the output file size after successful transmission of data at receiver's end. These parameters prove that the present system is energy-efficient, enhances data security, and reduces the possibility of data loss. According to the experimental observation during encryption and decryption process, it has been examined that the length of the plaintext at sender's end (which is converted from sample .wav audio) is equal to the length of the final plaintext at the receiver's end (which is further converted to .wav audio file). As a result, it can be concluded that the data loss ratio is near zero.

Figure 14.9 is created to illustrate the correlation between the sample file size and the encryption/decryption time for .docx files. As it is observed from the figure, there exists a linear relationship between the source file size and the encryption-decryption time.

Upon comprehensive analysis of the results presented in Section 14.5, a summary overview has been derived. The encryption and decryption time exhibit a linear correlation with the size of the source file. Remarkably, there is minimal disparity between the encryption and decryption times for a given file, indicating that the computational complexity of both processes is relatively similar. Furthermore, the Chi square value for the .docx file surpasses that of the .txt file. This observation suggests differences in statistical distribution or characteristics between the two file formats. In this proposed technique, audio data inputs in various languages, such as Greek, Bengali, Spanish, and Chinese, have been utilized, leveraging Unicode for

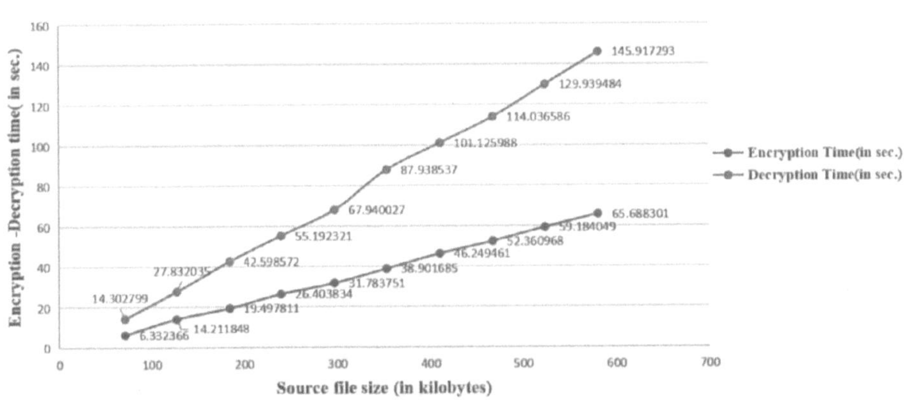

FIGURE 14.9 Relations between sample file size and encryption-decryption time for .docx files for the proposed technique.

representation. The input is extended in multiples of 8 bits to ensure easy scalability. This experiment also furnishes that for conversion of plaintext and ciphertext from the .wav file to .txt/.docx file, .txt file is more efficient as compared to .docx file, due to the requirement of less encryption and decryption time.

14.6 CONCLUSION AND FUTURE SCOPE

In conclusion, the present technique introduces a novel strategy for ensuring the secure transmission of audio data across WASN, employing the RKR algorithm and CNN model. Through the integration of dynamic key management functionalities within the communication protocol, it effectively tackles essential security challenges while also mitigating energy consumption concerns. The presented technique boasts simplicity and efficiency with minimal encoding and decoding times, despite its relatively high block length. Moreover, the encoded string generated by this method does not entail any overhead bits. Its straightforward implementation in various high-level programming languages renders it suitable for practical applications, thereby enhancing message transmission security. Since it is symmetric cryptographic algorithm, it does not generate any public key, it only generates private key, which is transferred to the user. In this technique, XOR operation has been performed which is easy to decrypt because if someone does repeated XOR operation, he/she gets back the same after same iteration. As a result, the complexity of the proposed technique is less. This system uses only hexadecimal digits as Unicode uses 0 to 9 and a/A to f/F, so attacker could guess to see these that there perform any hexadecimal operation. For this reason attacker may be use different types of hexadecimal techniques to decrypt these and they need not apply versatile technique to decode.

In future, this technique will be applicable in image cryptography, audio and video steganography as well as in multimedia to transmit information securely. Future research directions also include to refine the CNN model through the augmentation of layers and the fine-tuning of hyperparameters, thereby accelerating the encoding-decoding process and elevating security measures and further optimization of the proposed technique and its application in real-world scenarios.

REFERENCES

1. Anand, A., & Bhambri, P. (2018). Orientation, Scale and Location Invariant Character Recognition System using Neural Networks. International Journal of Theoretical & Applied Sciences, 10(1), 106–109. ISSN No. (Print): 0975-1718. ISSN No. (Online): 2249-3247. Published by Research Trend. Retrieved from www.researchtrend.net.
2. Bhambri, P. & Rani, S. (2024). Ethical Issues for Climate Change and Mental Health. In D. Samanta& M. Garg (Eds.), Impact of Climate Change on Mental Health and Well-Being (pp. 178–198). IGI Global. https://doi.org/10.4018/979-8-3693-2177-5.ch012
3. Ismail, S., Dawoud, D.W., Reza, H. (2023). Securing wireless sensor networks using machine learning and blockchain: A review. Future Internet, 15(6): 200.
4. Faris, M., Mahmud, M. N., Salleh, M.F.M., Alnoor, A. (2023). Wireless sensor network security: A recent review based on state-of-the-art works. International Journal of Engineering Business Management, 15: 18479790231157220.

5. Mugheri, A. A., Siddiqui, M.A., Khoso, M. (2018). Analysis on security methods of wireless sensor network (WSN). Sukkur IBA Journal of Computing and Mathematical Sciences, 2(1): 52–60.
6. Harn, L., Hsu, C.-F., Xia, Z., He, Z. (2021). Lightweight aggregated data encryption for wireless sensor networks (WSNs). IEEE Sensors Letters, 5(4): 1–4.
7. Dhabliya, D., Soundararajan, R., Selvarasu, P., Balasubramaniam, M.S., Rajawat, A.S., Goyal, S., Raboaca, M.S., Mihaltan, T.C., Verma, C., Suciu, G. (2022). Energy-efficient network protocols and resilient data transmission schemes for wireless sensor networks: An experimental survey. Energies, 15(23): 8883.
8. Li, X., Wang, C., Yang, Z., Yan, L., Han, S. (2018). Energy-efficient and secure transmission scheme based on chaotic compressive sensing in underwater wireless sensor networks. Digital Signal Processing, 81: 129–137.
9. Al Shehri, W. (2017). A survey on security in wireless sensor networks. International Journal of Network Security & Its Applications (IJNSA), 9(1): 25–32.
10. Fang, W., Zhang, W., Chen, W., Pan, T., Ni, Y., Yang, Y. (2020). Trust-based attack and defense in wireless sensor networks: a survey. Wireless Communications and Mobile Computing: 1–20.
11. Salau, A.O., Marriwala, N., Athaee, M. (2021). Data security in wireless sensor networks: Attacks and countermeasures. In: Mobile Radio Communications and 5G Networks: Proceedings of MRCN 2020: 173–186, Springer.
12. Khashan, O.A., Ahmad, R., Khafajah, N.M. (2021). An automated lightweight encryption scheme for secure and energy-efficient communication in wireless sensor networks. Ad Hoc Networks, 115: 102448.
13. Sadkhan, S.B., Salman, A.O. (2018). A survey on lightweight-cryptography status and future challenges. In: 2018 International Conference on Advance of Sustainable Engineering and Its Application (ICASEA): 105–108, IEEE.
14. Zhang, Y., Xiang, Y., Zhang, L. X., Rong, Y., Guo, S. (2018). Secure wireless communications based on compressive sensing: A survey. IEEE Communications Surveys & Tutorials, 21(2): 1093–1111.
15. Yousefpoor, M.S., Yousefpoor, E., Barati, H., Barati, A., Movaghar, A., Hossein zadeh, M. (2021). Secure data aggregation methods and countermeasures against various attacks in wireless sensor networks: A comprehensive review. Journal of Network and Computer Applications, 190: 103118.
16. Pritchard, S.W., Hancke, G. P., Abu-Mahfouz, A.M. (2017). Security in software-defined wireless sensor networks: Threats, challenges and potential solutions. In: 2017 IEEE 15th International Conference on Industrial Informatics (INDIN): 168–173, IEEE.
17. Shahraki, A., Taherkordi, A., Haugen, Ø., Eliassen, F. (2020). Clustering objectives in wireless sensor networks: A survey and research direction analysis. Computer Networks, 180: 107376.
18. Debnath, A., Mondal, U.K. (2023). Lossless audio codec based on CNN, weighted tree and arithmetic encoding (LACCWA). Multimedia Tools and Applications: 1–23.
19. Bhairnallykar, S.T., Narawade, V. (2023). Segmentation of MR images using DN convolutional neural network. International Journal of Information Technology, 1–12. https://doi.org/10.1007/s41870-023-01461-x.
20. Jintanachaiwat, W., Siriborvornratanakul, T. (2023). Vision-based image similarity measurement for image search similarity. International Journal of Information Technology, Springer: 1–6. https://doi.org/10.1007/s41870-023-01437-x.
21. Hemalatha, S., Mahalakshmi, M., Vignesh, V., Geethalakshmi, M., Balasubramaniam, D., Jose, A. A. (2023). Deep Learning Approaches for Intrusion Detection with Emerging Cybersecurity Challenges. In: International Conference on Sustainable Communication Networks and Application (ICSCNA). 1522–1529.

22. Reshma, R., Anand, A. J. (2023). Predictive and Comparative Analysis of LENET, ALEXNET and VGG-16 Network Architecture in Smart Behavior Monitoring. In: Seventh International Conference on Image Information Processing (ICIIP). 450–453.
23. Potluri, S., Tiwari, P. K., Bhambri, P., Obulesu, O., Naidu, P. A., Lakshmi, L., Kallam, S., Gupta, S., & Gupta, B. (2019). Invention titled "Method of Load Distribution Balancing for Fog Cloud Computing in IoT Environment" (Application No. 201941044511). Published on 29th November 2019 (Issue No. 48/2019, Page No. 56118). Retrieved from http://www.ipindia.nic.in/writereaddata/Portal/IPOJournal/1_4813_1/Part-1.pdf
24. Panda, S. K., Reddy, G. S. M., Goyal, S. B., Thirunavukkarasu, K., Bhambri, P., Rao, M. V., Singh, A. S., Fakih, A. H., Shukla, P. K., Shukla, P. K., & others. (2019). Invention titled "Method for Management of Scholarship of Large Number of Students based on Blockchain" (Application No. 201911034937). Published on 06th September 2019 (Issue No. 36/2019, Page No. 40572). Retrieved from http://ipindia.gov.in/writereaddata/Portal/IPOJournal/1_4785_1/Part-1.pdf
25. Efron, B., Hastie, T. (2021). Computer Age Statistical Inference, Student Edition: Algorithms, Evidence, and Data Science, 6. Cambridge University Press.

15 Enhancing Cybersecurity with Distributed Models and Sparse Mixture of Experts

Ashok J, Jayapratha T, Arunmozhi S A, Gayathri B, Karthick K, and Mayakannan S

15.1 INTRODUCTION

Data privacy and security are major concerns for traditional centralized machine learning (ML) models, especially when dealing with sensitive information in fields like healthcare and cybersecurity. Data leakage instances in the past few years have increased worries about data privacy, and the General Data Protection Regulation (GDPR) of the European Union (EU) places constraints on the gathering and sharing of personal information about EU citizens. Concurrently, there is a growing need for data and computational resources to power complex artificial intelligence (AI) models that analyze data. People and businesses alike are wary of sharing data, despite its obvious value. Moreover, operating expenses rise and environmental implications are substantial as a result of deep learning systems' computing needs, which cause high energy consumption and carbon emissions.

One potential approach to these problems is federated learning (FL), which allows for several entities to work together to develop a global model without actually sharing any data. While protecting the privacy of data, this paradigm aids in the discovery of correct models from dispersed data. When dealing with huge datasets or broad computer networks, FL can be taxing on resources like memory, computation power, energy, and network bandwidth. The increased computational resources needed to power AI's rapid advancement—fueled by increasingly extensive and computationally costly ML models—have resulted in a substantial increase in the technology's carbon footprint [1].

The communication and compute overheads associated with FL's ability to support complicated ML tasks in a dispersed and privacy-aware manner could result in higher carbon emissions compared to centralized systems. Studies show that FL model training can produce as much as 80 kg of CO_2e, which is more than the emissions from training bigger models in a centralized environment with AI accelerators. Training overheads across varied client hardware, increased communication costs, and sluggish convergence all contribute to this inefficiency. More widespread use of FL in industry and decentralization of ML activities are two factors that could

DOI: 10.1201/9781003521020-15

raise FL's global carbon footprint [2]. Sustainable FL is complicated by the fact that renewable electricity is not readily available in all areas. Therefore, encouraging more environmentally friendly ML applications requires maximizing FL efficiency.

Green methods to FL seek to balance energy efficiency with performance, in response to the pressing need to lessen AI's environmental effect [3]. In order to lessen the negative effects of artificial intelligence on the environment, "green AI" advocates for more energy-efficient algorithms, more eco-friendly hardware, and less carbon emissions from data centers. The reliance of end-user devices on local energy mixes makes it challenging to provide Florida with renewable energy, even though centralized AI can be powered by such energy. The majority of methods that aim to bridge the gap between green AI and FL concentrate on strategies for task assignment and energy-aware node selection, but they fail to tackle the complex new environment of vertical FLVFL.

Mixture of experts (MoE) designs have been in the spotlight as of late for their ability to strike a balance between model capacity, energy efficiency and computational cost. A classical MoE model's final prediction is the result of linearly combining the predictions made by a group of experts, who are homogeneous prediction sub-nets, and being weighted by a gate sub-net. Minimizing computing costs, Sparsely-Gated Mixture-of-Expert (SMoE) models choose a small number of experts according to the current instance of input data. When it comes to green FL applications, sparse MoE models are great because they decrease data transfer costs and protect local data from information leakage. This is accomplished by selecting a small number of expert sub-models for each data instance using the gate sub-model.

Energy consumption and carbon footprint analysis of FL applications, with a special emphasis on horizontal FL (HFL) environments has been the subject of recent study. Less research has focused on VFL applications, however, in which parties hold distinct feature spaces of overlapping real-world objects. To avoid data leakage, secure communication protocols have been suggested; however, these protocols incur substantial communication and computational costs due to their reliance on encryption-based data transformation and numerous peer-to-peer connections [4].

Since their inception, sparse MoE models for FL have mostly concentrated on the heterogeneous data distributions and model personalization scenarios with an emphasis on HFL. There have been proposals to use a MoE-based model in a VFL context, but this approach has complications with data leaking and high communication and computing requirements. In order to facilitate scalable calculations while maintaining data security, this study presents VFL_MoE, a new and efficient MoE-based method to VFL. To minimise communication overheads and guarantee privacy, the gate is trained using a lesser selection of clean data characteristics. To further alleviate computational demands, the method makes use of a data-reduction factor to regulate the data fraction utilized for each training epoch.

Here are the key points of this proposal:

- A model architecture for VFL similar to MoE that lets the coordinator get only the outputs from expert models per instance, thus minimizing the disclosure of private local data.

- A distributed algorithm that can train this model in a way that is both cost-effective and private.
- An experimental research and theoretical analysis demonstrating the model's efficiency and effectiveness.

15.2 BACKGROUND

15.2.1 Privacy and Vertical Federated Learning

The introduction of data-driven technology in the last several years has caused a sea change in the way data is handled, examined, and used in many different fields. Data privacy, security, and personal information sovereignty have become more important issues alongside this advancement [5]. The concept of data sovereignty centers on the idea that the rightful owner should have control over their data when it comes to digital information. It even goes as far as asking owners to choose which parts of their data to share and with whom. When it comes to digital data sovereignty in India, there are two main areas to focus on: cloud sovereignty, which is all about using federated cloud infrastructures and services that are in line with current regulations, and secure online data exchange among various consortium members or groups of companies [6].

There should also be business, legal, and cloud-based laws in place, as well as written contracts that control data usage and access and specify how data can be shared with other organizations or entities [7]. To provide an example, think about a situation where a healthcare operator is involved in the data cycle from start to finish, providing specific data and then making use of the results of analytics and ML processes. Not all data aspects may be accessible depending on the operator's function (patient, medical practitioner, or paramedical personnel) in this complex scenario, which requires a number of activities [8]. In addition, it may be necessary to anonymize specific data pieces before sending them, and there may be limitations on exporting data of the nation of origin or the European Community. In order to effectively manage this kind of situation, a compelling approach that aims to balance the advantages of data-driven perceptions with the need to protect information authority is the FL paradigm [9].

The dispersed ML method known as FL permits the training of models across numerous decentralized devices or servers that store local data, all without the need to exchange the raw data. Using FL, model training can take place locally on data-housing devices or servers, reducing the burden of raw data transport. Also, FL offers decentralized training, which means that models can be trained locally on distributed devices [10]. This keeps data private while they work together to improve the model's performance. With FL, the models from multiple sources were updated without revealing who did what, which ensures confidentiality and privacy. Differential privacy is another approach that may be used to create a strong privacy framework; it prevents specific data points from being identified by adding noise to individual updates.

There are still several issues with FL, such as communication overhead, data that is not equally distributed (non-IID), and security concerns, despite its immense potential. Recent developments, such as the implementation of VFL frameworks have prompted researchers to concentrate on finding solutions to these problems in the hopes of making FL even more useful and resilient in a variety of settings [11]. Data is partitioned across different parties in VFL according to features. The main goal is to make it possible for various groups to work together to build a prediction model without compromising any sensitive information. A more complex method for dissecting the loss function at each party is required in the VFL setting compared to HFL.

Traditional approaches take one of two forms: (a) all parties involved in the training have equal ownership of the model, or (b) the model is divided up among them [12]. As shown in Figure 15.1, in the second scenario, characteristic of a traditional neural network (NN) is for each node to convert the input data into a representation that can be processed further. The final data is relayed to the next level until the conclusion of either the inference or training phase. As the backpropagation process continues, the gradient is likewise sent out to all of the connecting nodes. In addition, that can help alleviate the computing load on each node, which can be quite challenging in real-world settings due to restricted computational resources.

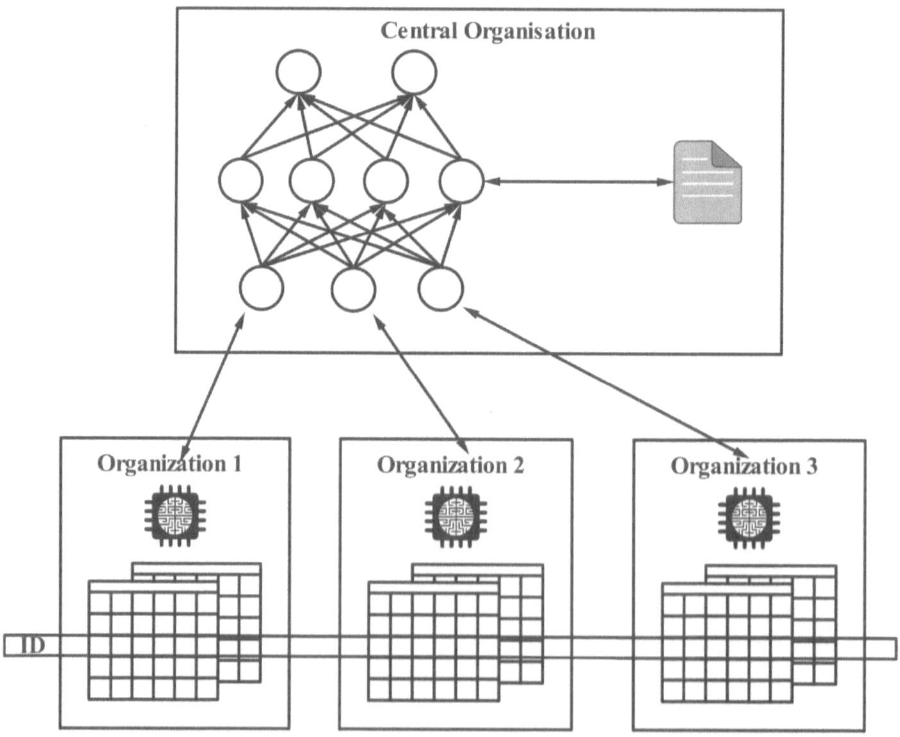

FIGURE 15.1 The data for training and the model that is to be trained are distributed between many businesses in a vertical federated learning scenario.

15.2.2 Mixture of Experts Classifiers

Model scaling is crucial for improving and deploying ML systems in the real world, as recent advances in the field have shown [13]. Extensive use of deep learning in domains such as audio analysis, computer vision, and natural language processing is largely attributable to the fact that training data and model sizes may be easily scaled up. Nevertheless, the computational cost is exponentially increasing as model size increases, which is outpacing the rate of hardware progress and creating problems with sustainability [14]. Hence, in order to capture the complication of real-world data, ML models are getting bigger and bigger. In the quest for computing efficiency, new designs have emerged that strike a compromise between the computational cost of the models and their capacity. The MoE paradigm allows model scaling without increasing computing, which is a plausible fix for this problem [15]. One way that MoE accomplishes this is by using a modular neural network architecture. In this architecture, some parts of network are activated dependent on the input. Figure 15.2 shows the architecture for a MoE with n experts based on the input element x, for each subnet Ei to evaluate the categorization function.

The SMoE and the MoE framework achieve this equilibrium by using conditional computation, a method that allows the active model components to be adjusted in real-time based on the input [16]. Conditional computing in MoE models enables the utilization of dynamic sparsity, as opposed to the fixed sparsity patterns generated by conventional weight pruning methods. A MoE model is a flexible model that can adapt to different data regimes because, unlike static techniques, it keeps all parameters but selectively activates sections of them, rather than permanently removing weights to decrease computation and parameters [17].

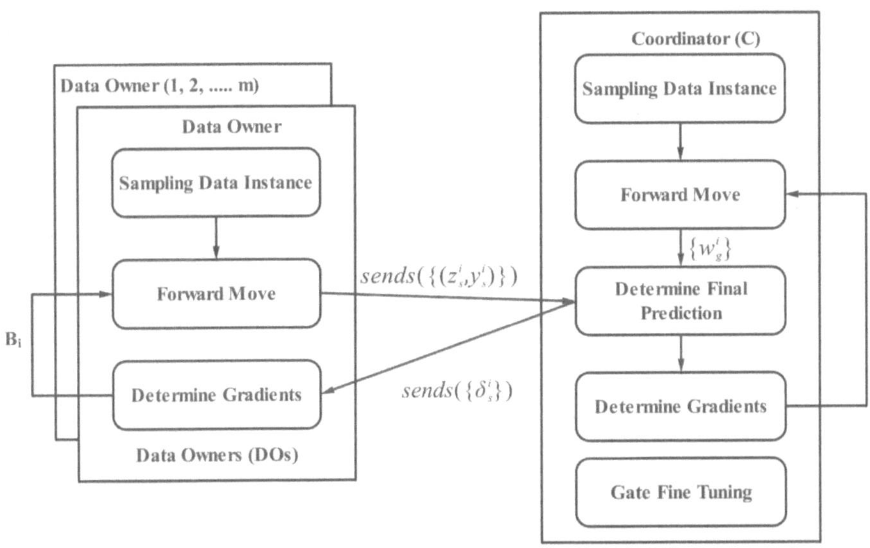

FIGURE 15.2 An example of the typical architecture for a MoE with n experts. Based on the input x, each subnet Ei, or expert, calculates a distinct categorization function.

Important to a MoE model is its gating mechanism, which implements this conditional computation paradigm. It theoretically expresses the process of directing input flow to the right specialists by means of distinct neural network modules that are qualified in various areas of the input space [18, 19]:

$$G(x) = \text{Softmax}\left(Z(x)\right) \tag{15.1}$$

Z is a multi-layer perceptron network that takes an input representation x and outputs a real-valued competency ratings that are not normalized. In these scores, authors can see the expert competency weights. An explanation of these gating scores using probability is guaranteed by the Softmax function, which is used to encourage a sparse activation pattern. For SMoE models in particular, the network G of sparse gates takes in tokens as input and calculates a dispersal across the expert networks, which is expressed as:

$$G(x) = \text{Softmax}\left(\text{Top}\left(Z(x), k\right) \cdot Z(x)\right) \tag{15.2}$$

The highest(v, k) function generates a v-dimensional vector with 0s otherwise and 1s at indices matching the k highest values in Z(x). In order to determine which experts are required to classify x, the gate can apply a sparse technique [20, 21]. Concerning the experts, the final output is influenced by the score for gating G(x)_i of each expert E_i, which is specified independently. The gating network G determines the weights, and the SMoE/MoE output is a weighted sum of expert outputs:

$$y = G(x)^t \left[E_1 \ldots E_m\right]^t = \sum_{i=1}^{m} G(x)_i \cdot E_i(x) \tag{15.3}$$

whereas superscript t indicates the vector/matrix transpose operator and m is number of experts.

The ability to scale model capacity efficiently while maintaining a constant computational budget has been shown by SMoE models, which take use of the sparsity in expert activations. For tasks where computing resources are limited, SMoE models are attractive because of their efficiency without sacrificing the network's representational power. Finally, MoE models, especially sparse ones, signify a sea change towards neural network topologies that are more computationally efficient and scalable. They allow for a manageable approach to handling model capacity and complexity, which in turn allows ML system performance to continue improving without significantly raising computing overhead.

15.2.3 The Proposed VDL Method: Structure of the Model and Training Algorithm

The following section introduces a new approach to solve the VFL problem that builds upon the preceding section's work. With this methodology, we hope to improve upon previous VFL approaches by striking a better balance between the competing goals of reducing computing costs and increasing model accuracy, all the while protecting the local raw data of data owners to an acceptable degree. Two parts of the methodology are described in the following sections: first, a distributed training algorithm; and second, the Federated Classification Model (FCM) classification model requires an architecture similar to that of the MoE.

15.3 MODEL ARCHITECTURE

Using a distributed training technique and a MoE-like model architecture, we train a Federated Classification Model in our method. It clearly specifies the suggested VFL_MoE model, which adapts the conventional abstract MoE model to the problem setting and effectively handles the scalability as well as privacy challenges that are common in real-world VFL systems. The subsequent functional elements, each executed as distinct segments of an all-encompassing neural network model, make up a VFL_MoE's architecture, which is conceptually comparable that of a conventional neural-net Mixture of Expert classifier: (i) The ability for each member of a set of "expert" classifiers E1,…, Em to evaluate fresh data instance x and make a classification prediction; (ii) A module for combiners that incorporates a static product-sum sub-network alongside an attainable gate sub-network G and a linear convex combination system. This scheme ensures that the overall prediction is influenced by the normalized competency scores assigned by G to each expert, with the more competent an expert leading the pack. Specifically, following the sparse-gating computation stated in Equation 15.2, Every weight that the gate returns is zero, with the exception of the top k. Using the best possible k value to return the x prediction simplifies the product-sum combination. This is similar to sparse MoEs.

Two important ways in which this work's suggested combination technique deviates from both traditional and sparse MoEs are:

- The gate employs a partial representation x0 of x, linked to the shared data structures of all nodes in the VFL network, to ascertain the competency weight of the experts. This helps to optimize computation while ensuring privacy. This is done for each data instance's vectorial representation $x \in X0 \times X1 \times \cdots \times Xm$.
- Authors can minimize computation and communication costs by utilizing an ad hoc training loss function to encourage the gate net to be as selective as feasible during training [22, 23].

In order to maintain the confidentiality of information and optimize computational performance, the VFL_Mixture of expert model is physically divided into multiple sub-networks that are assigned to different nodes in VFL network. This process is described in detail below.

- In training, each expert is bound to a specific node of the data owner (DO) and instructed to use just the features stored there. Instead of employing data embedding or encryption—which would increase computing costs and increase the risk of losing crucial data—this method allows the expert to be taught directly on raw data.

- Specifically, the coordinator node C maintains the ground-truth DI classifications in the gate sub-network G, which comprises the non-trainable product-sum module of the combiner sub-model in its entirety. The sub-vector x0i, which represents the mapping of xi onto the common characteristic sub-space X0, is extracted by the gate for each occurrence of input data xi ∊ D. The sub-vector for each data instance of D is replicated in all federation nodes, including node C. Therefore, no information on the private information of data owner/client nodes needs to be provided to the gate (coordinating party).

Following this, the authors prove that generalizing our study to multi-class classification is easy by concentrating on a binary classification context, where there are only two classes in C that need to be distinguished.

Based on this basic assumption, the suggested federated categorization model's physical and functional architecture can be formally described as follows.

15.3.1 The Proposed Training Method: Algorithm VFL_MoE

Using a new federated training algorithm, called the VFL_MoE algorithm from here on out, we aim to find a VFL_MoE classifier.

The user can regulate the computational and communication costs associated with executing the algorithm and implementing the VFL_Mixture of Expert classifier on new data instances by modifying three hyperparameters: the maximum number of training epochs, the expert-selection factor hyperparameter k, which shows the number of expert estimates taken into consideration for classifying data instances, and a batch-reduction variable r ∊ (0, 1), which establishes the percentage of total example batches used for training in accordance with the repeated random sampling (RRS) technique [24, 25]. In addition, the algorithm considers the size of the small batches, the rate of learning used to improve the gate parameters, and the hyper-parameters utilized for each expert, denoted as ηe. Figure 15.3 shows the flowchart for the VFL_MoE framework.

Algorithm 1: The distributed algorithm VFL_MoE pseudo-code.

Data: Tensors X and Y storing the input feature vectors $x_0^i, \ldots, x_m^i$ and class label $\hat{y}^i$ (for $i \in [1 \ldots N]$), respectively for all the data instances d^i of a distributed dataset D;

Requires: expert-selection factor k; max. number e of epochs; batch-reduction factor r; batch size b; learning rates η_g and η_e for the gate's and experts' parameters, respectively.

Result: A VFL_MoE model $\langle \mathcal{N}_g, E_1, \ldots, E_m \rangle$ with optimized parameters $\Theta = [\Theta_g \Theta_1 \ldots \Theta_m]^t$.

1. The coordinator node C and every data owner node DO_s (with $\in [1.m]$) concurrently initialize the parameters Θ_g and Θ_s of the gate sub-net $\mathcal{N}_g$ and of expert E_s, respectively;

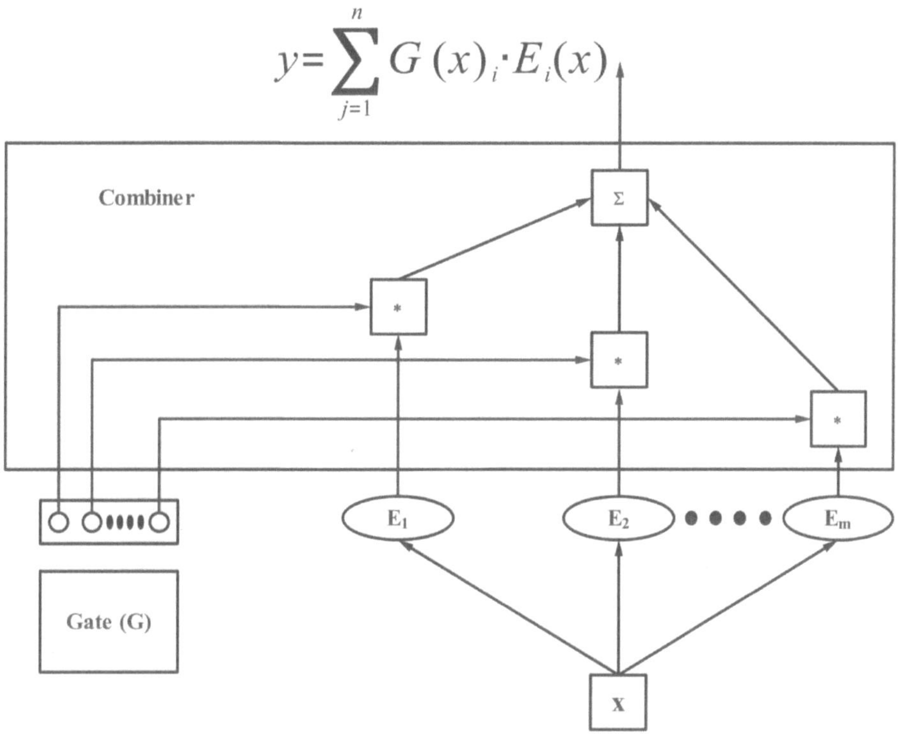

FIGURE 15.3 Flow chart for the VFL_MoE framework, which is based on MoE.

2. C generates a seed $\varepsilon \in \mathbb{N}$ to be used by all the nodes involved in the training for randomly sampling the same subset of data batches in every training epoch;
3. C sends the values of the seed ε and of the factor r to all the DO nodes $DO_1,\ldots,DO_m$;
4. foreach epoch $e \in [1\ldots e]$ do
5. All the nodes sample a vector $\mathcal{I}$ of $\lceil r \cdot N \rceil$ data instance indices using the same seed ε;
6. Let $B_1,\ldots,B_{n_b}$, with $n_b = \lceil \frac{r \cdot N}{b} \rceil$, denote the training batches, regarded as sets of (data instance) indexes s.t. $B_i = \{\mathcal{I}(j') | j' \in [(i-1)\cdot b+1\ldots i\cdot b]\}$ for any $i \in \{1,\ldots,n_b\}$; foreach batch B_i s.t. $i \in [1\ldots n_b]$ do
7. For each data-instance index $j \in B_i$ do in parallel
8. Every node DO_s performs a forward pass through its expert model E_s on the current data instance d^j (i.e. the one referred to by index j), to compute the respective logit $z_s^j = f_s(x_s^j;\Theta_s)$ and prediction $y_s^j = \sigma(z_s^j)$ (cf. Def. 2);
9. C performs a forward pass through the gate $\mathcal{N}_g$ on the current data instance d^j, to
10. Compute the respective vector $w_g^j = g(x_0^j;\Theta_g)$ of expert weights;

11. Every DO_s node sends the logit and output values it has computed for the instances of the current batch to C, as an ordered list of pairs $\left(z_s^J, y_s^J\right)$ for all j in B_i;
12. For each data-instance index $j \in B_i$ do
13. C computes the final prediction $\hat{y}^j = \left(w_g^j\right)^t \left[y_1^j \ldots y_m^j\right]^t$, the total per-instance loss $\mathcal{L}(x^j, \hat{y}^j; \Theta)$ (based on Eq. 3) and the partial derivative $\delta_s^j = \frac{\partial \mathcal{L}(x^j, \hat{y}^j; \Theta)}{\partial f_s\left(x_s^j\right)}$;
14. C sends the derivatives $\left\{\delta_s^j \middle| j \in B_i\right\}$ to each DO node DO_s as an ordered list;
15. do in parallel
16. C computes gradients $\{\nabla_{\Theta_g} \mathcal{L}(x^j, \hat{y}^j; \Theta) | j \in B_i\}$, aggregates them all into average one $\mathcal{G}_g$, and updates the parameters of gate $\mathcal{N}_g$ via $\Theta_g := \Theta_g - \eta_g \cdot \mathcal{G}_g$;
17. Every DO_s computes gradients $\{\nabla_{\Theta_s} \mathcal{L}(x^j, \hat{y}^j; \Theta) | j \in B_i\}$, averages them all into $\mathcal{G}_s$ and updates the parameters of E_s via $\Theta_s := \Theta_s - \eta_e \cdot \mathcal{G}_s$
18. All DO nodes DO_s apply their experts to the $N - \lceil r \cdot N \rceil$ instances that were not used in the last iteration of the main loop (Steps 4-17) and send the resulting logit and output values to C;
19. The parameters Θ_g of gate $\mathcal{N}_g$ are fine-tuned by making C execute the loop over Steps 4-17 in isolation again -i.e. skipping Steps 9,11,14,16 and 17 - using the logit and output values it gathered in Step 18 and in Step 11 of the last iteration of the previous complete run of the loop;
20. return the updated version of $VFL\_\text{MoE}\left\langle \mathcal{N}_g, E_1, \ldots, E_m \right\rangle$.

Core computation steps: There are three primary steps to the distributed training method that has been developed:

- In Step 1, the various nodes of VFL network—a coordinator C and multiple DOs DO1..., DOm—build and maintain a VFL_MoE model with random initializations, in accordance with the model architecture. In addition, before beginning training, the nodes establish a shared expectation for coordinated iteration across data instances by deciding on a random seed to specimen the similar proportion r of data samples at every training phase.
- Steps 4–17 make up the main loop, which is executed in its entirety during the second phase to train the VFL_MoE model. The VFL_MoE model is optimized from beginning to end using this loop, which conceptualizes a conventional mini-batch-based SGD-like technique. To be more precise, for every batch B_i that contains b DI: (i) To begin, in Steps 8–10, the DIs linked to B_i are forward-simulated using the expert and gate sub-models. In particular, each DO node calculates its own logit and prediction, and the coordinator determines the weights; (ii) Following the aggregate of the intermediate results w_g^j, y_1^j, ..., y_m^j from the sub-models into a single overall forecast y^j for every j in B_i, per-instance losses are calculated in Steps 12–13, as functions of the model parameters Θ; (iii) The parameters are lastly revised in Steps 15–17 by averaging the ^6 back-propagation-generated per-instance gradients of all model variables in Θ.

Following the loss function definition, this subsection concludes with a more in-depth description of this stage.

- Step 19 summarizes the final core computation, which involves adjusting the gate sub-net's parameters Θ_g while maintaining a frozen state for the experts' parameters. This is achieved by re-executing the training loop that covers Steps 3 to 17 on the solitary coordinator node C, while DO nodes perform new calculations to skip those steps for efficiency.

Communication steps: The technique involves extensive communication between the data owner nodes and the coordination node C, in addition to the previously specified fundamental computational processes. The algorithm illustrates these talks as instances of the general data-transfer action dispatch for visualization purposes. For the ensure clarity, let us assume that all of these messages are straightforward and to the point, even though some could be implemented using more effective group communication methods. For instance, when exchanging data between the coordinator and the DO nodes, we could use broadcast, scatter, gather, etc. operations. For the time being, let's go over the first communication action that happens in Step 3, during startup, as we've already established that the majority of data-exchange activities are executed once for every optimization step, or mini-batch of training instances. So, each batch-wise communication only transfers a little amount of data: 1 scalar in the other direction (Step 14) and 2 scalars from each DO to C (Step 11).

Since the lone coordinator C performs the fine-tuning method in Step 19 independently of DO nodes, communications are not involved in the process per se. For this to be feasible, C must ascertain the output (logit) produced by the final experts for every scenario in dataset D, which is gathered at the conclusion of training. A collection of predictions for all the DI that were not used in the final training period are gathered by C in Step 18 from all the nodes that are linked experts. This is done to do this.

Loss function and details about the optimization steps: Allow us to specify the particular loss function employed throughout the training phase to optimize the model variables. Authors utilize s for all local expert Es and g for the gate sub-net, in order to ensure technical comprehensiveness [26, 27].

To summarize, for every training instance $(x,\hat{y})$ of every mini-batch taken into account in each epoch, the optimization algorithm executes the following key actions in order to discover a set of variables for VFL_mixture of expert model that reduces this function of loss:

- After applying each expert to their assigned local portrayal of x, DO nodes compute the predictions of all researchers for x during the forward pass.
- To get a general class prediction and calculate the loss value $\mathcal{L}(x,\hat{y};\Theta)$ according to Eq. 3, the output $g\left(x_0;\Theta_g\right)$ from the gate is combined with all the expert forecasts in coordinator node C. Computing the gradient $\nabla_{g\left(x_0;\Theta_g\right)}\mathcal{L}(x,\hat{y},\Theta)$ of the gate's output layer and the partial derivatives $\frac{\partial\mathcal{L}(x,\hat{y},\Theta)}{\partial Q(x,\hat{y},\Theta_S)}$ for each expert's logit nodes are also responsibilities of the coordinator. I am E_S.

- C can optimize the gate parameters greedily by obtaining batch-wise aggregated gradient $\mathcal{G}_g = \text{avg}_{(x,\hat{y})} (\nabla_\Theta \mathcal{L}(x, \hat{y}; \Theta))$ through back-propagating the gradient $\nabla_{g(x_0;\Theta_g)} \mathcal{L}(x, \hat{y}; \Theta)$.
- Alternatively, after being given $\frac{\partial \mathcal{L}(x,\hat{y},\Theta)}{\partial Q(x,\hat{y},\Theta_s)}$ each DO_s can use backpropagation to obtain batch-wise average gradient $\mathcal{G}_g = \text{avg}_{(x,\hat{y})} (\nabla_{\Theta_g} \mathcal{L}(x, \hat{y}, \Theta))$ which is then used to optimize the expert's parameters E_s.

15.3.2 The Algorithm's Cost-Benefit Analysis

Computation costs: A two-layer feed-forward gate $\mathcal{N}b$ and m expert algorithms *E*1.,...,*Em* make up each instance of the small and shallow sub-networks that comprise the proposed VFL_MoE federated classification model [28, 29].

The neurons of number in the second layer is represented by, *d*0 = *X*0 and *d*0 plus d0(d0+1). An m-dimensional output is generated by an m-parameter gate sub-net. With a 1D output, each expert sub-net really has $d_s + 1$ parameters, where $d_s = |X_s|$. This means that there is a total of m·$\Sigma_s(d_s + 1) = m \cdot d$ variables in all specialists. Here, d=|X| is sum of all the input features (after numbering them) plus $|X_0| \cdot (m - 1)$, and each expert sub-net has $d_s + 1$ parameters.

Each of these sub-networks contains fewer than 4 × 10^4 floating-point integers, therefore they require very little main/GPU memory to be processed and saved, given that d_s is less than $\max_{s=0}^{m} d_s < d \ll 10^4$ and that $m \ll 100$, as in the instance research mentioned in Section 5.

As a stand-in for the overall energy cost of the process, let's concentrate on the entire quantity of floating-point operations (FLOPs) executed in this case for the purpose of clarity and consistency. This metric, which represents the amount of basic mathematical operations executed in the computation, will enable us to measure the effectiveness of suggested algorithm regardless of software and hardware platform it runs on. This will enable us to compare it more broadly with other solutions, both current and future [30].

The forward pass through every layer of the gate and experts' subnet, with d_IN and d_"OUT" neurons, needs $d_{OUT} \cdot (d_{IN} + 2)$ floating-point operations, where non-linearity calculations and bias addition are taken into account using the latter value, given the easy feed-forward design of two networks. During the overall model training, m·$(d_0' + d_i + 4) + d_0' \cdot (d_0 + 2)$ floating-point operations are executed, whereas the gate fine-tuning necessitates C =·$(2d_0' + d_i + 6) + 4d_0' \cdot (d_0 + 2)$, add d(d0+2) and multiply by m. After estimating the cost of each computation step in the back-propagation and gradient-related techniques as 2·C_"for", as well as taking into account that the overall quantity of training procedures is = to e·r·N (in the D training dataset, where N is the overall number of occurrences), we obtain the following cost estimate for the first algorithm: Where P is the overall number of parameters in the model, the equation can be expressed as $3 \cdot e \cdot r \cdot N \cdot m \cdot (2d_0' + d_i + 6) + 4d_0' \cdot (d_0 + 2) = \Theta(e \cdot r \cdot N \cdot P)$.

This computation just needs $(d_0' + d_i + 4) + d_0' \cdot (d_0 + 2)$ floating-point operations per test instance, and it can be done quickly and efficiently with the algorithm and a

VFL_MoE by just execution a forward pass-through gate and k experts selected for the prediction.

The communication cost is equal to the overall quantity of communications done in Algorithm 1's main loop, which is $e \cdot m \cdot \left\lfloor \frac{r \cdot N}{b} \right\rfloor$ assuming that all communications are conveyed using point-to-point communications in pairs. The most data that can be sent in each of these messages is 2·h floating-point numbers. The transmission of 1 and $2 \cdot (N - \lceil r \cdot N \rceil)$ floating-point numbers is a part of Step 2 and Step 18 of m communications.

The total number of messages sent and received by an algorithm is $2 \cdot m + e \cdot m \cdot \left\lfloor \frac{r \cdot N}{b} \right\rfloor$ and the total number of floating-point integers substituted is Θ (e·m·r·N), presuming that e·r≥1.

Transferring all local raw data in embedded versions from DOs to the coordinator is a requirement of both the approach proposed and the more traditional FL methods [31, 32]. Sharing elevated dimensional parameters/gradients at every optimization step (per mini-batch) also leads to larger data exchanges.

Below the modest assumption that e·r≥1, the processing as well as communication costs of proposed training method VFL_Mixture of expert (as shown in method 1) scale linearly with the total number ⌈r·N⌋ of occurrences handled over all training periods. By adjusting hyper-parameter r, it is possible to simply manage these expenses.

15.4 EXPERIMENTAL SETUP

15.4.1 Dataset Preparation

A training, validation, and test subset was created from each dataset. Of the total data set, 80% was utilized for training purposes, while the remaining 20% was utilized for validation purposes. We saved 20% of the dataset for testing purposes.

The public component of the KronoDroid dataset contained all features from the Intents group and five features from the Permissions group (WRITE EXTERNAL STORAGE, RECEIVE BOOT_COMPLETED, RECEIVE SMS, READ SMS, GET_TASKS). The coordinating node and all of the local nodes shared these traits. All of the local nodes kept their private characteristics.

Four sections were randomly created from the Adult dataset's characteristics. In order to distribute instances to different experts, the gate relied on one component that had common properties. For the purpose of group study, the remaining three components were split evenly among three separate companies.

15.4.2 Model Variants

In order to assess the VFL_MoE (VM) model, we contrasted it with Centr_MoE (CM), a centralized version. By leveraging all raw data and concentrating the MoE on the Coordinator node, Centr_MoE functions as a theoretical upper bound, ignoring privacy standards commonly found in VFL environments. It is a perfect example of how to measure things like communication costs, privacy, and correctness.

15.4.3 Parameters

In order to evaluate VFL_MoE's performance, we used a two-pronged approach to parameter setting, changing two key parameters, kkk and rrr:

- kkk: From one to three experts were utilized by the MoE model.
- rrr: An increment of 0.125 was used to vary the data/compute reduction factor from 0.075 to 1 for the number of batches each training period.

Fixed hyperparameters included:

- Maximum training epochs (eee) = 40
- Learning rates: ηg=10^{-3}\eta_g = 10^{-3}ηg=10^{-3} for the gate, ηe=10^{-4}\eta_e = 10^{-4}ηe=10^{-4} for experts
- Batch size (bbb) = 64
- There was just one linear layer in each expert model, however two internal linear layers totaling 512 neurons were present in the coordinator model.

15.4.4 Effectiveness Metrics

The purpose of this research is to identify malware by solving a binary classification problem that divides data into two categories: "normality" (non-malicious) and "attack" (malicious). In order to assess the efficacy of the model, we take four important measures: the accuracy score, the false positive rate (FPR), the area under the curve (AUC), and the F1 score.

15.4.5 Accuracy Score (ACC)

ACC=TP+TNTP+TN+FP+FN\text{ACC} =\frac{\text{TP} +\text{TN}}{\text{TP} +\text{TN} +\text{FP} +\text{FN}}ACC=TP+TN+FP+FNTP+TN Measures the proportion of correct predictions out of all predictions. While a high accuracy score is indicative of a trustworthy model, it might be deceiving when dealing with datasets that are imbalanced.

15.4.6 Area Under the ROC Curve (AUC)

AUC\text{AUC}AUC is calculated by plotting True Positive Rate (TPR) against FPR at various thresholds. It evaluates the model's capacity to differentiate between classes, which is essential for datasets that are imbalanced.

15.4.7 F1 Score (F1)

F1=2·Precision·RecallPrecision+Recall\text{F1} = 2\cdot\frac{\text{Precision}\cdot\text{Recall}}{\text{Precision} + \text{Recall}} F1=2·Precision+RecallPrecision·Recall Balances Precision and Recall, important for minimizing both false positives and false negatives. It is particularly relevant when both types of errors have significant consequences.

15.4.8 False Positive Rate

FPR = FPFP + TN\text{FPR} = \frac{\text{FP}}{\text{FP} + \text{TN}} FPR=FP + TNFP Critical in environments where false alarms are costly, as a high FPR can lead to unnecessary resource allocation and reduced trust in the detection system. The model's performance can be better understood with the help of each metric. In order to create a strong, trustworthy, and effective malware detection system, a thorough review takes into account the contexts and trade-offs.

15.5 EXPERIMENTAL RESULTS

Using the recently compiled benchmark dataset KronoDroid, which contains a large collection of malware specimens influencing Android operating system-based from 2008 to 2020, this segment objects to assess the ability of proposed algorithm VFL_ Mixture of expert in exactly identifying malicious behaviors. To further explore VFL_MoE's performance in a different application scenario, authors also included the publicly available Adult dataset in our experimental study [33, 34].

15.5.1 Datasets

From 2008 all the way up to 2020, the KronoDroid dataset contains samples of both good and bad Android applications. The timestamped data samples that make up this collection span a large amount of time. A total of 489 features are used to characterize each sample; 289 of these features are dynamic while the others are static. This dataset has become a standard in the cybersecurity field, especially for research on how Android malware has changed over time and how detection methods have improved.

There are 36,755 safe apps and 41,382 malicious ones in KronoDroid, representing 240 different malware families [35]. As far as Android-centric hybrid feature datasets go, this one is the most comprehensive. There are two separate sub-datasets in the dataset, one for actual devices and one for emulators. This separation makes it easier to conduct analyses in a variety of runtime contexts. We were able to successfully test our VFL system by vertically separating the dataset into 4 sections—System Calls, Permissions, Intents, and Others—due to this categorization. Attributes belonging to these categories are as follows: 289, 173, 7, and 8, correspondingly. Following dataset normalization, all features that were determined to be non-contributory or had a strong association with Malware label (e.g., Detection Ratio) were eliminated.

Data collected from various households' census forms makes up the Adult dataset. It captures several socioeconomic characteristics with its fourteen unique qualities [36, 37]. Predicting whether a specific household has an income above 50,000 is the goal of this dataset. There are a total of fourteen features in the first Adult dataset, with eight being categorical and six being continuous. An important step is the discretization of continuous features into quantiles; a binary feature is then used to represent each quantile. In a similar vein, m binary features are generated from m category features, which are defined by m separate categories.

15.5.2 Competitor and Baseline Approaches

With precision, privacy protection, and reduced communication and computing requirements as our primary goals, we conducted a thorough evaluation of VFL method, VFL_MoE, in this research. Authors put it up against two standards, SVFL and Baseline. SVFL utilizes the top-performing expert for classification, taking use of large data sets but at the expense of greater communication and computing expenses; it is the sole rival known to integrate a MoE into a VFL environment. Conversely, VFL_MoE enhances privacy by minimizing the transfer of embedded data and reducing communication costs by combining predictions from different experts. Compared head-to-head, VFL_MoE uses less resources and lessens the likelihood of data leaking, although it might miss some supervision signals while training with fewer instances. At its most basic level, baseline depicts a scenario in which nodes do not cooperate with one another and local models run autonomously, providing maximum privacy with minimum communication costs. We can see that VFL_MoE strikes a good mix between accuracy, privacy, and efficiency in this comparison, and we can see that SVFL and Baseline are good metrics to use to judge how relevant VFL_MoE is.

15.5.3 Performance Analysis Using the KronoDroid Dataset

In Table 15.1, authors thoroughly examine the VFL method, VFL_Mixture of expert, compares to its ideal upper bound version, Centre Mixture of expert, for different values of the variables $k \in \{1,2,3\}$ and $r \in \{0.075,\ldots,1.0\}$.

The quantity of expert estimates during inference (variable $k \in \{1, 2, 3\}$), the number of training batch repetitions ($r \in \{0.075,\ldots,1.0\}$), and various combinations of these three variables are examined in this table.

One thing that stands out from the results in Table 15.1 is how well Centr_MoE handles changes in k and r. We can probably thank Centr_MoE's extensive public and non-public feature set for MoE gate for this resilience [38]. There may be less need to combine many predictions to make up for possible mistakes in the expert selection process now that the gate has access to so much data. Figure 15.4 shows that Centr_MoE's performance curve begins to flatten at about $r = 0.375$, which lends more credence to this theory. This early flattening indicates that Centr_MoE's MoE gate can improve its selection of expert with less training data if it has access to all unmasked data and can thus better find patterns in the input data. Again, though, we must not forget that Centr_MoE is an unrealistic ideal model in real-world VFL settings because of how it is designed. Because of this, its main purpose is to provide a theoretical upper limit that our federated solution, VFL_MoE, may be compared to.

With the exception of baseline, all of these approaches were evaluated using an expert-selection factor, k, which can take values from 1 to 2. Performance metrics for

TABLE 15.1
Thorough Assessment of the VFL-based Method's Effectiveness Upper Bound Form of VFL_MoE and Its KronoDroid Dataset's Centr_MoE

K	Techniques	r	ACC (↑)	AUC (↑)	F1 (↑)	FPR (↓)
1	Centr_MoE (CM)	0.075	0.902	0.968	0.901	0.047
		0.125	0.932	0.98	0.933	0.048
		0.250	0.943	0.986	0.946	0.052
		0.375	0.944	0.987	0.946	0.052
		0.500	0.944	0.986	0.947	0.053
		0.625	0.949	0.989	0.951	0.053
		0.750	0.949	0.99	0.951	0.056
		0.875	0.947	0.988	0.949	0.055
		1.000	0.947	0.987	0.949	0.052
	VFL_MoE (VM)	0.075	0.864	0.934	0.864	0.095
		0.125	0.888	0.949	0.89	0.088
		0.250	0.912	0.964	0.915	0.087
		0.375	0.917	0.968	0.921	0.091
		0.500	0.921	0.968	0.924	0.087
		0.625	0.922	0.969	0.926	0.084
		0.750	0.923	0.971	0.927	0.087
		0.875	0.925	0.973	0.929	0.088
		1.000	0.923	0.971	0.927	0.088
2	Centr_MoE (CM)	0.075	0.909	0.975	0.908	0.046
		0.125	0.931	0.98	0.933	0.045
		0.250	0.941	0.984	0.943	0.047
		0.375	0.945	0.988	0.947	0.052
		0.500	0.947	0.989	0.95	0.055
		0.625	0.948	0.988	0.95	0.051
		0.750	0.946	0.987	0.948	0.053
		0.875	0.948	0.989	0.95	0.056
		1.000	0.95	0.99	0.952	0.055
	VFL_MoE (VM)	0.075	0.888	0.961	0.887	0.069
		0.125	0.91	0.967	0.912	0.068
		0.250	0.922	0.974	0.924	0.065
		0.375	0.925	0.976	0.928	0.068
		0.500	0.93	0.98	0.933	0.072
		0.625	0.934	0.982	0.937	0.071
		0.750	0.932	0.98	0.935	0.07
		0.875	0.932	0.98	0.935	0.069
		1.000	0.934	0.983	0.937	0.072
3	Centr_MoE (CM)	0.075	0.902	0.968	0.901	0.047
		0.125	0.932	0.98	0.933	0.048
		0.250	0.943	0.986	0.946	0.052
		0.375	0.944	0.987	0.946	0.052
		0.500	0.944	0.986	0.947	0.053

(*Continued*)

TABLE 15.1 (*Continued*)
Thorough Assessment of the VFL-based Method's Effectiveness Upper Bound Form of VFL_MoE and Its KronoDroid Dataset's Centr_MoE

K	Techniques	r	ACC (↑)	AUC (↑)	F1 (↑)	FPR (↓)
		0.625	0.949	0.989	0.951	0.053
		0.750	0.949	0.99	0.951	0.056
		0.875	0.947	0.988	0.949	0.055
		1.000	0.947	0.987	0.949	0.052
	VFL_MoE (VM)	0.075	0.864	0.934	0.864	0.095
		0.125	0.888	0.949	0.89	0.088
		0.250	0.912	0.964	0.915	0.087
		0.375	0.917	0.968	0.921	0.091
		0.500	0.921	0.968	0.924	0.087
		0.625	0.922	0.969	0.926	0.084
		0.750	0.923	0.971	0.927	0.087
		0.875	0.925	0.973	0.929	0.088
		1.000	0.923	0.971	0.927	0.088

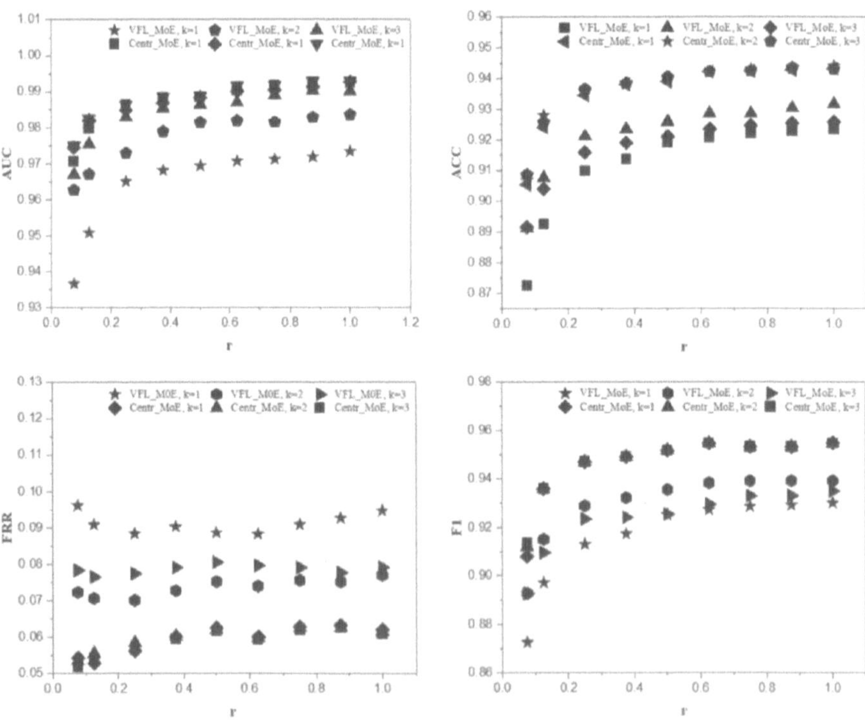

FIGURE 15.4 Comparative study of several VFL_MoE configurations and their optimal upper-bound variation, Centr_MoE, using the KronoDroid dataset, covering a range of parameter k values ∈ {1, 2, 3}. During the training phase, the comparison considers several values of batch-decrease factor r ∈ {0.075,..., 1}.

Centr_MoE (CM) and VFL_MoE (VM) are also provided for values between 0.25 and 0.75 for the batch-reduction factor hyperparameter. Keep in mind that r = 1.0 for both SVFL and baseline because neither of them uses a technique to decrease the quantity of training batches.

The performance of VFL_MoE is not noticeably worse than that of perfect model Centr_Mixture of expert for maximum accuracy metrics across multiple values of kkk and rrr, as can be shown in Tables 15.1 and 15.2. The only metric where there is a more noticeable difference is FPR. The performance disparity gets smaller as kkk and rrr get higher. For example, VFL_MoE demonstrates a -2.1% performance disparity in AUC, -3.2% gap in ACC, -83.7% gap in FPR, and -3.2% gap in F1 when k = 1k = 1k=1 and r=0.25r = 0.25r=0.25. By raising rrr to 0.75, the disparity is narrowed to -1.8%, -2.7%, -68.1%, and -2.4%. The gap is even narrower for larger kkk values. Beyond k=2k = 2k=2, increasing rrr produces diminishing results; thus, the optimal performance-efficiency trade-off appears to be k=2k = 2k=2 with r=0.25r = 0.25r=0.25.

The VFL_MoE, SVFL, and Baseline techniques are compared in Table 15.2. Even though fewer training batches are used by VFL_MoE, its performance is almost identical to that of SVFL when k=1k = 1k=1 and r≥0.5r\geq 0.5r≥0.5. Compared to SVFL, VFL_MoE performs marginally worse with r=0.25r=0.25, yet it uses just a quarter as much computing power. The AUC and FPR metrics, in particular, show that VFL_MoE outperforms SVFL when k=2k=2k=2. Also, in every setup, VFL_MoE beats the Baseline method. This shows that standalone local models can't cut it when it comes to classification, and that advanced federated methods are needed, such as VFL_MoE, which combine different data views from different nodes.

TABLE 15.2
Comparative Evaluation of Several Models Using the KronoDroid Dataset: The Suggested Approach Its Optimum Upper-Bound Variation, VM Baseline, CM, and Rival SVFL

K	Techniques	r	ACC (↑)	AUC (↑)	F1 (↑)	FPR (↓)
-	Baseline	1.0	0.873	0.946	0.884	0.118
1	VM		0.91	0.962	0.921	0.087
	CM	0.25	0.941	0.984	0.948	0.05
	VM		0.921	0.97	0.936	0.083
	CM	0.50	0.946	0.988	0.957	0.052
	VM		0.921	0.969	0.933	0.087
	CM	0.75	0.947	0.988	0.953	0.054
2	VM		0.922	0.976	0.936	0.061
	CM	0.25	0.943	0.986	0.953	0.046
	VM		0.928	0.978	0.939	0.072
	CM	0.50	0.945	0.987	0.952	0.053
	VM		0.932	0.982	0.947	0.066
	Centr_MoE	0.75	0.948	0.989	0.958	0.052

15.5.4 Examination of the Adult Dataset's Performance

The methodology utilized for the assessment of VFL_MoE on the KronoDroid dataset is also applied to the Adult dataset. Table 15.3 provides a full overview of outputs, comparing VFL_MoE to other models and demonstrating its performance on various datasets.

Every approach except baseline was evaluated using an expert-selection factor, where k is a number between 1 and 2. The performance results for Centr_Mixture of expert and VFL_ Mixture of expert are presented across in-between values of batch-decrease variable r ∈ {0.25, 0.50, 0.75} in order to make the results more readable. Keep in mind that r = 1.0 for both SVFL and baseline because neither of them uses a technique to decrease the quantity of training batches.

Table 15.3 displays results of the VFL_MoE study on the Adult dataset, which are consistent with the tendencies seen in the KronoDroid dataset. In particular, for most accuracy measures across all possible combinations of k and r, VFL_MoE performs quite similarly to the ideal model Centr_MoE. The projected performance disparity among VFL_MoE and Centr_MoE is greatest at k=1 and r=0.25, as one would expect. As k and r increase in value, this disparity shrinks, becoming negligible when VFL_ Mixture of expert is activated with k=2 and r=0.75. When using half or more of the training groups that SVFL usages (r≥0.5), VFL_MoE outperforms SVFL in a head-to-head comparison with Table 15.3, and this holds true regardless of the values of k. Importantly, even with smallest quantity of training data (r=0.25), VFL_MoE outperforms SVFL on the F1 metric. The exact percentage of improvement in F1 score for VFL_Mixture of expert at k=1 and 2 where r=0.52 is 8.3% and 8.8%, respectively. To beat SVFL in the FPR metric, VFL MOE needs

TABLE 15.3
Comparative Evaluation of Various Models Using the Adult Dataset: The Suggested Methodology Its Optimum Upper-Bound Variation, VM Baseline, CM, and Rival SVFL

k	Techniques	r	ACC (↑)	AUC (↑)	F1 (↑)	FPR (↓)
-	Baseline	1.0	0.821	0.86	0.596	0.103
1	VM		0.816	0.88	0.653	0.167
	CM	0.25	0.823	0.903	0.677	0.182
	VM		0.847	0.899	0.659	0.097
	CM	0.50	0.848	0.907	0.69	0.119
	VM		0.848	0.899	0.668	0.085
	CM	0.75	0.854	0.907	0.686	0.093
2	VM		0.827	0.896	0.658	0.154
	CM	0.25	0.823	0.905	0.682	0.184
	VM		0.848	0.9	0.671	0.098
	CM	0.50	0.847	0.906	0.685	0.121
	VM		0.851	0.903	0.676	0.074
	CM	0.75	0.855	0.908	0.691	0.091

additional data (r=0.75). These results demonstrate how versatile and effective VFL MOE is in situations where training data is limited. Consistent with the results from the KronoDroid dataset, VFL_MoE beats the Baseline model on the Adult dataset for practically all values of k and r. Under the strictest criteria (k = 1 and r = 0.25), the only time this does not hold true is when Baseline comes out on top in ACC by a hair's breadth and in FPR by a substantial margin of almost 40%. Regardless, VFL_MoE outperforms baseline in terms of AUC and F1 score, increasing by 2.3% and 9.8%, respectively, even under these limited conditions.

15.5.5 Ablation Study

To assess the distinct effects of different setups within our methodology, an ablation experiment was carried out. The reference approach for this research is the different of VFL_MoE that strikes a good compromise between performance and efficiency, with two experts for every prediction (k = 2). Three more straightforward versions were contrasted with this approach:

1. A Random Ensemble (k=1) is a modification of the MoE algorithm, where the data-driven gating technique is substituted with a random ensemble mechanism. In this mechanism, for every data example, a single expert is randomly selected to categorize the tuple. It is important to note that this reduced version of the proposed methodology is really same as the reference Baseline technique that has been evaluated in the experimental research [39, 40].
2. A Random Ensemble (k=2) is a method in which two experts are selected randomly for each data instance. The classification of the instance is determined by taking the average of the probability offered by the two experts.
3. With VFL_MoE (k=1), we use the MoE's gate function to pick a single expert for every data instance.

Table 15.5 presents the outcomes for the Adult dataset, while Table 15.4 presents the outcomes for the KronoDroid dataset, respectively, from the ablation study. In all trials, there was no batch decrease factor applied, and each variant processed the whole set of training batches (r = 1).

Incorporating the MoE mechanism outperforms Random Ensemble's purely random method on the KronoDroid dataset, as anticipated. Results across all versions

TABLE 15.4
The KronoDroid Dataset Ablation Research

Techniques	k	ACC (↑)	AUC (↑)	F1 (↑)	FPR (↓)
VM	1	0.922	0.97	0.934	0.089
VM	2	0.933	0.982	0.94	0.071
Random Ensemble	1	0.875	0.951	0.891	0.113
	2	0.91	0.969	0.921	0.091

TABLE 15.5
Results from the Adult Dataset's Ablation Analysis

Techniques		ACC (↑)	AUC (↑)	F1 (↑)	FPR (↓)
VM	1	0.85	0.901	0.671	0.08
VM	2	0.852	0.903	0.67	0.073
Random Ensemble	1	0.822	0.861	0.601	0.098
	2	0.835	0.881	0.616	0.085

and performance indicators are even better when going from one expert to two. Using two experts with the MoE technique improves FPR by about 25% compared to a random selection, which is remarkable.

In the Adult dataset, we saw a similar trend of improvement across the board, but with smaller disparities. For this particular instance, the FPR shows an improvement of approximately 18% when examined under identical circumstances.

15.5.6 Green FL

The concept of "green FL" emerged from the growing body of literature that examines the environmental impact and energy efficiency of FL systems; this approach seeks to lessen emissions without sacrificing model accuracy. While earlier research evaluated FL's carbon footprint using analytical or simulation approaches, more current work uses data-driven methodologies to evaluate FL in the actual world. Resource management systems, optimizing bandwidth and task allocation among heterogeneous devices, controlling energy usage through modifying accuracy targets, and balancing training time and energy consumption are key areas of research. Alternative strategies for lowering carbon emissions have investigated ways to compress models and enhance communication efficiency. The majority of efficiency-focused research has concentrated on HFL models, although the less-explored VFL models have higher communication costs owing to the requirement for tight cooperation between participants. Secure communication protocols and encryption methods are solutions that try to stop information from leaking, but they usually cause a lot of extra work when it comes to processing and data sharing. New developments such as Hybrid FL (HBFL) and the FedHD algorithm improve communication efficiency by resolving data splits in feature and sample spaces, enabling clients to execute several local updates while tracking global gradient information.

15.5.7 Mixture of Experts in FL

For effective FL methods in a green AI setting, an architectural concept known as a MoE can be used to distribute the learning process among numerous specialized models. By efficiently communicating only pertinent expert updates between nodes and the central server, MoEs can improve overall model correctness and handle data heterogeneity. This is achieved while decreasing communication overheads. The adaptability of MoEs to different FL contexts (HFL and VFL) is made possible by their modular design, which also optimizes resource utilization and offers the ability for personalized learning. While there has been some research into using MoE in HFL situations, current methods such as FedMix and personalized FL (PFL) improve FL's accuracy and handle data heterogeneity, but they frequently overlook energy efficiency. One example is PFL, which trains a public, universal model first, and then lets each client build their own model with their own private data, integrating the results through MoE. By training a combination of user-specific and generalized models, FedMix and related methods increase performance on non-IID data. Neither energy efficiency nor a VFL architecture are priorities for these approaches.

There has been very little research on using MoE models to VFL contexts; our exploratory work is the sole prior effort. Our present concept dramatically outperforms the framework in terms of reducing computational resources, improving communication efficiency, and protecting user privacy. Using masked (embedded) data shards was the method, which increased the likelihood of "inversion attacks" that might jeopardize data privacy and resulted in significant communication and computing costs. Reduced communication overheads and improved privacy are the results of present framework's usage of an ad hoc MoE architecture, which supplies the gate with a subset of low sensitive data that is already accessible to all nodes. We also regulate the amount of data groups per training epoch using RRS approach, which helps us achieve a balance between computational efficiency and efficacy. As part of our research, we conducted a comprehensive study of the impact of data reduction factor as well as the quantity of experts selected on accuracy, providing valuable insights into how the model performed in various contexts. This method vastly improves upon earlier ones by striking a better balance between privacy, efficiency, and classification accuracy. As far as we are aware, our approach is the first to find a neural categorization model in a VFL situation that prioritizes privacy while still meeting efficiency requirements; this model is based on MoE.

15.6 CONCLUSION

A VFL model was introduced in this study which utilizes a neural architecture that is influenced by the MoE paradigm. This approach is created with the express purpose of reducing the cost of computing and communication in a multi-party, distributed, privacy-conscious context. With a particular emphasis on a malware detection case study in cybersecurity, we tested the suggested methodology on real-world datasets. Approaches using the full training dataset resulted in an FPR reduction of 18.2% and 16.9%, respectively, while our method still achieved better performance across all accuracy metrics when given 50% and 75% of the training instances, respectively. These results from the experiments show that our VFL framework can maintain privacy while balancing data and computational efficiency with accuracy. This method is well-suited for VFL environments that need precise models, energy efficiency, and stringent privacy adherence because each node processes its own private data internally and only shares the final master output with coordinating node. This paradigm holds great promise for improving cybersecurity threat detection and prevention through safe collaboration. It might be especially helpful in healthcare, where many organizations have access to private patient data that cannot be shared owing to privacy concerns.

By selectively activating the highest k experts per instance, our VFL_MoE model's sparse routing method decreases computing and communication costs during inference. On the other hand, evaluating training losses and gradients need the output of all experts, which adds another layer of dense computation to the learning process. It is still difficult to implement a sparse MoE training scheme since the discrete output of the gate needs to be approximated using differentiable methods. Routing based on heuristics, estimators that pass straight through, and systems similar to REINFORCE all have drawbacks, such as inefficiency or slow convergence. Our proposed solution reduces the impact of these problems by using a data sampling technique to decrease the number of mini-batches utilized for each training session. By utilizing various neural architectures for the expert and gate sub-models, or by transforming non-tabular data into vectorial form, the framework can be adjusted to accommodate diverse types of data. In addition, it may be easily modified to handle situations involving multiple classes of data. To further improve learning efficiency, future work will investigate hierarchical MoE designs and integrate advanced privacy-preserving approaches while minimizing the cost of computation and communication.

REFERENCES

1. S. Chen, Z. Wu, and P. D. Christofides, "Cyber-Security of Decentralized and Distributed Control Architectures with Machine-Learning Detectors for Nonlinear Processes," in *Proceedings of the American Control Conference*, 2021, pp. 3273–3280.
2. J. Anand, J. J. Tamilselvi, and S. Janakiraman, "Analyzing the Performance of Diverse LEACH Algorithms for Wireless Sensor Networks," *Int. J. Advanced Networking and Applications (IJANA)*, vol. 4, no. 03, pp. 1610–1615, 2012.
3. T. Arauz, P. Chanfreut, and J. M. Maestre, "Cyber-Security in Networked and Distributed Model Predictive Control," *Annu Rev Control*, vol. 53, pp. 338–355, 2022.
4. J. Anand, and K. Sivachandar, "Performance Analysis of ACO-Based IP Traceback," *Int J Comput Appl*, vol. 59, no. 1, 2012.
5. S. Chen, Z. Wu, and P. D. Christofides, "Cyber-Security of Centralized, Decentralized, and Distributed Control-Detector Architectures for Nonlinear Processes," *Chemical Engineering Research and Design*, vol. 165, pp. 25–39, 2021.
6. S. Gaba, I. Budhiraja, V. Kumar, S. Garg, and M. M. Hassan, "An Innovative Multi-Agent Approach for Robust Cyber–Physical Systems Using Vertical Federated Learning," *Ad Hoc Networks*, vol. 163, 2024.
7. M. M. Salim, Y. Sangthong, X. Deng, and J. H. Park, "Federated Learning-Enabled Zero-Day DDoS Attack Detection Scheme in Healthcare 4.0," *Human-Centric Computing and Information Sciences*, vol. 14, 2024.
8. R. Vaishnavi, J. Anand, and R. Janarthanan, "Efficient security for Desktop Data Grid using cryptographic protocol," in *2009 International Conference on Control, Automation, Communication and Energy Conservation*, IEEE, 2009, pp. 1–6.
9. W. Alghamdi, S. Mayakannan, G. A. Sivasankar, J. Singh, B. R. Naik, and C. Venkata Krishna Reddy, "Turbulence Modeling Through Deep Learning: An In-Depth Study of Wasserstein GANs," in *Proceedings of the 4th International Conference on Smart Electronics and Communication, ICOSEC 2023*, 2023, pp. 793–797.
10. J. Anand, D. Srinath, R. Janarthanan, and C. Uthayakumar, "Efficient Security for Desktop Data Grid Using Fault Resilient Content Distribution," *International Journal of Engineering Research and Industrial Applications*, vol. 2, no. VII, pp. 301–313, 2009.

11. D. D. Bikila, "DEEP LEARNING FOR CYBER SECURITY IN THE INTERNET OF THINGS (IOT) NETWORK," in *IDIMT 2023: New Challenges for ICT and Management - 31st Interdisciplinary Information Management Talks*, 2023, pp. 391–398.
12. V. Ravi, T. D. Pham, and M. Alazab, "Attention-Based Multidimensional Deep Learning Approach for Cross-Architecture IoMT Malware Detection and Classification in Healthcare Cyber-Physical Systems," *IEEE Trans Comput Soc Syst*, vol. 10, no. 4, pp. 1597–1606, 2023.
13. A. M. Gejea, S. Mayakannan, R. M. Palacios, A. A. Hamad, B. Sundaram, and W. Alghamdi, "A Novel Approach to Grover's Quantum Algorithm Simulation: Cloud-Based Parallel Computing Enhancements," in *Proceedings of the 4th International Conference on Smart Electronics and Communication, ICOSEC 2023*, 2023, pp. 1740–1745.
14. P. P. Kumar, K. Duraiswamy, and A. J. Anand, "An Optimized Device Sizing of Analog Circuits Using Genetic Algorithm," *European Journal of Scientific Research*, vol. 69, no. 3, pp. 441–448, 2012.
15. A. K. Abasi, M. Aloqaily, B. Ouni, and M. Hamdi, "Optimization of CNN-Based Federated Learning for Cyber-Physical Detection," in Proceedings - IEEE Consumer Communications and Networking Conference, *CCNC*, 2023.
16. A. U. Karimy, and P. C. Reddy, "Enhancing IoT Security: A Novel Approach with Federated Learning and Differential Privacy Integration," *International Journal of Computer Networks and Communications*, vol. 16, no. 4, 2024.
17. E. M. Campos, P. F. Saura, A. González-Vidal, J. L. Hernández-Ramos, J. B. Bernabe, and G. Baldini, "Evaluating Federated Learning for Intrusion Detection in Internet of Things: Review and Challenges," *Computer Networks*, vol. 203, 2022.
18. T. V. Khoa, Y. M. Saputra, and D. T. Hoang, "Collaborative Learning Model for Cyberattack Detection Systems in IoT Industry 4.0," in IEEE Wireless Communications and Networking Conference, *WCNC*, 2020.
19. Y. Djenouri, and A. N. Belbachir, "Empowering Urban Connectivity in Smart Cities using Federated Intrusion Detection," in *2023 IEEE 10th International Conference on Data Science and Advanced Analytics, DSAA 2023 - Proceedings*, 2023.
20. S. Mayakannan, M. Saravanan, R. Arunbharathi, V. P. Srinivasan, S. V. Prabhu, and R. K. Maurya, *Navigating Ethical and Legal Challenges in Smart Agriculture: Insights from Farmers*. 2023.
21. R. Girimurugan, C. Shilaja, A. Ranjithkumar, R. Karthikeyan, and S. Mayakannan, "Numerical Analysis of Exhaust Gases Characteristics in Three-Way Catalytic Convertor Using CFD," in *AIP Conference Proceedings*, 2023.
22. U. Zukaib, X. Cui, C. Zheng, D. Liang, and S. U. Din, "Meta-Fed IDS: Meta-Learning and Federated Learning Based Fog-Cloud Approach to Detect Known and Zero-Day Cyber Attacks in IoMT Networks," *J Parallel Distrib Comput*, vol. 192, 2024.
23. R. Myrzashova, S. H. Alsamhi, A. Hawbani, E. Curry, M. Guizani, and X. Wei, "Safeguarding Patient Data-Sharing: Blockchain-Enabled Federated Learning in Medical Diagnostics," *IEEE Transactions on Sustainable Computing*, pp. 1–15, 2024.
24. K. Vijayalakshmi, P. M. Sitharselvam, I. Thamarai, J. Ashok, G. Sathish, and S. Mayakannan, *Secure and Private Federated Learning through Encrypted Parameter Aggregation*. 2024.
25. S. Mayakannan, N. Krishnamurthy, K. V. Devi, R. Deepalakshmi, S. Rani, and A. J. Anand, *Navigating the Complexity of Macro-Tasks: Federated Learning as a Catalyst for Effective Crowd Coordination*. 2024.
26. Y. A. Kadakia, F. Abdullah, A. Alnajdi, and P. D. Christofides, "Encrypted Distributed Model Predictive Control of Nonlinear Processes," *Control Eng Pract*, vol. 145, 2024.
27. R. A. Nogueira, R. Bourdais, S. Leglaive, and H. Guéguen, "Expectation-Maximization Based Defense Mechanism for Distributed Model Predictive Control," in *IFAC-PapersOnLine*, 2022, pp. 73–78.

28. M. Zuppelli, A. Carrega, and M. Repetto, "An Effective and Efficient Approach to Improve Visibility Over Network Communications," *J Wirel Mob Netw Ubiquitous Comput Dependable Appl*, vol. 12, no. 4, pp. 89–108, 2021.
29. F. Ullah, L. mostarda, D. Cacciagrano, C. Chen, and S. Kumari, "Semantic-Based Federated Defense for Distributed Malicious Attacks," *IEEE Consumer Electronics Magazine*, pp. 1–9, 2024.
30. S. Zhao, R. Bharati, C. Borcea, and Y. Chen, "Privacy-Aware Federated Learning for Page Recommendation," in *Proceedings - 2020 IEEE International Conference on Big Data, Big Data 2020*, 2020, pp. 1071–1080.
31. M. A. Ferrag, O. Friha, L. Maglaras, H. Janicke, and L. Shu, "Federated Deep Learning for Cyber Security in the Internet of Things: Concepts, Applications, and Experimental Analysis," *IEEE Access*, vol. 9, pp. 138509–138542, 2021.
32. L. Yuan, L. Su, and Z. Wang, "Federated Transfer-Ordered-Personalized Learning for Driver Monitoring Application," *IEEE Internet Things J*, vol. 10, no. 20, pp. 18292–18301, 2023.
33. M. Akter, N. Moustafa, and B. Turnbull, "SPEI-FL: Serverless Privacy Edge Intelligence-Enabled Federated Learning in Smart Healthcare Systems," *Cognit Comput*, vol. 16, no. 5, pp. 2626–2641, 2024.
34. M. Itani, H. Basheer, and F. Eddine, "Attack Detection in IoT-Based Healthcare Networks Using Hybrid Federated Learning," in *2023 International Conference on Smart Applications, Communications and Networking, SmartNets 2023*, 2023.
35. Z. Qin, L. Yao, D. Chen, Y. Li, B. Ding, and M. Cheng, "Revisiting Personalized Federated Learning: Robustness Against Backdoor Attacks," in *Proceedings of the ACM SIGKDD International Conference on Knowledge Discovery and Data Mining*, 2023, pp. 4743–4755.
36. O. K. Chandraumakantham, S. Gajendran, and S. Marappan, "Enhancing Intrusion Detection Through Federated Learning with Enhanced Ghost_BiNet and Homomorphic Encryption," *IEEE Access*, vol. 12, pp. 24879–24893, 2024.
37. S. S. Li, X. Zhang, S. Zhou, H. Shu, and R. Liang, "PQLM - Multilingual Decentralized Portable Quantum Language Model," in *ICASSP, IEEE International Conference on Acoustics, Speech and Signal Processing - Proceedings*, 2023.
38. S. Lu, Z. Gao, Q. Xu, C. Jiang, A. Zhang, and X. Wang, "Class-Imbalance Privacy-Preserving Federated Learning for Decentralized Fault Diagnosis With Biometric Authentication," *IEEE Trans Industr Inform*, vol. 18, no. 12, pp. 9101–9111, 2022.
39. N. Subramanian, L. Ravi, M. Shaan, M. Devarajan, T. Choudhury, K. Kotecha, and S. Vairavasundaram, "Securing Mobile Devices from Malware: A Faceoff Between Federated Learning and Deep Learning Models for Android Malware Classification," *Journal of Computer Science*, vol. 20, no. 3, pp. 254–264, 2024.
40. A. Chaudhuri, A. Nandi, and B. Pradhan, "A Dynamic Weighted Federated Learning for Android Malware Classification," in *Lecture Notes in Networks and Systems*, 2023, pp. 147–159.

16 Anomaly Detection in SIEM Data

User Behavior Analysis with Artificial Intelligence

Vedat Önal, Halil Arslan, and Özkan Canay

16.1 INTRODUCTION

The continuously evolving and increasingly uncontrolled surge of cyber threats targeting organizations is becoming more sophisticated, leaving institutions vulnerable to next-generation threats. As artificial intelligence (AI)-driven threat detection and prevention techniques gain prominence, organizations must develop secure IT infrastructures and plan robust cybersecurity strategies to safeguard their information assets, detect attacks, and mitigate their impacts. In this context, the role of security information and event management (SIEM) systems is becoming a critical component for the security infrastructures of modern organizations [1, 2]. SIEM systems gather security data from various sources, such as devices, applications, and other security tools operating within an organization's IT infrastructure, and process them through analysis. This process allows for real-time detection of potential threats against monitored systems, enabling effective and timely responses to incidents while minimizing their impact. At the core of SIEM functionality lies anomaly detection, a critical process to identify deviations from normal behavior that may indicate security breaches. A recent study demonstrated how automatic email categorization based on content can improve security monitoring and event management, particularly in detecting suspicious email activities [3].

SIEM systems operate based on several fundamental principles. First, they can collect and standardize data from various sources. This standardization process simplifies the interpretation and analysis of data for security teams, allowing them to conduct the analysis more efficiently. In addition, they convert diverse types of collected data into a consistent and standardized format. Second, SIEM systems employ advanced correlation and rule-based engines to detect patterns, relationships between data, and anomalies within the collected information. These systems apply predefined rules and heuristic methods to identify suspicious activities, policy violations, and potential threats, simultaneously alerting security teams to help them take necessary actions [4].

As cybersecurity data's volume, velocity, and complexity increase, the need for robust and high-impact anomaly detection methods becomes paramount. Although traditionally used rule-based or signature-based approaches are valuable to

DOI: 10.1201/9781003521020-16

organizations, they fail to address cyber attackers' continuously evolving strategies and tactics [5]. This gap has led to the growing importance of user behavior analysis (UBA), which provides a more explainable, interpretable, and holistic understanding of security-related organizational activities. UBA examines and analyzes the patterns, trends, and deviations in how users interact with digital systems and networks, such as computers, servers, firewalls, cloud services, and virtual machines. By recognizing standard employee behavior, security teams can more effectively identify anomalies that may signal malicious activities such as unauthorized access, data breaches, and other insider threats. The approach improves threat detection accuracy and helps establish a proactive security framework, enabling organizations to mitigate the impact of potentially risky incidents before they occur.

The integration of AI and machine learning (ML) techniques represents a pivotal development in detecting user behavior anomalies within SIEM systems, providing an effective solution to address emerging cyber threats. Advanced analytics methods not only uncover complex patterns, detect subtle deviations, and adapt more effectively to changing user behaviors compared to traditional rule-based approaches, but they also enable security teams to predict and prevent evolving cyber threats with greater accuracy and speed. SIEM systems that leverage the analytical power of AI and ML can become more logical, responsive, and effective in safeguarding critical information assets, such as customer/employee data and financial systems, while ensuring the integrity of their infrastructures. The growing importance of SIEM systems and anomaly detection further highlights the essential contribution of AI/ML methods in enhancing UBA capabilities and shaping the future of cybersecurity.

16.1.1 Importance of User Behavior Analysis

UBA has emerged as a critical aspect of modern cybersecurity, particularly in SIEM systems. Security teams aim to effectively identify deviations from standard patterns and trends in user interactions with systems and networks that may indicate potential security breaches. UBA provides a detailed and comprehensive understanding of security-related activities within an organization. This approach extends the detection capabilities of organizations beyond traditional rule-based or signature-based methods, particularly in addressing complex and next-generation tactical attacks. By analyzing user activity patterns, SIEM systems establish expected behavior as a baseline criterion, enabling the detection of anomalies that may signal malicious activities such as unauthorized access, data breaches, or insider threats.

In the current cyber era, the increasing volume and velocity of security incidents and the complex tactics continuously developed by attackers can render traditional SIEM approaches inadequate. The importance of UBA within SIEM systems lies in its significant role in accurately and timely detecting threats to organizations and facilitating the development of necessary security configurations. By examining the behavior of a typical user, security teams can more effectively differentiate between legitimate activities and security incidents that carry potential threats. This improved differentiation reduces the likelihood of false positives, allowing teams to focus on more targeted and impactful responses [6]. UBA weakens the impact of attacks before they escalate into significant incidents and will enable organizations to update their security policies and preventive strategies proactively.

Advanced SIEM systems can detect anomalies—behaviors that deviate from normal user behavior—enabling early detection of potential threats and preventing the spread of malicious activities within an organization. In an era where cybersecurity is rapidly evolving, developing proactive approaches such as these is critical, as they bridge the gap between defensive capabilities and identifying breaches that could lead to destructive incidents. The integration of UBA models into SIEM systems allows organizations to expand their security frameworks and facilitate the detection of high-impact potential threats. SIEM solutions, therefore, provide a more comprehensive perspective by correlating user activities with additional security-related information, including application logs, network activity, and intelligence from threat analysis. The improved capacity allows security teams to inform their decisions better, focus on risk reduction efforts, and deploy more effective security strategies in response to potential attacks [7, 8].

Moreover, the role of UBA models in anomaly detection continuously enhances and strengthens SIEM systems, keeping them vigilant and responsive. These models, when integrated into security systems, analyze historical user behavior activities collected by SIEM, allowing the identification of emerging threat trends. Strengthening the threat detection algorithms within the UBA model helps organizations update their security policies and strategies, thereby improving the overall efficiency of SIEM solutions. In the face of evolving tactics and strategic scenarios employed by cyber attackers, iterative learning and adaptation processes are paramount. Integrating models like UBA becomes essential in the struggle to protect critical information assets and ensure the reliability of IT infrastructures. These models, providing a deeper contextual insight into user activities, improve both the precision and effectiveness of identifying and responding to threats, thus simplifying the task for security teams in managing and reducing risks.

16.2 ARTIFICIAL INTELLIGENCE IN ANOMALY DETECTION

Anomaly detection aims to identify deviations from the norm or unexpected events within a dataset. This process finds extensive use in fraud detection in cybersecurity, enhancing customer satisfaction on e-commerce platforms, sentiment analysis on social media, and quality control in healthcare services. AI has transformed anomaly detection by introducing methods that substantially boost the precision and effectiveness of detection. This revolution is especially significant in cybersecurity, where the sheer volume of security data, its increasing velocity, and the complexity of next-generation attacks often render traditional rule-based or signature-based SIEM systems insufficient. This situation increasingly motivates security teams to integrate AI/ML models into SIEM systems.

AI-driven anomaly detection approaches utilize various techniques, each with its strengths and applications [9, 10]. ML models, including supervised and unsupervised methods, can effectively detect trends and abnormal behaviors in network traffic and user activities. Recent advancements in deep learning (DL) approaches, such as using neural networks for intrusion detection, have shown significant potential in addressing emerging cybersecurity challenges [11]. The behavioral patterns of a standard user interacting with networks and systems form the foundation for these models. Essentially, these models are trained on historical data to identify deviations that could indicate potential security breaches [12].

DL, a subset of ML, holds significant promise in anomaly detection. For instance, architectures like long short-term memory (LSTM) and recurrent neural networks (RNNs) have demonstrated high performance in classifying network traffic as normal or abnormal, capturing complex patterns in user behavior over time. In addition, convolutional neural networks (CNNs) are used for malware detection, leveraging their ability to recognize the structure of events to identify suspicious software behaviors [13]. Recently, ensemble methods, which combine multiple AI models under a single framework, have gained popularity in anomaly detection due to their enhanced detection accuracy. Ensemble approaches offer greater robustness and flexibility in detecting complex security threats by compensating for the weaknesses of individual algorithms while reinforcing their strengths.

The integration of AI-driven anomaly detection models into SIEM systems provides substantial benefits. SIEM systems utilizing advanced techniques better detect subtle deviations and highly complex patterns that traditional SIEM systems might overlook. AI-based SIEM solutions create a proactive security framework by offering more robust and effective adaptability to the ever-evolving threat landscape and the behavioral variability rooted in human psychology. These systems help reduce cyber risks by automating workflows related to threat detection and incident response, thereby significantly improving the efficiency of security operations. AI-driven solutions can filter security incidents within workflows, focus the attention of security teams on high-risk activities, reduce manual workload, and enable faster incident response [14, 15].

The advantages of AI-based anomaly detection systems over traditional techniques are numerous and substantial. AI methods improve accuracy in detecting and classifying normal and abnormal behaviors, with lower false favorable rates and the ability to handle multi-class classification problems. Moreover, AI-powered systems bring a more resilient and dynamic approach to cybersecurity by adapting more effectively to changes in user behavior and evolving threat models. However, integrating AI into anomaly detection also introduces unique challenges and obstacles. Issues such as explainability, interpretability, and the potential for attacks targeting AI models must be carefully considered. Researchers must address these challenges carefully to ensure that AI-based security solutions are effectively integrated into SIEM systems. The advancements in AI-driven anomaly detection significantly enhance the capabilities of SIEM systems, enabling more accurate, efficient, and proactive security measures. As the cybersecurity landscape continues to evolve, integrating such innovative techniques is crucial for organizations to stay ahead of sophisticated cyber threats and protect their critical assets [16–20].

16.2.1 Artificial Intelligence Techniques and Algorithms

Anomaly detection in SIEM systems using AI techniques has emerged as a significant advancement. AI techniques offer advanced analytical methods that detect abnormalities deviating from normal behavior more effectively than traditional approaches, with better adaptability to changing user behaviors. Various AI-based approaches have been researched and applied for anomaly detection in SIEM systems, including supervised, unsupervised, and semi-supervised learning algorithms, as well as DL techniques.

16.2.1.1 Supervised Anomaly Detection

Supervised anomaly detection techniques use labeled training datasets, including regular and abnormal data, to perform anomaly detection. This approach creates separate prediction models for normal and abnormal classes and then compares these models to detect anomalies [21]. Supervised algorithms that have shown significant success include models such as K-nearest neighbors (KNN), Random Forests (RF), Decision Trees (DT), Support Vector Machines (SVM), and Neural Networks. These algorithms learn from labeled historical data to build predictive models that classify new incoming data as normal or abnormal. Accordingly, the system is designed to evaluate the structure of incoming data by utilizing patterns learned from prior examples. This process enables the rapid detection of outliers in any dataset [21]. However, two challenges arise: anomalies tend to be far less frequent in training data than typical examples, and it can be difficult to define precise and representative labels, particularly for the anomaly class.

16.2.1.2 Unsupervised Anomaly Detection

Unsupervised anomaly detection techniques are commonly applied to unlabeled datasets. This approach detects anomalies by analyzing structural and distributional differences in data without predefined labels. Anomalies are identified by examining the inherent structure of the data, which allows for the detection of outliers without the need for predefined labels. Popular unsupervised algorithms used for anomaly detection include clustering algorithms, isolation forests, and one-class SVMs. These algorithms analyze the structural characteristics and distributions within the data to detect anomalies. Unsupervised methods are typically used to understand the internal structure of datasets and define abnormal behaviors. However, if the underlying assumptions fail, unsupervised techniques can suffer from high false alarm rates [21].

16.2.1.3 Semi-Supervised Anomaly Detection

Semi-supervised anomaly detection techniques are employed in scenarios where only a tiny portion of the dataset is labeled. Since data labeling is time-consuming and costly, labeling every piece of data is often not feasible. In this context, semi-supervised learning comes into play. These methods attempt to identify anomalies in unlabeled data by leveraging the information provided by the labeled data [22]. Anomalies are generally characterized as points deviating from the overall structures and distributions of the dataset. Detecting these deviations is critical in many sectors, such as fraud detection in finance, malicious activity detection in software, or disease diagnosis in healthcare. Semi-supervised methods can be supported by various ML algorithms, such as SVMs, DL methods (especially deep neural networks), and graph-based approaches. These algorithms learn the characteristics of labeled data and assess whether the unlabeled data conforms to those structures [21, 22].

16.2.1.4 Deep Learning-Based Anomaly Detection

DL has garnered significant attention for anomaly detection in SIEM systems. DL has advanced anomaly detection in complex and high-dimensional datasets. DL-based models can automatically uncover essential features within the data and understand its complex structure. With their multi-layered architectures, these models can extract meaningful information from raw time series data. This capability allows them to learn complex relationships and undefined rules that traditional methods struggle to identify, achieving highly accurate anomaly detection. In this regard, DL models demonstrate exceptional sensitivity and recall in detecting anomalies [23]. DL-based models typically extract behavioral information from historical data and can suggest potential unfavorable changes in the future. For this reason, simply predicting distribution is limited in identifying contextual and collective anomalies. CNNs, RNNs, and LSTMs have been employed to identify complex patterns and detect temporal dependencies in data, facilitating effective anomaly detection. In addition, hybrid modeling with different window sizes on time series data provides a powerful and effective anomaly detection capability [24].

16.2.1.5 Ensemble Learning-Based Anomaly Detection

Ensemble methods, which integrate various ML and DL models, have been created to boost anomaly detection capabilities. Such techniques effectively enhance anomaly detection's precision, robustness, and dependability in SIEM systems by leveraging the strengths of diverse algorithms and methodologies. Ensemble learning algorithms contribute to higher detection scores by increasing diversity and reducing feature redundancy, particularly in irregular and imbalanced datasets. Approaches such as XGBoost, gradient boosting machines (GBM), random forests (RF), and generalized linear models (GLM) have proven to be highly effective in anomaly detection scenarios, leading to the creation of robust and reliable model ensembles. Combining ensemble learning approaches represents a significant advancement in data analytics and ML applications, crucial in improving data security and quality [25, 26].

In addition to these techniques, natural language processing (NLP) and other analysis techniques compatible with SIEM systems have become integral to anomaly detection developments. NLP can be applied to extract and analyze contextual information out of non-structured data sources, such as system logs and security documents, to detect potential anomalies. On the other hand, graph-based approaches utilize relationships and connections between entities, events, and activities to detect complex and multi-layered attacks. Applying these AI techniques in SIEM systems has led to significant advancements in detecting high-impact incidents with unknown attack techniques. SIEM solutions using these techniques can detect behaviors that deviate from standard user behavior with greater accuracy, reduce false positive rates, and provide timely, effective responses to emerging threats.

Advanced anomaly detection systems have become indispensable in addressing cybersecurity vulnerabilities that exceed traditional SIEM products in organizations, particularly amid data's growing scale, speed, and intricacy. Furthermore, advancements in AI technology enhance the detection capabilities of SIEM products and enable more proactive and automated security measures. AI-based SIEM solutions can predict potential threats, prioritize mitigation efforts, and facilitate the timely and efficient application of appropriate security controls by continuously learning and adapting to behavioral events collected from users. As a result, AI-driven anomaly detection mechanisms strengthen organizations' defense frameworks, allowing security teams to stay ahead of sophisticated cyber attackers and protect critical assets.

16.2.2 Comparison with Traditional Methods

Compared to conventional rule or signature-based methods, AI and ML/DL-integrated SIEM solutions offer significant advantages, focusing on improving anomaly detection capabilities in SIEM systems. This AI-driven approach outperforms traditional SIEM approaches regarding anomaly detection accuracy, flexibility with varying datasets, and the ability to learn from large, complex data structures. Traditional anomaly detection methods rely on predefined rule templates or signature information in data collection systems to detect user behaviors. While these approaches are practical for identifying historical cyber threats, they struggle to adapt to cyber attackers' new tactics, scenarios, and strategies. Maintaining an extensive rule set capable of predicting all scenarios is challenging and often limited for rule-based SIEM systems. As the cybersecurity environment becomes more flexible, traditional SIEM systems increasingly lose effectiveness compared to modern solutions, which are better equipped to handle evolving threats and reduce false-positive rates, leaving traditional systems more prone to undetected threats and inefficiencies [10, 27].

In contrast, AI-driven anomaly detection methods utilize advanced algorithms and statistical models to uncover complex data structures and deviations in user behavior and network interaction activities. AI-based approaches are better equipped to adapt to the constantly changing user behaviors and can detect anomalies in real time with higher accuracy. By employing ML, DL, and ensemble learning methods, SIEM systems can more detail relationships in standard user behaviors, detecting even minor deviations.

AI-driven anomaly detection methods offer significant advantages. The first is their ability to handle the continuously growing volume of security data, the uncontrolled increase in data flow velocity, and the various tactics developed by attackers. Traditional SIEM systems often struggle to process and analyze datasets at increasing volumes and speeds, leading to delayed threat detection and incident response. In contrast, AI-based SIEM systems can leverage techniques like ML, DL, NLP, and neural networks to detect the complex structures of potential threats and provide real-time notifications and incident response [24, 28].

Another benefit of AI-based anomaly detection systems is their enhanced capacity to adjust to shifting user behaviors and newly arising threats, in contrast to traditional SIEM systems. AI-based SIEM systems continuously learn from newly collected user behavior data and update their models, allowing them to swiftly detect emerging attack vectors and respond effectively, which helps establish a more robust and resilient defense framework against cyber threats. This flexibility is vital in the constantly shifting cybersecurity environment, where cyber attackers persistently devise new strategies to bypass conventional security measures.

While rule-based and signature-based approaches still exist in today's SIEM systems, the growing acceptance of AI-driven anomaly detection techniques is inevitable. SIEM systems incorporating user behavior analytics (UBA), ML, DL, NLP, and neural networks are better equipped to accurately detect abnormal behaviors, minimize false alarms, and deliver valuable insights to help security teams mitigate risks more effectively. Next-generation SIEM systems offer a more holistic, flexible, and efficient protection against diverse cyber threats by utilizing the capabilities of advanced analytical techniques, thereby enhancing an organization's overall security posture [10, 14].

16.3 METHODOLOGIES FOR USER BEHAVIOR ANALYSIS

Understanding user behavior has become a critical area that spans various applications, from social platforms to e-commerce, healthcare services, and cybersecurity. Research in this field and advancements in software provide valuable insights by analyzing how users interact with digital platforms, systems, and servers. These insights foster innovation, improve user experiences, and enhance decision-making processes. In the context of SIEM systems, UBA involves a set of methodologies to identify patterns, trends, and anomalies in user interactions. These methodologies often rely on the meticulous collection, preprocessing, and analysis of data such as event logs, browsing history, sensor data, and network flow data using ML, DL, and statistical approaches.

One prominent approach in UBA is using statistical solutions designed to detect patterns, trends, and deviations in user activities by applying traditional statistical methods. Methods such as time series forecasting, regression analysis, and anomaly detection are employed to establish a typical user behavior baseline and detect deviations from this norm. Statistical methods are often valued for their interpretability and ability to provide insights into the underlying factors influencing user behavior. On the other hand, ML models offer a more advanced and adaptable approach to UBA. These models are designed to detect complex anomalies that may escape traditional statistical techniques by learning and recognizing the structural features of data gathered from users' historical interactions.

Algorithms like random forests, decision trees, and support vector machines (SVMs) commonly applied in supervised learning are often employed in UBA to categorize user activities as either normal (0) or abnormal (1) using pre-labeled data. Unsupervised learning methods, such as clustering algorithms and anomaly detection methods, can detect outlier data structures and anomalies without relying on pre-labeled data. Additionally, the emergence of DL, a subset of ML, has further extended UBA capabilities in SIEM systems. Architectures such as LSTM and CNN have shown remarkable performance in modeling data's temporal and structural features within user activities. DL approaches are adept at capturing the complexity and dynamism of user behavior, allowing for the highly accurate detection of anomalies and the identification of elusive threats.

An effective UBA system consists of three core components, regardless of the chosen methodological approach. These components comprise data preprocessing, feature engineering, model training, validation, and practical case studies and applications. The topic of practical examples and real-world applications is covered in detail in section 4. In the data preprocessing and feature engineering phase, tasks such as data cleaning, normalization, and extracting features that effectively capture the nuances of user behavior are performed.

Feature selection and design are essential for optimizing the effectiveness of the selected analytical models, highlighting the importance of the expertise of security personnel and a deep understanding of user activities within organizations. In addition, training and validating UBA models is crucial for ensuring their reliability and generalizability. Techniques such as cross-validation, holdout testing, and benchmark datasets may be employed to accurately evaluate models' ability to detect anomalies under various potential attack scenarios. It is also essential for models to continuously monitor collected data and self-improve. Indeed, user behaviors and possible cyber threats evolve continuously, and SIEM systems are expected to adapt and advance over time.

The choice of methodologies for UBA in SIEM systems depends on specific requirements, data availability, and the organization's overall cybersecurity strategy. Hybrid structures that combine multiple techniques can effectively address the multidimensional nature of the data collected from users by compensating for each method's weaknesses, thus improving the precision and resilience of anomaly detection. As the SIEM and UBA fields continue to evolve, integrating these methodologies alongside emerging technologies like edge computing and federated learning will become increasingly crucial for developing more intelligent, adaptable, and privacy-focused security solutions.

16.3.1 Data Preprocessing and Feature Engineering

Data analysis is critical in many areas, such as understanding user behavior on digital platforms, providing personalized recommendations, and supporting cybersecurity. Raw user data collected from various sources—such as web server logs, Windows event viewer data, Linux and MacOS system logs, sensor data, and network flow data—are often unstructured and noisy, making them inadequate for serial analysis models. Therefore, data preprocessing and feature engineering processes are essential for transforming this raw data into a format that can be effectively used for modeling and predicting user behavior. This step is crucial for enhancing the performance of models used in UBA.

One of the primary obstacles in UBA is managing the extensive and varied data types that are collected. SIEM systems typically aggregate data from numerous sources, spanning both structured and unstructured formats, making it challenging to filter out the most relevant features for detecting anomalous user behavior. The preprocessing phase emphasizes data cleaning, standardization, and dimensionality reduction to remove extraneous or insignificant features from the user data.

Data cleaning improves the accuracy and reliability of input data by identifying and removing errors, incomplete information, or duplicates. Standardization methods, such as min-max scaling or z-score normalization, help unify data formats from different sources, enhancing data quality and better model training. Dimensionality reduction techniques, like Principal Component Analysis (PCA) and t-distributed Stochastic Neighbor Embedding (t-SNE), highlight the most critical features, minimizing data complexity without sacrificing the integrity of underlying models.

In the context of UBA, feature engineering is crucial for enabling the model to differentiate between normal and abnormal user activities. Therefore, feeding the system with new and meaningful features from the input data is of utmost importance. This process may involve extracting timestamps and behavioral attributes from raw data for various activities, such as login/logout events, file access frequencies, and network access events. By incorporating domain-specific attributes, UBA models can more accurately capture nuances in user behavior and improve the precision of anomaly detection.

Additionally, integrating contextual information, such as organizational hierarchy, departmental structure, access privilege levels, or permissions, can tailor the interpretation of detected anomalies and prioritize them based on their potential impact. This contextualization allows security teams and analysts to understand high-impact potential threats better before an incident occurs, enabling incident response mechanisms to be planned with more informed strategies.

16.3.2 Model Training and Validation

After raw log data collected from various channels on the systems and platforms used by users undergoes preprocessing and feature engineering, the model training and validation process is critical to enable UBA models used in SIEM systems to detect security threats in dynamic environments more effectively. This process is carried out by analyzing data formats, demographic structures of the population, and the relationships or dependencies between actions. The selection of the most suitable algorithm and model optimization are the fundamental pillars of this process.

Selecting the suitable ML algorithm for UBA is vital during the model training phase. SIEM systems can employ various AI techniques, such as supervised learning methods (e.g., RF, SVM), unsupervised learning approaches (e.g., K-means clustering, autoencoders), and DL models (e.g., RNNs, CNNs). The specific requirements of the anomaly detection task, the characteristics of the available data, and the need for model interpretability influence the choice of algorithm. Supervised methods such as regression, classification, sliding window, or time series analyses are often based on a user's past actions. These models are trained on categorized datasets according to input attributes and data variability, allowing the algorithms to learn the relationships between features and categorized data and increase prediction accuracy. For instance, in e-commerce platforms, supervised learning models can recommend products by analyzing past purchase behaviors and demographic data. Similarly, in security models, this approach can detect abnormal behaviors based on actions like system activity, network traffic flow, and file access.

On the other hand, unsupervised algorithms uncover hidden patterns, segments, behavioral characteristics, and anomalies in user data without labeled data. These approaches reveal the underlying structure within the data, making them useful in scenarios where data structures are unclear or the goal is to identify unknown user behaviors or subgroups. For instance, clustering algorithms group users with similar behaviors, enabling personalized recommendations, targeted marketing, or early detection of suspicious activities. Anomaly detection methods using unsupervised models can assist in identifying outliers or unusual user behaviors that may signal potential security threats, fraud, or other issues.

Tuning hyperparameters is an essential aspect of the training phase, as it aims to maximize the selected algorithms' performance on validation data. Approaches such as grid search, random search, or advanced techniques like Bayesian optimization are employed to identify the best hyperparameter combinations. Proper model optimization ensures that algorithms can effectively capture the underlying patterns of user behavior and accurately detect anomalies. Additionally, employing a robust validation strategy is essential for assessing the generalization capability of trained models. These validation strategies are critical for ensuring that the models developed for SIEM systems can accurately detect anomalies in user behavior.

Techniques such as cross-validation (CV), holdout validation, and time series validation are used to assess models' reliability and performance resilience. In CV, the dataset is split into multiple subsets, with one subset used for model training and another for performance validation, repeating this process numerous times across different data segments. Holdout validation, on the other hand, separates the dataset into distinct training and testing groups, utilizing the first for model training and the second for evaluation purposes. The choice of validation method is influenced by factors such as the dataset's size, characteristics, and the complexity of the models employed in UBA.

After the training phase, various performance metrics evaluate the model's effectiveness based on the specific problem and the desired outcomes. These metrics are vital for assessing the alignment of model predictions with actual data and overall performance. Commonly employed metrics in UBA include accuracy, precision, sensitivity, F1-score, ROC curve, and mean squared error (MSE). These measures capture the model's capacity to identify user behaviors and anomalies and make relevant predictions or recommendations. For instance, in an e-commerce platform, performance metrics for a model that predicts user churn can be assessed by measuring the percentage of correct predictions and the model's ability to minimize false positives [29–32]. In the security domain, a model targeting detecting abnormal behavior in employees should exhibit high precision and sensitivity, ensuring the model correctly identifies potential attacks while minimizing false alarms. Such accuracy allows the security team to avoid unnecessary time spent on false alarms.

SIEM systems can develop accurate and reliable AI-driven anomaly detection capabilities by following rigorous model training and validation procedures. Additionally, the interpretability and explainability of models are essential for security analysts to understand the reasoning behind anomaly detection decisions and take appropriate actions. Together with performance metric evaluation, model interpretability, data preprocessing, feature engineering, model selection, and the training-validation phases enhance the efficiency and effectiveness of UBA methodologies in SIEM systems.

16.4 CASE STUDIES AND APPLICATIONS

16.4.1 Real-World Examples and Applications

The practical effectiveness and evaluation results of UBA methods developed based on real-world attack scenarios are crucial. Researchers and developers across various sectors provide detailed and insightful information about UBA's application strategies, the challenges encountered, and the systems' shortcomings through analyses of actual incidents. The integration of UBA into SIEM systems has shown promising results across various sectors, including finance, healthcare, retail, telecommunications, and energy, demonstrating its effectiveness in achieving more robust security outcomes.

A notable case study in the finance sector by Kotagiri presents a system for detecting and preventing fraud by applying AI-powered UBA in U.S. banks. The banking sector must cope with high volumes of security alerts and many false positives, leading to significant operational burdens and delayed incident response times. In this sector, using AI to monitor normal user behavior in real time and detect fraud through behavior analysis and adaptive learning mechanisms is a cornerstone of security incident management. The core of the proposed system lies in using ML algorithms, which were developed based on the accuracy and sensitivity required to detect fraudulent transactions. The algorithms performed with a commendable accuracy of 85%, a recall of 90%, and an F1 score of 87% [33].

Another fraud-related case study by Rieke and colleagues focuses on AI-powered UBA implementation in a large financial institution. By integrating an AI-supported SIEM system, the institution established a more apparent baseline for normal user behavior through user behavior analytics. The system detected subtle anomalies such as unusual login patterns, abnormal file access, and suspicious data transfers, often indicative of potential threats or unauthorized access attempts. The results of this case study are impressive: the AI-powered SIEM system reduced the institution's manual incident response workload by 86.76%. Additionally, the system classified potential threats with a misclassification risk of less than 0.001%, ensuring that genuine security incidents were not overlooked. The improved accuracy and efficiency allowed the institution to respond to emerging threats promptly and effectively, reducing the overall risk of data breaches and reputational damage [34].

UBA was also applied in another case study that involved e-commerce platforms. The research conducted by Rimakka and Aras on Amazon's shopping platform focused on how web usage mining techniques were used to analyze customer browsing patterns, purchase habits, and product interactions. The researchers preprocessed web log data to extract features such as cart addition behavior, product review frequency, and the helpfulness rating of reviews. Using this data, they developed predictive models that accurately forecast customer churn and identified cross-selling opportunities. Insights from this analysis allowed the retailer to personalize product recommendations, optimize website layout, and tailor marketing campaigns, significantly increasing customer loyalty and revenue [35].

Similarly, UBA has proven to be highly beneficial in the healthcare sector. In R&D efforts within healthcare services, analysis of sensor device data, electronic health records, and patient-generated results has provided a better understanding of addiction levels, disease progression, and factors affecting overall health. An example is the AI-powered e-health monitoring study by Fang and colleagues. In this study, dynamic AI/ML models were developed based on real-time behavioral and health data collected from the user's phone. By using deep reinforcement learning techniques, users could have a more personalized exercise experience, and the system encouraged them to adopt healthier lifestyles. An essential aspect of this study was boosting users' motivation and evaluating the negative impacts encountered when they failed to meet daily exercise goals [36].

There has been a noticeable rise in cyberattacks targeting critical infrastructures such as national energy grids and transportation systems in recent years. UBA techniques integrated with SIEM systems are crucial in detecting abnormal behaviors in this context. A study by González-Granadillo and colleagues examined the role of SIEM systems in critical infrastructures and their development in detail. The study highlighted the effectiveness of SIEM solutions in detecting and managing cyberattacks and emphasized the importance of integrating these systems with big data analytics in the future. Furthermore, integrating AI/ML techniques with SIEM systems was identified as a significant innovation in anomaly detection. These technologies enable the rapid detection of potential threats by identifying anomalies in network traffic and system behaviors. Overall, the study comprehensively addressed how SIEM systems contribute to security management in critical infrastructures and explored future development areas [2].

Various field case studies demonstrate that integrating UBA into SIEM systems provides significant advantages in combating cyber incidents. Organizations leveraging AI and ML techniques have enhanced their security infrastructures, improved anomaly detection accuracy, and responded to emerging threats promptly and effectively. Establishing a baseline for normal user behavior and quickly identifying deviations has proven to be a powerful strategy for combating sophisticated cyberattacks and insider threats. As organizations continue to face evolving cybersecurity challenges, the lessons learned from these case studies will lead to prioritizing UBA as a critical component of security strategies and the development and implementation of advanced SIEM solutions.

16.4.2 Results and Outcomes

Applying UBA in SIEM systems has yielded significant findings and promising results, especially in improving anomaly detection and reducing false positives, as demonstrated through extensive research and real-world cyber incident applications. The analysis of findings from studies on user behavior has highlighted the profound impact of UBA methodologies on decision-making processes and organizational security strategies. Simultaneously, these studies have underscored the lessons learned in this dynamic research area and pointed to future directions. One of the critical benefits of UBA is its ability to enable users to make more informed, data-driven decisions. Security teams within organizations benefit from insights gained by analyzing standard user activity data collected from various endpoints, including servers, cloud services, employee computers, and sensor devices. These insights help develop mechanisms that understand the relationships between user activities, allowing the organization's IT infrastructure to more effectively detect deviations that could indicate potential security incidents and inform strategic decision-making accordingly.

A key advantage of integrating UBA into SIEM solutions is the significant enhancement in anomaly detection accuracy based on user data. Studies have shown that by establishing a baseline of expected user behavior, SIEM systems can more effectively distinguish legitimate activities from suspicious incidents, resulting in a marked reduction in false alarms. This reduction permits security teams to concentrate on investigating genuine threats and reacting more effectively rather than being overwhelmed by high volumes of false positives. Additionally, using UBA has enhanced the proactive capabilities of SIEM systems, enabling organizations to mitigate risks before they escalate into major incidents. SIEM solutions can detect subtle deviations from normal activities, allowing security teams to identify potential threats early and prevent malicious activities from spreading. This proactive approach has proven invaluable within the quickly changing cybersecurity environment, in which the capability to anticipate threats is essential to protecting corporate assets.

Integrating UBA techniques into SIEM systems also strengthens an organization's security framework, giving it a significant advantage against cyber attackers. SIEM solutions offer a comprehensive perspective on potential security vulnerabilities and risks by correlating user activities with additional security-related data, including application logs, network traffic, and threat intelligence. This contextual awareness empowers security teams to make informed decisions, prioritize mitigation efforts, and implement more effective security strategies. The iterative learning and adaptation processes enabled by UBA techniques can contribute to continuous improvement and development in SIEM systems. By analyzing historical user behavior data, security teams can identify emerging trends and potential weaknesses, allowing for improvements in detection algorithms, updates to security policies, and an enhancement of the SIEM solution. This adaptive approach is crucial for staying ahead of the evolving next-generation tactics of cyber attackers.

16.5 CHALLENGES AND FUTURE DIRECTIONS

As the UBA field in SIEM systems continues to evolve, several key challenges and future directions must be carefully considered. These factors play a crucial role in developing next-generation anomaly detection capabilities and ensuring the long-term effectiveness of SIEM solutions in countering new and sophisticated cyber-attack techniques.

One of the primary challenges lies in data quality and privacy. Effective UBA relies on the availability of high-quality, comprehensive datasets that accurately capture user activities and interactions. However, such data must be collected and processed to respect individual privacy and comply with relevant data protection regulations. Striking the right balance between data's utility and privacy preservation is a delicate and ongoing challenge for security professionals. Additionally, the dynamic nature of user behavior poses a significant challenge for SIEM systems. As users' interaction patterns with digital systems evolve, the models and algorithms used for anomaly detection must be continuously updated to maintain their accuracy.

It is essential to continuously improve and adapt these models to stay ahead of constantly changing tactics employed by cyber attackers.

Another challenge is the need for extensive and diverse datasets to train and validate AI-powered anomaly detection models. While significant progress has been made in developing AI methods for UBA, the availability of high-quality, representative datasets remains a persistent barrier. Security researchers and developers must collaborate to create and maintain standardized, publicly available datasets that capture the nuances of user behavior across different sectors and domains. Moreover, the ethical implications of UBA must be carefully examined. The collection and analysis of user data, even for security purposes, raise concerns about individual privacy, authority, and potential misuse. SIEM system developers and security teams must implement robust governance frameworks, transparency measures, and user-centered controls to ensure that UBA is conducted ethically and with all parties involved understanding and consent.

Numerous emerging trends and innovations are set to influence UBA within SIEM systems. Progress in AI and ML, especially in deep reinforcement and federated learning, is expected to enhance anomaly detection models' adaptability, interpretability, and scalability. Furthermore, integrating SIEM systems with advanced security solutions, such as user and entity behavior analytics (UEBA) and platforms designed for security orchestration, automation, and response (SOAR), can improve their capacity to identify, analyze, and mitigate complex security challenges. The increasing use of Internet of Things (IoT) devices and the growing importance of cloud-based environments also introduce new challenges and opportunities for UBA within SIEM systems. Adapting anomaly detection methodologies to these new and evolving digital ecosystems will require innovative approaches to handle user interactions' scale, diversity, and dynamic nature.

As organizations continue to face the complexities of the cybersecurity landscape, the importance of UBA in SIEM systems will only increase. By addressing challenges related to data quality, privacy, model adaptation, and ethics while embracing new technologies and emerging trends, security professionals can further enhance the effectiveness of SIEM solutions in defending organizations against sophisticated cyber threats.

16.5.1 Data Quality and Ethical Concerns

Ensuring the quality and reliability of data is a critical aspect of UBA in SIEM systems. As organizations strive to leverage the power of AI/ML to enhance anomaly detection processes, the integrity and accuracy of the data become increasingly important. Poorly regulated or biased datasets can lead to flawed models, resulting in false positives, missed anomalies, and potentially catastrophic security breaches. One of the core challenges in UBA is the inherent complexity and variability of user actions and interactions. The data gathered from different origins, including systems, networks, and applications, can be fragmented, inconsistent, or noisy [16–20].

Addressing data quality issues requires robust data preprocessing and feature engineering techniques to ensure models are trained on clean, representative, and contextually relevant information. Effective preprocessing includes data cleaning, normalization, and feature selection, which help reduce the impact of outliers, missing values, and irrelevant attributes. As a result, the model's capacity to precisely identify patterns in user behavior is greatly enhanced. On the other hand, feature engineering focuses on identifying and extracting the most valuable features from raw data, allowing AI models to better distinguish between normal and abnormal user activities.

In addition to data quality, ethical concerns related to UBA in SIEM systems must also be addressed. The gathering, storage, and examination of user data raise issues concerning privacy, authority, and the possible abuse of sensitive information. Organizations must establish clear policies and procedures to handle user data carefully and comply with relevant privacy regulations and industry standards.

One way to mitigate these ethical challenges is by using privacy-preserving methods like data anonymization, differential privacy, and secure multi-party computation. These techniques allow valuable insights to be derived from user data while minimizing the risk of identifying individuals and preserving privacy. Additionally, organizations should consider obtaining explicit user consent to collect and analyze data, fostering transparency and trust.

Ethical concerns go beyond data privacy to encompass the responsible development and deployment of AI-powered anomaly detection systems. Developers and security teams must ensure these systems are designed reasonably, accountable, and transparent, avoiding potential biases and unintended consequences that may disproportionately affect specific user groups. The interpretability and transparency of AI models are essential in helping security professionals comprehend the system's decisions and take necessary actions.

As organizations continue to benefit from UBA and AI-powered anomaly detection in SIEM systems, ensuring data quality and addressing ethical concerns remain crucial. By proactively tackling these issues, security teams can build reliable, efficient SIEM solutions that protect organizational assets while upholding the rights and privacy of employees and customers. Ongoing research and collaboration between academia, industry, and regulatory bodies will be critical in shaping the future of UBA and AI-driven cybersecurity.

16.5.2 Emerging Trends and Innovations

As anomaly detection in SIEM systems continues to evolve, several emerging trends and innovations shape this critical cybersecurity area's future. Integrating advanced AI techniques, including user feedback in model development, and enhancing real-time monitoring capabilities are among the key advancements driving the progress of anomaly detection methodologies.

In recent years, the use of AI and ML in anomaly detection has gained significant attention. Researchers and developers are increasingly exploring the potential of advanced AI algorithms to detect complex patterns in emerging cyber threats, recognize subtle deviations, and respond to evolving user behaviors more efficiently than traditional rule-based approaches. For instance, DL techniques, such as RNNs and long short-term memory (LSTM) models, have demonstrated exceptional accuracy in classifying network traffic and detecting anomalies. These AI models can capture time-dependent and sequential patterns in user activities, offering more precise and context-aware anomaly detection.

In addition to applying standalone AI models, hybrid approaches that combine the advantages of multiple techniques are gaining popularity. For example, when combined with traditional rule-based or signature-based detection methods, an AI-driven UBA system can utilize both strategies' strengths, resulting in more comprehensive and robust anomaly detection. Hybrid systems that integrate AI's pattern recognition abilities with the established rule sets and expertise embedded in SIEM systems can tackle a broader spectrum of security threats and adapt more effectively to emerging attack vectors.

Another emerging trend in anomaly detection is incorporating user insights and human-in-the-loop mechanisms. Researchers and security experts are exploring methods to combine the expertise of security analysts and contextual understanding to improve the accuracy and reliability of anomaly detection models. This human-machine collaboration can boost detection algorithms, reduce false positives, and integrate real-world insights that purely data-driven approaches may miss. By fostering this symbiotic relationship between humans and AI, SIEM systems can better adapt to the dynamic nature of user behavior and evolving security threats.

In addition, improving real-time monitoring and incident response is becoming increasingly significant in the SIEM domain. SIEM systems can now detect anomalies and react to them with unmatched speed and efficiency through advances in stream processing, in-memory computing, and edge computing. This near-instant analysis and decision-making significantly shorten the duration of cyber incidents, allowing security teams to implement pre-planned response protocols before new threats can escalate. Integrating real-time monitoring with predictive analytics and threat intelligence strengthens organizations' proactive security stance, enabling them to anticipate and counteract evolving cyber tactics attackers use.

The emerging trends and innovations in anomaly detection within SIEM systems highlight the continuous efforts to enhance critical security solutions' accuracy, adaptability, and responsiveness. As organizations navigate the challenges of the modern digital era, the ability to efficiently detect and mitigate security risks using advanced AI techniques, user-centric strategies, and real-time monitoring will be crucial for safeguarding strategic assets and maintaining the resilience of IT infrastructures.

16.6 CONCLUSION

A thorough examination of the shifting dynamics of SIEM systems and anomaly detection underscores the essential role of UBA in bolstering cybersecurity efforts. As organizations contend with the increasing volume, velocity, and complexity of security-related data, the demand for more sophisticated and adaptive threat detection approaches has become more urgent. Examining user behavior patterns and trends has emerged as a fundamental component of modern SIEM systems, enabling security teams to gain a more detailed and comprehensive understanding of potential security incidents. SIEM solutions can more effectively identify anomalies that may indicate the presence of malicious actors or insider threats by establishing a baseline for standard user activities. This proactive approach to security monitoring improves the accuracy of threat detection and enables organizations to mitigate risks before they escalate into significant breaches.

Integrating AI and ML techniques has further strengthened the capabilities of SIEM systems in the UBA domain. Advanced analytics methods now offer the potential to uncover complex patterns, detect subtle deviations, and adapt more effectively to changing user behaviors than traditional rule-based approaches. SIEM solutions that leverage the power of AI/ML can become more thoughtful, more responsive, and more effective in protecting corporate assets and ensuring the integrity of critical data and systems. As advancements in detecting anomalies from security data collected through AI/ML continue, the value of UBA is expected to grow exponentially. The precise identification and response to deviations from standard user behavior will be critical in defending against advanced cyber threats.

In this context, addressing the challenges and emerging trends outlined in the literature review will be essential. The accuracy and dependability of the data used in UBA will play a critical role, as the integrity of foundational data directly impacts the effectiveness of anomaly detection. Moreover, developing adaptive and explainable AI dynamics will be essential to increasing the trustworthiness and interpretability of these systems, mainly as they are utilized in high-risk decision-making mechanisms within cybersecurity.

The ethical implications of UBA, particularly concerning privacy and data protection, will require more careful consideration than is currently given to similar data protection practices, especially as these techniques become more widespread and their potential risks become more pronounced. Striking the right balance between security and individual privacy will be a delicate challenge that organizations must overcome by implementing privacy-preserving methods and resilient governance frameworks. Additionally, the ongoing evolution of cyber threats, driven by the continuous adaptation of tactics and techniques by attackers, will require the constant development and refinement of UBA methodologies. SIEM systems must be flexible and adaptable, capable of learning from new data and patterns, staying ahead of trends, and effectively protecting against emerging threats.

REFERENCES

1. M. Goldstein, S. Asanger, M. Reif, and A. Hutchison. 2013. Enhancing Security Event Management Systems with Unsupervised Anomaly Detection. *International Conference on Pattern Recognition Applications and Methods (ICPRAM)*, 530–538. https://doi.org/10.5220/0004230105300538
2. G. González-Granadillo, S. González-Zarzosa, and R. Diaz. 2021. Security information and event management (SIEM): Analysis, trends, and usage in critical infrastructures. *Sensors*, 21(14), 4759. https://doi.org/10.3390/S21144759
3. T. Anuradha, J. J. Sharmila, K. Kanimozhiraman, K. Kalaiselvi, and C. K. Shruthi, and J. A. A. 2024. Automatic categorization of emails into folders based on the content of the messages. *Proceedings of the 2024 Third International Conference on Intelligent Techniques in Control, Optimization and Signal Processing (INCOS)*, 1–6. https://doi.org/10.1109/INCOS59338.2024.10527463
4. T. Laue, T. Klecker, C. Kleiner, and K.-O. Detken. 2022. A SIEM architecture for advanced anomaly detection. *Open Journal of Big Data*, 6(1), 26–42. https://doi.org/10.25968/OPUS-2321

5. R. Zuech, T. M. Khoshgoftaar, and R. Wald. 2015. Intrusion detection and big heterogeneous data: A survey. *Journal of Big Data*, 2(1), 1–41. https://doi.org/10.1186/S40537-015-0013-4
6. I. Kotenko, D. Gaifulina, and I. Zelichenok. 2022. Systematic literature review of security event correlation methods. *IEEE Access*, 10, 43387–43420. https://doi.org/10.1109/ACCESS.2022.3168976
7. G. Desetty. 2024. Unveiling hidden threats with ML-powered user and entity behavior analytics (UEBA). *Turkish Journal of Computer and Mathematics Education (TURCOMAT)*, 15(1), 44–50. https://doi.org/10.61841/turcomat.v15i1.14394
8. T. Wu, H. He, X. Gu, Y. Peng, Y. Zhang, Y. Zhou, and S. Xu. 2013. An intelligent network user behavior analysis system based on collaborative Markov model and distributed data processing. *Proceedings of the 2013 IEEE 17th International Conference on Computer Supported Cooperative Work in Design*, 221–228. https://doi.org/10.1109/CSCWD.2013.6580966
9. Aakash Aluwala. 2024. AI-driven anomaly detection in network monitoring techniques and tools. *Journal of Artificial Intelligence & Cloud Computing*, 3(3), 1–6. https://doi.org/10.47363/JAICC/2024(3)310
10. D. Sugumaran, Y. M. Mahaboob John, J. S. Mary C, K. Joshi, G. Manikandan, and G. Jakka. 2023. Cyber defence based on artificial intelligence and neural network model in cybersecurity. *Proceedings of 8th IEEE International Conference on Science, Technology, Engineering and Mathematics (ICONSTEM 2023).* https://doi.org/10.1109/ICONSTEM56934.2023.10142590
11. S. Hemalatha, M. Mahalakshmi, V. Vignesh, M. Geethalakshmi, D. Balasubramanian, and J. A. A. 2023. Deep learning approaches for intrusion detection with emerging cybersecurity challenges. *Proceedings of the 2023 International Conference on Sustainable Communication Networks and Application (ICSCNA)*, 1522–1529. https://doi.org/10.1109/ICSCNA58489.2023.10370556
12. H. Oikawa, T. Nishida, R. Sakamoto, H. Matsutani, and M. Kondo. 2020. Fast semi-supervised anomaly detection of drivers' behavior using online sequential extreme learning machine. *Proceedings of the 2020 IEEE 23rd International Conference on Intelligent Transportation Systems (ITSC 2020).* https://doi.org/10.1109/ITSC45102.2020.9294659
13. P. S. Muhuri, P. Chatterjee, X. Yuan, K. Roy, and A. Esterline. 2020. Using a long short-term memory recurrent neural network (LSTM-RNN) to classify network attacks. *Information*, 11(5), 243. https://doi.org/10.3390/INFO11050243
14. S. O. Olabanji, Y. A. Marquis, C. S. Adigwe, A. Ajayi, T. O. Oladoyinbo, and O. O. Olaniyi. 2024. AI-driven cloud security: Examining the impact of user behavior analysis on threat detection. *Asian Journal of Research in Computer Science*, 17(3), 57–74. https://doi.org/10.9734/AJRCOS/2024/V17I3424
15. L. Benova, and L. Hudec. 2023. Using web server logs to identify and comprehend anomalous user activity. *Proceedings of the 17th International Conference on Telecommunications (ConTEL 2023).* https://doi.org/10.1109/CONTEL58387.2023.10199092
16. R. Rana, P. Bhambri, and Y. Chhabra. 2024. Evolution and the future of industrial engineering with the IoT and AI. In *Integration of AI-Based Manufacturing and Industrial Engineering Systems With the Internet of Things* (pp. 19–37) CRC Press.
17. P. Bhambri, and A. Khang. 2024. Ethical and privacy considerations in AEI deployment. In N. Kumar; S. K. Pal; P. Agarwal; J. Rosak-Szyrocka; & V. Jain (Eds.), *Human-Machine Collaboration and Emotional Intelligence in Industry 5.0* (pp. 386–404). IGI Global. https://doi.org/10.4018/979-8-3693-6806-0.ch021
18. P. Bhambri, and A. Khang. 2024. New theoretical paradigms in cognitive psychology. In N. Kumar; S. K. Pal; P. Agarwal; J. Rosak-Szyrocka; & V. Jain (Eds.), *Harnessing*

Artificial Emotional Intelligence for Improved Human-Computer Interactions (pp. 13–32). IGI Global. https://doi.org/10.4018/979-8-3693-2794-4.ch002

19. P. Bhambri, S. Rani, and P. K. Pareek. 2024. Paperless paradigm: Intelligent automation in document and record management. In D. Darwish (Ed.), Hyperautomation in Business and Society (pp. 164–183), IGI Global Publication, USA. https://www.igi-global.com/book/hyperautomation-business-society/335477
20. P. Bhambri, S. Rani, and P. K. Pareek. 2024. Financial innovations: Intelligent automation in finance and insurance sectors. In D. Darwish (Ed.), Hyperautomation in Business and Society (pp. 226–243), IGI Global Publication, USA. https://doi.org/10.4018/979-8-3693-3354-9.ch012
21. B. Nassif, M. A. Talib, Q. Nasir, and F. M. Dakalbab. 2021. Machine learning for anomaly detection: A systematic review. *IEEE Access*, 9, 78658–78700. https://doi.org/10.1109/ACCESS.2021.3083060
22. F. Zhou, G. Wang, K. Zhang, S. Liu, and T. Zhong. 2023. Semi-supervised anomaly detection via neural process. *IEEE Transactions on Knowledge and Data Engineering*, 35(10), 10423–10435. https://doi.org/10.1109/TKDE.2023.3266755
23. K. Choi, J. Yi, C. Park, and S. Yoon. 2021. Deep learning for anomaly detection in time-series data: Review, analysis, and guidelines. *IEEE Access*, 9, 120043–120065. https://doi.org/10.1109/ACCESS.2021.3107975
24. M. Carratù, V. Gallo, A. Pietrosanto, P. Sommella, G. Patrizi, A. Bartolini, L. Ciani, M. Catelani, and F. Grasso. 2023. Anomaly detection on industrial electrical systems using deep learning. *Conference Record - IEEE Instrumentation and Measurement Technology Conference (2023-May)*. https://doi.org/10.1109/I2MTC53148.2023.10175908
25. Mosavi, F. Sajedi Hosseini, B. Choubin, M. Goodarzi, A. A. Dineva, and E. Rafiei Sardooi. 2021. Ensemble boosting and bagging based machine learning models for groundwater potential prediction. *Water Resources Management*, 35, 23–37. https://doi.org/10.1007/s11269-020-02704-3
26. T. Zhang, D. Du, G. Zhao, Q. Gao, W. Zhao, and S. Zhang. 2018. Method and system for detecting anomalous user behaviors: An ensemble approach. *Proceedings of the International Conference on Software Engineering and Knowledge Engineering*, 263–269. https://doi.org/10.18293/SEKE2018-036
27. V. S. Stency. 2024. Redefining cyber defense: The evolution of threat detection with artificial intelligence. *Management and Commerce*, 5(2). https://doi.org/10.46632/rmc/5/2/7
28. S. Gadal, R. Mokhtar, M. Abdelhaq, R. Alsaqour, E. S. Ali, and R. Saeed. 2022. Machine learning-based anomaly detection using k-mean array and sequential minimal optimization. *Electronics*, 11(14), 2158. https://doi.org/10.3390/ELECTRONICS11142158
29. K. Suguna, and K. Nandhini. 2017. Frequent pattern mining of web log files working principles. *International Journal of Computer Applications*, 157(3), 975–8887.
30. R. Krueger, T. Tremel, and D. Thom. 2017. VESPa 2.0: Data-driven behavior models for visual analytics of movement sequences. *Proceedings of the 2017 International Symposium on Big Data Visual Analytics (BDVA 2017)*. https://doi.org/10.1109/BDVA.2017.8114626.
31. C. Guo, and J. Xie. 2024. Machine learning builds embedded interaction model to guide knocking behavior in 3-6 year olds. *Design for Inclusion*, 128, 1–10. https://doi.org/10.54941/AHFE1004802.
32. S. Zhou, and N. S. Hudin. 2024. Advancing e-commerce user purchase prediction: Integration of time-series attention with event-based timestamp encoding and graph neural network-enhanced user profiling. *PLoS One*, 19(4), e0299087. https://doi.org/10.1371/JOURNAL.PONE.0299087.
33. Kotagiri. 2023. Mastering fraudulent schemes: A unified framework for AI-driven US banking fraud detection and prevention. *International Transactions in Artificial Intelligence*, 7(7), 1–19. Retrieved from https://isjr.co.in/index.php/ITAI/article/view/197.

34. R. Rieke, M. Zhdanova, J. Repp, R. Giot, and C. Gaber. 2013. Fraud detection in mobile payments utilizing process behavior analysis. *Proceedings of the 2013 International Conference on Availability, Reliability and Security (ARES 2013)*, 662–669. https://doi.org/10.1109/ARES.2013.87
35. D. Rimakka, and R. A. Aras. 2023. User segmentation based on purchasing habits and preferences on the amazon platform using k-means clustering. *Inspiration: Jurnal Teknologi Informasi Dan Komunikasi*, 13(2), 103–109. https://doi.org/10.35585/INSPIR.V13I2.63
36. J. Fang, V. C. S. Lee, H. Ji, and H. Wang. 2024. Enhancing digital health services: A machine learning approach to personalized exercise goal setting. *Digital Health*, 10, 1–10. https://doi.org/10.1177/20552076241233247

17 AI-Driven Security System for Biometric Surveillance

Özgür Önday

17.1 INTRODUCTION

Considerable concerns have been raised regarding the potential adverse effects of artificial intelligence (AI) on privacy [1]. It is crucial to acknowledge that not all AI applications depend on personal data; therefore, some implementations may not pose any privacy issues. Nevertheless, utilizing extensive datasets for the training and validation of machine learning models can give rise to various complications. Privacy, a multifaceted concept that will be examined in greater detail later, is integral to the discussion surrounding ethics and privacy within the context of AI. A significant component of this discourse involves the apprehension that the deployment of AI technologies might violate data protection principles, which could result in harm to particular individuals or groups whose data is subjected to analysis by AI systems [2].

Concerns regarding privacy and ethics are pertinent to a wide array of digital technologies, including AI. In the absence of suitable safeguards, there exists a significant risk that personal data may be exploited in ways that contravene data protection principles or infringe upon legitimate privacy preferences. A crucial legal acknowledgment of the "right", the initial expression of a "right to privacy," grounded in legitimate privacy preferences, emerged in the nineteenth century [3]. Warren and Brandeis articulated the defined "right to be let alone," which was influenced by a significant technological advancement of that era: the capability to photograph individuals. This innovation prompted anxieties that had not been relevant when the process of capturing a person's likeness necessitated that they remain seated for prolonged durations before a painter [4].

The evolution of legislation and regulation related to data protection has mirrored technological advancements and the subsequent threats to privacy since the nineteenth century. The advent of electronic computers, which significantly improved data processing capabilities, sparked extensive academic discourse on this subject, ultimately leading to the establishment of principles termed fair information practices [5]. These principles, which originated in the United States in 1973, continue to play a vital role in contemporary discussions surrounding data protection.

Beginning in the 1970s and 1980s, these principles have significantly impacted both legislation and its content. The introduction of Directive 95/46/EC in 1995 at the European level established a unified framework that included particular data

DOI: 10.1201/9781003521020-17

protection principles. Subsequently, this directive was succeeded by the General Data Protection Regulation (GDPR) [6] which came into force in 2018.

Since AI is not the inaugural potential danger to privacy or ethical standards, it is imperative to investigate the reasons behind the frequent perception of AI technologies as significant ethical issues regarding privacy [7]. A key element of this comprehension lies in the capacity of machine learning to facilitate the creation of complex data classifications, which can subsequently be employed to categorize and profile individuals. This form of profiling may indeed serve as a deliberate objective in the use of AI, particularly when an organization aims to pinpoint prospective clients for targeted marketing initiatives [8].

The utilization of personal data by AI can significantly enhance surveillance capabilities beyond what was possible prior to the advent of AI technology. This includes the automated monitoring of individuals through their biometric information, such as facial recognition, which will be elaborated upon in the subsequent examples. While there may be valid justifications for the creation and implementation of such surveillance—along with morally commendable outcomes like the prevention of gender-based violence—it is essential to acknowledge that AI-driven surveillance can also lead to unintended consequences [9]. A primary challenge lies in the necessity to balance data protection, a moral imperative, against other competing moral values. This consideration is crucial from an ethical standpoint, particularly given that data protection is subject to stringent regulations, whereas other ethical concerns and potential moral benefits generally do not face equivalent levels of oversight. The ensuing cases of privacy infringements facilitated by AI serve to illustrate this argument [10].

17.2 INCIDENTS OF PRIVACY BREACHES RESULTING FROM ARTIFICIAL INTELLIGENCE

17.2.1 Case 1: Utilization of Personal Information by Authoritarian Governments

China has risen to prominence as a significant global force in the realm of AI development. The country skillfully leverages the vast amounts of data it collects from its populace, as demonstrated by its social credit scoring system. This system assesses trustworthiness scores for each individual based on a wide array of data points, which include social media interactions, local government records, and personal behaviors. To aggregate this information, various data platforms are utilized, forming what has been characterized as "a state surveillance infrastructure" [11]. Citizens who achieve high scores benefit from perks such as lower utility costs and improved booking conditions, whereas those with lower scores may experience the withdrawal of certain services [12]. Within the context of China, this system receives substantial backing, as Chinese citizens "interpret it through frames of benefit-generation and promoting honest dealings in society and the economy instead of privacy-violation" [13].

Every state gathers data regarding its citizens for various objectives. Certain objectives may receive considerable public backing, such as the distribution of healthcare services or financial aid, whereas others, including tax collection, might encounter a more tepid reception [14]. Furthermore, authoritarian regimes have the capacity to exploit citizen information to reinforce their power structures [15]. A pertinent example is China, where studies indicate that citizens often assess the system according to its perceived advantages [16].

It has been contended that China possesses robust data protection legislation. Nevertheless, these laws are not applicable to state entities [17], which means that government utilization of data for initiatives like social credit scoring remains unregulated. This stands in contrast to the European context, where data protection laws are obligatory for both governments and state organizations. The practice of social credit scoring is a subject of considerable debate; however, it bears similarities to methods such as "nudging" employed by democratic governments to promote healthy behaviors, such as quitting smoking or engaging in physical exercise [18].

The topics of social credit scoring and nudging continue to generate considerable discourse; nonetheless, there exist arguments supporting their adoption. In contrast, the employment of AI for the repression of citizens exceeds the implications of these practices. Reports suggest that China employs AI to surveil behaviors deemed suspicious, with a particular emphasis on religious expressions among its Uighur population [19]. Individuals within the Uighur community in the Xinjiang region endure extensive data collection and analysis, which evaluates not only their interactions with religious texts but also their residential addresses, movement patterns, pregnancy statuses, and various other personal details. The overarching goal of this data acquisition seems to be the augmentation of state authority over the Uighur populace. China's human rights record, particularly regarding the Uighurs, indicates that such uses of data and AI analysis are likely to result in heightened restrictions on freedoms and human rights [20]. By utilizing AI, authoritarian regimes may increasingly find it manageable to analyze large volumes of data, including social media information, thus streamlining the identification of content that may incite governmental response [21–24].

17.2.2 Case 2: Genetic Privacy

Numerous initiatives in genetics are recognized for their profound influence on medical progress, particularly through personalized medicine and the detection of hereditary conditions. A prominent illustration of this is the Saudi Human Genome Program (SHGP), launched by the King of Saudi Arabia in 2013 to achieve these aims [25]. Findings from research indicated, "90.7% of [Saudi] participants agreed that AI could be used in the SHGP" [25]. However, the same study also uncovered "a low level of knowledge … regarding sharing and privacy of genetic data" [25], underscoring a potential gap between the understanding of the benefits and the associated risks of AI-driven genetic research.

Genetic information offers profound insights into various medical conditions, along with potential risks and predispositions to diseases, surpassing the insights provided by other data types. Consequently, it bears resemblances to medical data and often falls under stricter data protection regulations, thereby categorizing it as a unique form of data within numerous legal frameworks. However, the implications and opportunities associated with genetic data extend beyond its applications in medicine. An individual's genetic profile can reveal details regarding their ancestry, heritage, and lineage [12, 26–29]. Thus, access to genetic information can yield both advantages and disadvantages, raising a multitude of ethical challenges. For example, while genomic databases may facilitate advancements in research related to cancer and rare diseases, the risk of reidentifying even anonymized datasets presents considerable privacy concerns for the families involved [30].

As the expenses associated with gene sequencing persist in their decline, it is reasonable to anticipate that genetic information will integrate into standard healthcare practices within the next ten years [31]. This development prompts inquiries regarding the governance, storage, and security of such data. To ensure the viability of this genetic information and to yield pertinent scientific or diagnostic insights, it necessitates the application of Big Data analytics methodologies, which are generally founded on various forms of AI [32].

In conjunction with the use of genetic information within the healthcare sector, a growing number of private organizations, such as 23andMe, Ancestry, and Veritas Genetics [33], are offering gene sequencing services in a commercial framework. This evolution raises further questions concerning data ownership and the security protocols implemented by these companies, as well as instilling concerns regarding the possible use of data should one of these entities experience bankruptcy or undergo acquisition [34–36].

Addressing ethical dilemmas can result in unexpected revelations, such as when a genetic assessment contradicts established familial relationships, revealing that an individual's ancestry is not as previously believed. While some may respond to this information with humor or mild embarrassment, in other instances—especially where lineage is vital to the validation of a social standing—the implications of such evidence can be significantly detrimental. It could be contended that these repercussions are inherent to genetic data and should be managed through suitable information and consent protocols. Nevertheless, genetic data inherently relates to multiple individuals. For instance, if a sibling undergoes genetic testing, many of the results will have implications for other family members. Should such an analysis indicate that a parent carries a gene associated with a particular disease, it is likely that the risk for other siblings to develop this disease would be heightened, despite their not having undergone genetic testing themselves. This scenario illustrates the potential conflicts that may arise from the possession and dissemination of such information.

The examination of genetic data through AI has the potential to yield significant medical insights. This premise underpins the operational framework of private gene-sequencing companies. These organizations operate on the belief that the aggregation of extensive genetic information, complemented by additional data supplied by their clients, will enable them to discern genetic patterns that may aid in the prediction or elucidation of diseases. Consequently, this approach paves the way for advancements in medical research and the discovery of cures, which could represent a highly profitable venture.

From an ethical perspective, this scenario poses considerable difficulties, primarily because the companies engaged in the data analysis are generally the main beneficiaries, whereas individual data subjects or contributors may only receive updates concerning the insights generated from their data inputs. Furthermore, there exists apprehension that these analyses might facilitate the forecasting of disease progressions without the ability to offer interventions or treatments [37]. This situation could force patients to face daunting choices based on complex probabilities, a challenge that most laypersons are ill-equipped to manage.

A significant concern is the occurrence of mission creep, which refers to the blurring of the original objective of data collection due to a shift in intent or the emergence of an entirely new purpose. An illustrative example of this phenomenon is the growing inclination among law enforcement agencies to obtain wider access to genetic data, thereby enabling the identification of offenders through techniques such as genetic fingerprinting. It is essential to understand that once data is digitized, it becomes increasingly difficult to manage. Moor (2000) elucidates this concept using the analogy of grease in an internal combustion engine; in a similar vein, data stored in an electronically accessible format proves remarkably challenging to eradicate. Just as lubricants can infiltrate unexpected areas within an engine, making removal efforts potentially futile, genetic information raises concerns regarding its future applications and currently unanticipated uses. Given the highly personal nature of such data, the implications could lead to substantial and unpredictable consequences.

The foundation of the Saudi case rests on the assumption that the sharing of genetic information will produce beneficial outcomes. Nonetheless, there exists an insufficient amount of data to demonstrate whether ethical dilemmas have arisen or are expected to arise. A major concern in this regard is the tendency for data to be easily compromised, indicating that postponing action until ethical issues have been thoroughly resolved may not be wise. It is unlikely that simply appearing in the presence of these issues will be adequate. By that time, the proverbial genie may have already escaped the bottle, and the "greased" data could become unmanageable.

17.2.3 Case 3: Biometric Surveillance

"Nijeer Parks represents the third individual known to have been apprehended for a crime he did not commit due to an inaccurate facial recognition match" [38]. Parks faced false allegations of theft and attempting to strike a police officer with his vehicle, despite being located 30 miles away at the time of the incident. "Facial recognition … [is] highly effective with white males, significantly less accurate with Black females, and even suboptimal with white females" [39]. This issue becomes especially concerning when "law enforcement places greater trust in facial recognition technology than in the individual" [39].

Biometric surveillance pertains to the utilization of data associated with human physiology for the accurate monitoring or tracking of individuals. A notable instance of this practice is the employment of facial recognition technology for the identification and tracking of a person. Broadly defined, biometric surveillance encompasses any direct observation of an individual, including those suspected of criminal activity. The primary rationale for incorporating biometric surveillance into the discourse surrounding privacy issues lies in the capacity of AI systems to significantly amplify the extent of such monitoring initiatives. In contrast to earlier times when a single observer could monitor only one person or a limited group, the emergence of machine learning, image recognition technologies, and the extensive use of closed-circuit television cameras facilitates comprehensive surveillance within communities. While automatic facial recognition and tracking constitute only one facet of biometric surveillance, they exemplify the most advanced iteration and elicit substantial public concern regarding privacy, as evidenced by the aforementioned case.

Biometric surveillance is considered ethically contentious for a variety of reasons. Its implementation can occur without the knowledge of the individuals whose data is being collected, thereby fostering both the potential and perception of omnipresent surveillance. While some may perceive such extensive monitoring as beneficial for enhancing security and decreasing criminal activity, it has been compellingly argued that exposure to such surveillance can result in considerable harm. Brown [40], referencing Giddens [41] and other scholars, posits that individuals require a "protective cocoon" to shield themselves from external observation. This protective environment is essential for cultivating a sense of "ontological security," which is vital for psychological and mental well-being. In light of this reasoning, the ethical concerns surrounding pervasive surveillance primarily stem from the psychological repercussions it may inflict simply through its existence. Such surveillance practices can induce self-censorship and contribute to what is termed "social cooling" [42], which refers to alterations in social interactions prompted by the fear of potential repercussions. The introduction of AI-driven large-scale biometric surveillance is likely to exacerbate this phenomenon.

17.3 WHAT IS THE IMPORTANCE OF EXAMINING THE ETHICS OF ARTIFICIAL INTELLIGENCE?

AIs are fundamentally constructs—entities devoid of ethical considerations, or, in other terms, they function within a domain of ethical neutrality. It is crucial that we refrain from ascribing agency to these constructs, especially when engaging in discussions about the potential for conferring legal personhood upon AIs. The discourse surrounding ethics pertains specifically to human ethics—the moral frameworks of those who design, develop, implement, and utilize AI systems. The inquiry into ethical dilemmas has been a topic of philosophical investigation since at least the era of Aristotle, whose text Nicomachean Ethics was composed in 350 BCE, and has also been extensively examined in ancient literature such as the Hebrew Bible, the Upanishads, and various other historical documents. Recognizing ethical challenges within society, particularly concerning information technology, is certainly not a contemporary issue.

What makes the exploration of the ethical implications surrounding AI crucial? This question is driven not only by legitimate philosophical concerns but also by practical factors that demand consideration. When unethical practices undermine trust within a system, the potential benefits may be lost. Historical examples highlight this assertion. For instance, although no scientific evidence suggests inherent problems with genetically modified foods, public confidence in these products notably diminished between 2003 and 2004, particularly within populations in Britain and Europe, leading to their widespread rejection. This erosion of trust in the UK occurred despite a declaration by Margaret Beckett MP, who was then the Secretary of State for Environment, Food and Rural Affairs, stating that "There was no scientific case for ruling out all GM crops or products." Additionally, the discredited claims made by former physician Andrew Wakefield, which falsely associated the MMR vaccine with autism, have resulted in declining vaccination rates for measles, mumps, and rubella in several nations, some of which are now facing alarmingly low immunization levels, thereby contributing to a rise in measles-related deaths, particularly among children.

As stated by the EU AI High Level Expert Group, "Trustworthiness is essential for individuals and societies to create, implement, and utilize AI systems. If AI systems—and the individuals responsible for them—do not clearly demonstrate their trustworthiness, adverse outcomes may arise, which could hinder their acceptance and obstruct the achievement of the significant social and economic advantages they have the potential to deliver.

17.3.1 Contemporary Ethical Frameworks

An effective method to showcase ethical principles and cultivate trust is through the publication of an ethical charter. Numerous ethical charters for AI currently exist in the market. There is, in fact, a potential risk that corporations may engage in "charter shopping" until they identify a framework that aligns with their objectives. The OECD has established and disseminated The Principles of AI, which serve as the foundation for the regulation and the safe, appropriate advancement of AI. A total of 44 governments, encompassing all members of the G20 as well as several nations outside of the OECD, have endorsed these principles. While they do not possess legal authority, their impact is significant. The principles are outlined below.

These principles are commendable and ought to guide regulation while serving as the foundation for all individuals involved in AI, whether they are developers or users. Nevertheless, this is not always the case.

17.3.2 Principal Ethical Risks and Related AI Ethics Cases

The principal ethical issues and potentially associated risks and cases related to those risks which we will discuss are listed below.

17.3.2.1 Bias

What accounts for the bias present in AI systems? The answer lies in our own biases, which exist within each of us. While we recognize some of these biases, others remain unnoticed. It is important to note that not all biases are detrimental. We often gravitate towards newspapers that align with our perspectives and favor individuals who share similarities with us. Additionally, many individuals exhibit biases against those who differ from themselves, including foreigners, immigrants, and individuals of varying colors or religions. Numerous other instances of bias can also be identified.

Bias is integrated into AI systems through various mechanisms. A notable example is the demographic composition of AI engineers, which predominantly consists of young, white males. This group may not recognize that the systems they develop exhibit biases, potentially functioning more effectively for white males compared to black females. AI systems inherit biases from the prejudiced data present in their training datasets. If the data mirrors the biases found within the broader population or specific segments thereof, it becomes biased, resulting in AI systems that perpetuate these biases during their operational phases. Due to their rapid deployment and widespread use, such biases are disseminated quickly and extensively.

Is this significant? Not necessarily. In the context of machine translation, the focus lies on the quality of the output in the target language. Gender bias can also infiltrate this process. Turkish employs gender-neutral pronouns. Certain automated translation systems render "o bir mühendis" as "he is an engineer," "o bir doktor" as "he is a doctor," "o bir hemşire" as "she is a nurse," and "o bir aşçı" as "she is a cook." This may be viewed as more offensive than fundamentally problematic.

Biases associated with gender and race are of considerable significance in numerous contexts. A prominent example can be seen in the difficulties encountered by facial recognition technology (FRT), which often grapples with these biases. In the United Kingdom, law enforcement agencies such as the Metropolitan Police and South Wales Police have extensively integrated FRT into their operations. Nevertheless, the use of this technology remains contentious due to substantial inaccuracies, particularly regarding certain racial demographics, as well as concerns about privacy. A relevant instance that highlights this issue is the legal action taken by Ed Bridges against the South Wales Police in the High Court of England. Although Bridges did not succeed in his lawsuit, the court acknowledged, among other conclusions, that the existing legal framework is insufficient. To guarantee the appropriate and non-arbitrary implementation of FRT, it is crucial to establish adequate measures. This claim was contested by the Information Commissioner, who expressed apprehensions about the sufficiency of the current legal structure. Recently, Lord Clement-Jones, the Chairman of the Lords Select Committee responsible for the report titled "AI in the UK: Ready, Willing and Able?" has introduced a Private Member's Bill in the House of Lords. The purpose of this legislation is to designate the application of Facial Recognition Technology (FRT) for overt surveillance in public spaces as a criminal offense, while also requiring the government to undertake a review of its usage within one year. While it is uncommon for such measures to be formalized into law, this specific proposal has the potential to compel the government to act.

Numerous additional instances of gender and racial bias exist. In 2019, Amazon discontinued a project involving an AI-driven human resources system due to its reinforcement of male gender bias, which resulted from being trained on the company's recruitment records.

In various American states, judges utilize an AI system known as the Correctional Offender Management Profiling for Alternative Sanctions tool (Compass) to assess the appropriateness of granting bail to accused individuals. In Wisconsin, this system also aids judges in determining the duration of sentences. The tool is based on several indicators, notably excluding race from its calculations. However, it does consider the residential location of the alleged offender, and due to the racial demographics prevalent in American urban areas, geography inadvertently serves as a proxy for racial factors. Consequently, a black individual facing accusations, who may very well not pose a risk of re-offending based on their history, is statistically more prone to being denied bail compared to a white individual with a similar record. The algorithm known as Compass is proprietary, and its developer, Equitant, refuses to disclose its operational mechanisms, claiming it as a trade secret. It is possible that they are unable to reveal the basis for its conclusions due to a lack of understanding of the process themselves.

A notable application of facial recognition technology (FRT) can be observed in the actions of the Chinese Communist Party within Xinjiang, a region in Western China, where it has been utilized to identify and detain approximately 1.8 million Uighur individuals in what are termed re-education camps. The precision of this technology is not especially pertinent, considering that Uighurs, a Turkic ethnic group, possess physical attributes that significantly contrast with those of the Han Chinese. This policy has attracted considerable scrutiny from Western media and has prompted American sanctions against the companies supplying the FRT; nonetheless, these actions have yet to yield any significant effect on the ongoing implementation of the policy.

17.3.2.2 Explainability

In contrast to conventional software applications, neural network-based AIs are unable to elucidate the rationale behind their conclusions, nor can their creators provide such explanations. If a financial institution employs AI to assess your eligibility for a loan and subsequently denies your application without offering a justification, this situation is both unjust and unethical, as it leaves you unaware of the necessary criteria for qualification. The same principle applies in the realm of insurance; if an insurer refuses coverage without providing an explanation, it raises similar concerns. The issue of explainability stands out as a distinct ethical challenge associated with AI, while discussions regarding other ethical dilemmas often trace their roots back to the philosophies of Aristotle.

Upon revisiting Compass, the case of Loomis v. Wisconsin centered around a six-year sentence handed down to Eric Loomis for his involvement in a drive-by shooting, which was partly influenced by the "high risk" score generated by Compass. Loomis contested his sentence, asserting that he had not been given the opportunity to evaluate the algorithm. The state Supreme Court ultimately ruled in favor of the state, concluding that awareness of the algorithm's output provided an adequate level of transparency. This situation raises significant ethical concerns, as it is a fundamental principle of Common Law that judges are required to provide a justification for their rulings.

Significant efforts are underway to address this issue. The most promising method seems to involve an external audit strategy, akin to that of a financial audit. However, at this moment, a comprehensive solution has yet to be established.

17.3.2.3 Liability for Failure

What occurs when situations take a turn for the worse? This inquiry is most commonly associated with automated vehicles (AVs), including self-driving cars and commercial transport. However, this question extends beyond just these modes of transportation.

Regarding automated vehicles, the question arises: should liability rest with the AV itself or the 'driver'? In the United Kingdom, a definitive answer exists.

According to the Automated & Electrical Vehicles Act 2018 (AEVA), liability falls upon the insurer—or the owner in the absence of insurance—if the AV is responsible for causing damage, injury, or death. Subsequently, the insurer possesses the right to seek recompense from either the vehicle's manufacturer or the developer of the defective component. Typically, the injured party requires prompt compensation, while insurers are able to wait for the completion of the post-crash investigation to ascertain the underlying cause before recovering their costs when appropriate.

Insurance coverage could be rendered void if the owner has tampered with the system or failed to update critical safety-related software. Nevertheless, what are the consequences if the software of the vehicle has been breached through hacking? Additionally, in the context of a fleet of vehicles, who is accountable for guaranteeing that software updates are conducted? In situations where the insurer refuses coverage, who is liable? Is it the individual 'driver'? This domain raises a multitude of unresolved questions, and presently, there exists an absence of case law to elucidate these issues.

Numerous applications of AI present similar concerns regarding liability in the event of failures. For example, should an AI-operated medical device that has been implanted within a patient's body malfunction, the question arises as to who bears responsibility. Is it the surgeon who performed the implantation, the hospital, or another party? What of manufacturers? How do off-road vehicles, such as tractors, fit into this discussion? The list continues to expand. Each of these liability issues presents both ethical and legal implications. There exists a lack of case law on the matter. In the context of human resource issues within the UK, the Equality Act 2010 will have significant implications, and comparable legislation exists in other nations. In other circumstances, it is probable that suppliers will attempt to evade the repercussions of failure through contractual means, although typically they cannot absolve themselves of responsibility in cases involving death or injury.

Two critical facets of these risks that cannot be overstated include the necessity of safeguarding such AIs against cybersecurity failures and the need for thorough testing. For example, should several automated vehicles fall victim to hacking, they could be transformed into formidable weapons—cars, buses, and trucks have previously been utilized as instruments of harm in numerous urban areas when operated by human terrorists.

Testing AI presents a notably challenging endeavor. This difficulty arises primarily from the fact that AI systems usually consist of numerous software components that may not have been evaluated in conjunction, despite the individual components having undergone testing. Additionally, these systems frequently incorporate publicly accessible open-source code, the testing status of which may remain ambiguous. Furthermore, the variety of use cases necessitating the development of test scenarios can be extensive, particularly in the case of autonomous vehicles. When manufacturers assert that an automated vehicle has been operated for several million miles, this statement does not provide any insight into the efficacy of the testing process. If an AI system is subjected to insufficient or flawed testing, its deployment would be deemed unethical.

17.3.2.4 Harmlessness

AIs ought to be designed to be non-threatening. In his 1942 work, I Robot, Isaac Asimov established his Three Laws of Robotics, predating Turing's contributions. The initial law stated, "A robot may not injure a human being or, through inaction, allow a human being to come to harm." Subsequently, he introduced a fourth law: "A robot may not harm humanity, or, by inaction, allow humanity to come to harm."

Currently, there exist two primary methods by which these laws are being violated. The first involves the harmful application of AI. AI, akin to any other instrument, possesses an ethically neutral nature and is subject to dual usage. Just as a knife may serve to slice a cake or inflict harm, it can be employed for both beneficial and detrimental purposes. Thus, one must question the rationale behind assuming such risks. Committing burglary may now be achieved through technological means, allowing individuals to steal "from the comfort of your own home," as the clichéd marketing slogan suggests. The implementation of AI can enhance the efficiency of such criminal activities by lowering expenses and amplifying the frequency of spear phishing attacks. In this process, comprehensive details regarding the victim, obtained from various sources, are utilized to establish trust, thereby facilitating the importation of a virus or trojan. The collection of such information is both costly and time-consuming; however, the implementation of AI significantly diminishes both the expenses and the required effort.

A comprehensive analysis of various threats linked to the malicious application of AI was presented in a report published in 2018. This document underscored the potential for AI systems to launch attacks against other AI systems, as well as the capacity of artificial intelligence to facilitate swifter, more cost-effective, and more frequent assaults on an array of systems, which include automated vehicles and utility services. Additionally, it emphasized the necessity for strategic planning and the formulation of countermeasures. Currently, AI is extensively employed in both offensive and defensive operations, including those that incorporate AI systems.

The alternative method involves the utilization of Lethal Autonomous Weapons Systems (LAWS). Drones serve as valuable instruments, applicable in various contexts such as crop assessment, delivering assistance to individuals affected by disasters, and locating submerged aircraft. Nonetheless, their potential applications extend beyond these humanitarian uses. When weaponized and arranged in swarms, these systems transform into a notably more potent offensive weaponry. The deployment of autonomous drones, specifically those guided by AI, raises considerable ethical concerns and poses legal challenges under international law. Such actions are explicitly forbidden. According to the Geneva Convention, a human operator possesses the ability to respond to evolving situations concerning the target—such as relocating to a hospital or mingling with a group of children—and subsequently abort the mission. Is it feasible for an AI to execute such a nuanced evaluation?

The British Government has resolved against the development or deployment of lethal autonomous weapon systems (LAWS), regardless of whether the adversary chooses to pursue such technologies. However, this policy may be subject to change.

Consider the possibility that the armed forces of the United States, Russia, China, and potentially Israel are actively engaged in the development of such technologies and may be ready to deploy them. In light of the effective utilization of human-operated drones—not only by the United States Air Force in regions such as Pakistan and Afghanistan but also by rival militias involved in the civil wars of Libya and Syria—what is the anticipated timeline for the emergence of lethal autonomous weapon systems (LAWS)?

17.3.2.5 The Ethical Use of Data

The functionality of all AI applications is fundamentally reliant on extensive datasets, which consequently raises concerns regarding privacy. A notable tension exists, as well as a trade-off, between the utilization of medical data for societal benefit and the safeguarding of individual privacy. Data within a dataset can be easily anonymized by eliminating identifiable information such as names, addresses, and other distinguishing features. Nevertheless, research indicates that when two datasets of a comparable nature are present, an overlap of less than 20% can facilitate the de-anonymization of the information contained within them.

As early as 2009, Netflix uncovered this issue when it published anonymized film reviews written by its subscribers. By cross-referencing these excerpts with reviews found on another platform, data analysts demonstrated that they could pinpoint individual subscribers and their viewing histories. This led to a lawsuit for breach of privacy filed by a gay customer, which resulted in a settlement by Netflix.

Solutions are emerging to address these challenges. Synthetic data is created through the artificial generation process, typically involving the application of an algorithm that introduces noise to real-world data, thereby forming a new dataset devoid of personal information. The resultant dataset retains the statistical characteristics of the original data while avoiding direct replication. This newly constructed dataset can subsequently be utilized for training the AI or as the foundational data on which it will function.

Additional issues arise regarding the unethical utilization of data by companies. For example, the 23andMe company employs Crispr technology to offer kits that enable users to send saliva samples for genetic analysis, aimed at uncovering insights into both their ancestry and potential future health. This practice prompts concerns about the potential loss of control over deeply personal information and the possibility of uncovering distressing family secrets. The field of genetics has a troubling history marked by abuses committed by eugenicists, who were fixated on the notion of eliminating "inferior" intelligence or maintaining racial "purity." As the advent of Crispr technology paves the way for the editing of embryos, the handling of genetic data necessitates heightened caution. While 23andMe has never claimed to identify intelligence within individuals' genetic makeup, companies like Gene Plaza permit users to upload their genetic information and assert their relative intelligence levels. Concurrently, members of the alt-right in the United States have publicly shared their 23andMe results on social media, taking pride in their white European heritage.

From a societal viewpoint, a significant issue emerges within this framework: the risk that the benefits of AI may be distributed unequally, favoring a privileged group of affluent individuals with technological expertise while disadvantaging the wider population. This concern has been articulated by Prof. Shoshana Zuboff. In her work, The Age of Surveillance Capitalism, she asserts that prominent information technology firms, including Google, leverage AI to create a new variant of capitalism, which she designates as 'surveillance Capitalism.' This paradigm is marked by individuals who, frequently without awareness, surrender their rights to personal data. She posits that individuals are generally predisposed to exchange their private information for perceived benefits, such as convenience, assistance with navigation, and connections with acquaintances and information. The ability to actively shape our futures is fundamentally compromised by predictive, data-driven AI systems. Interaction with the construct of surveillance capitalism and acquiescing to its demands for increasingly invasive access to everyday life involves more than simply sharing information; it requires the delegation of one's entire life course and the shaping of one's trajectory to market regulation and oversight. This situation is reminiscent of Pokémon Go participants being directed, illuminated by their devices, into establishments they had not previously contemplated visiting, as the company auctions virtual spaces to the highest bidders, including well-known brands like McDonald's and Starbucks.

17.3.2.6 Should AI's Have Legal Personality?

The shortcomings of AIs that do not fulfill expectations or lead to incidents present a substantial inquiry. The matter of liability stemming from such failures raises the question of whether AIs should be granted legal personality. In 2017, Saudi Arabia awarded legal personality to Sophia, a robot characterized as "female." Nevertheless, this precedent has not yet been embraced by any other jurisdiction.

In legal systems influenced by Common Law, Roman law, and various others, the framework recognizes two distinct classifications of entities: natural persons, which denote real individuals, and corporate persons, which include limited companies, partnerships, and governmental bodies. The latter category possesses legal personality, allowing them to participate in legal proceedings as both plaintiffs and defendants. The underlying principle is that these legal entities are ultimately directed by natural persons. Thus, one may question whether it is suitable to extend legal personality to machines, robots, and AIs.

The circumstances surrounding animals offer significant parallels. In 2015, a New York court examined the issue of legal personhood for chimpanzees in the case of Nonhuman Rights Project, Inc. v. Stanley. In this case, the Nonhuman Rights Project, an organization that operates independently of the government, submitted a writ of habeas corpus seeking the liberation of Hercules and Leo, two chimpanzees confined within a laboratory at Stony Brook University.

The NGO argued that the legislation fails to sufficiently define the term "person" within the context of habeas corpus. Given the lack of legal precedent regarding the application of habeas corpus to entities that are not human, the court decided to consider the issue. An amicus curiae brief was submitted by the Center for the Study of The Great Ideas, which contended that, under New York law, legal personality is exclusively assigned to humans and specific public and private entities, while also examining the notion of legal personality itself. The rationale for bestowing rights upon non-human entities is founded on their possession of human-like characteristics. Therefore, the notion of personhood should not be broadened to encompass animals. In its decision, the court rejected the classification of chimpanzees as persons, concluding that they do not possess the capacity to bear legal responsibility for their actions and are incapable of fulfilling obligations. Moreover, the court underscored that it is the ability to hold rights and obligations—not merely the physical resemblance to humans—that is essential for acknowledging a being's legal personality.

For precisely the same reasons, it is untenable to assert that a robot endowed with AI possesses free will that could result in the execution of forbidden actions to fulfill its own objectives. Consequently, it cannot be attributed any level of culpability, including negligence or recklessness. Furthermore, it is not feasible to hold such a robot accountable for damages arising from its mistakes, as exemplified by incidents involving autonomous vehicles or errors committed by surgical robots.

In conclusion, bestowing legal personhood is ill-advised: "My AI has caused you harm. How unfortunate; feel free to initiate legal action against it." Corporations possess capital and are thus capable of compensating damages in the event of a legal loss. When engaging with a company that has limited financial resources, one must exercise caution. Conversely, robots lack any financial assets.

17.4 CONCLUSION

This chapter has shed light on the field of AI ethics through the exploration of case studies that underscore the ethical challenges linked to AI, along with an array of strategies and tools aimed at addressing these concerns. It is crucial to recognize that AI ethics seldom presents clear-cut or unequivocal situations. While certain instances distinctly demonstrate unethical conduct, they often revolve around the reliability of the technology in question. For instance, it is vital that AI-driven robots do not introduce health, safety, or security hazards to users, such as the risk of a passenger's fatality in a self-driving car or vulnerabilities within a smart-home system that could facilitate a man-in-the-middle attack. More intricate are those scenarios where evaluating the ethical pros and cons does not lead to an immediate determination of the most appropriate course of action. A relevant example is the deployment of robots in elder care, which alleviates the strain on overburdened staff while concurrently reducing vital human interaction.

Upon examining the various case studies presented in this chapter across different example domains, several general observations can be made. The initial observation pertains to the context in which AI is applied. The case studies have been designed to be rooted in realistic and existing AI technologies, particularly those pertaining to currently pertinent machine learning. The application and integration of machine learning tools into broader systems is nearly always associated with ethical concerns. Consequently, these concerns do not center on AI itself, but rather on how AI is utilized and the implications that arise from such usage. For example, both the case regarding unfair dismissal and the case concerning gender bias pertain to the utilization of AI. The termination of employees without any human involvement in the process, as well as the training of AI systems on CVs that exhibit gender bias, illustrate issues related to AI application. It is important to clarify that this does not imply AI operates as an ethically neutral instrument; instead, it emphasizes that the wider context of AI usage—including prevailing moral values, societal practices, and formal regulations—must be considered in any ethical reflection and analysis.

This prompts an inquiry: in what ways do cases concerning AI ethics diverge from other instances of technology ethics? At a preliminary level, it is likely accurate to assert that they typically do not exhibit significant differences. Numerous ethical case studies presented herein are not intrinsically innovative, nor do we introduce matters that have not been previously contemplated. For example, the digital divide has been a topic of discussion and debate for many years. Nevertheless, the implementation of AI has the potential to intensify pre-existing issues and amplify established challenges.

In its most common manifestation, AI exhibits unique traits that set it apart from other technological innovations, especially in the realm of machine learning. Its ability to classify various phenomena allows it to make or suggest decisions. For example, an autonomous vehicle can identify the need to stop by recognizing an object as an obstacle on the road, while a law enforcement system may label an individual as likely to reoffend, irrespective of previous rehabilitation attempts. Such examples are often interpreted as signs of AI autonomy. However, it is essential to understand that this autonomy is not intrinsic. The functional capabilities of a machine learning model are influenced not only by its design but also by its integration into a broader socio-technical context, which may or may not permit these classifications to affect social realities. Therefore, autonomy should be viewed not as an inherent quality of AI but as a consequence of its application and incorporation within various systems.

What overarching conclusions can be derived from this compilation of cases involving ethically contentious applications of AI, along with the diverse interpretations of these challenges and the suggested responses? A significant point to emphasize is that ethical inquiries often arise from human interactions. Although the incorporation of AI into these interactions may alter the specific ethical dilemmas faced, it will not eliminate all ethical concerns, nor will it introduce entirely unforeseen issues. Engaging with ethical considerations regarding the actions we can and should undertake, the rationale behind our decisions, and how we assess the ethical implications of our actions and their outcomes is an intrinsic aspect of human nature. While Immanuel Kant asserted that good will is the sole ethical entity in existence, the mere possession of good will is inadequate when faced with complex consequences that may not be immediately apparent. For example, in the context of AI for Good, the most vulnerable communities might experience exacerbated challenges from climate change rather than receiving assistance due to the deployment of AI-driven systems. The situations faced by small-scale farmers in Brazil and Zimbabwe exemplify this point effectively; these farmers were refused credit access by bank managers who relied on seasonal climate forecasts to tackle the difficulties brought about by climate change. Likewise, seasonal workers in Peru encountered early terminations of employment based on predictions made from seasonal climate forecasts. In such instances, it is essential to avoid helicopter research aimed at assisting vulnerable groups in resource-constrained environments, as local partners typically possess a deeper comprehension of the impacts on these communities.

In practice, our examination of the responses to various cases has revealed a considerable array of initiatives that hold promise for enhancing our comprehension of AI ethics and tackling ethical dilemmas. These initiatives encompass individual awareness, assessments of AI impact, ethics-by-design methodologies, engagement with local partners in resource-constrained environments, and technical solutions related to AI explainability. Additionally, they include legal remedies, liability frameworks, and the establishment of new regulatory bodies. While none of these solutions serves as a comprehensive remedy capable of addressing the entirety of AI ethics independently, their collective implementation presents a viable opportunity to mitigate significant ethical challenges and avert potentially catastrophic outcomes. As systems ethics, AI ethics provides a framework of ethical responses. A principal challenge we currently encounter is the effective orchestration of existing ethical strategies to maximize societal benefits.

Addressing the challenges associated with AI ethics is a complex endeavor, and it is unrealistic to anticipate a resolution for all ethical dilemmas. It is essential to acknowledge that engaging with ethical considerations is inherently a human activity, and the incorporation of technology can introduce additional layers of complexity to established ethical queries. Moreover, it is important to understand that AI ethics frequently overlaps with the ethics of technology broadly, as well as with ethical concerns pertinent to both digital and non-digital technologies. Nonetheless, it is crucial to recognize the unique characteristics of AI ethics that warrant careful examination.

In this chapter, our objective has been to promote contemplation regarding various intriguing instances related to AI privacy and ethics. It is our aspiration that the reader has acquired valuable insights into addressing these challenges and recognizes the necessity of reflecting on and diligently pursuing the ethical implications of technology, as long as humanity continues to engage with it.

REFERENCES

1. EDPS. 2020. EDPS opinion on the European Commission's white paper on artificial intelligence: a European approach to excellence and trust (Opinion 4/2020). European Data Protection Supervisor, Brussels. https://edps.europa.eu/data-protection/our-work/publications/opinions/edps-opinion-european-commissions-white-paper_en. Accessed 29 September 2024.
2. Anuradha T., Jasphin Jeni Sharmila P., Kanimozhiraman, K. Kalaiselvi, C. K. Shruthi and Jose A. A., "Automatic Categorization of Emails into Folders Based on the Content of the Messages," 2024 Third International Conference on Intelligent Techniques in Control, Optimization and Signal Processing (INCOS), Krishnankoil, Virudhunagar district, Tamil Nadu, India, 2024, pp. 1–6.
3. Veale M, Binns R, Edwards L. 2018. "Algorithms that remember: model inversion attacks and data protection law". Phil Trans R Soc A 376:20180083. https://doi.org/10.1098/rsta.2018.0083.
4. Becker HA, Vanclay F (Eds) 2003. The international handbook of social impact assessment: conceptual and methodological advances. Edward Elgar Publishing, Cheltenham.
5. Sen A. 1988. On ethics and economics, 1st edn. Wiley-Blackwell, Oxford.
6. European Parliament, Council of the EU. 2016. "regulation (EU) 2016/679 of the European parliament and of the council of 27 April 2016 on the protection of natural persons with regard to the processing of personal data and on the free movement of such data, and repealing directive 95/46/EC (General data protection regulation)". Official Journal of European Union L119(11):1–88. https://eur-lex.europa.eu/legal-content/EN/TXT/PDF/?uri=CELEX:32016R0679&from=EN. Accessed 30 September 2024.
7. Buttarelli G. 2017. "Privacy matters: updating human rights for the digital society". Health Technol 7:325–328. https://doi.org/10.1007/s12553-017-0198-y.
8. Clarke R. 2009. "Privacy impact assessment: its origins and development". Comput Law Secur Rev 25:123–135. https://doi.org/10.1016/j.clsr.2009.02.002.

9. Culnan M. 1993. "How did they get my name?" "An exploratory investigation of consumer attitudes toward secondary information use". MIS Q 17(3):341–363. https://doi.org/10.2307/249775.
10. Dignum V. 2019. Responsible artificial intelligence: how to develop and use AI in a responsible way. Springer Nature Switzerland AG, Cham.
11. Liang F, Das V, Kostyuk N, Hussain MM. 2018. "Constructing a data-driven society: China's social credit system as a state surveillance infrastructure". Policy Internet 10:415–453. https://doi.org/10.1002/poi3.183
12. Raso FA, Hilligoss H, Krishnamurthy V et al. 2018. "Artificial intelligence & human rights: opportunities & risks". Berkman Klein Center Research Publication No. 2018-6. http://dx.doi.org/10.2139/ssrn.3259344
13. Kostka G. 2019. "China's social credit systems and public opinion: explaining high levels of approval". New Media Soc 21:1565–1593. https://doi.org/10.1177/1461444819826402.
14. ECP. 2019. "Artificial intelligence impact assessment. ECP Platform for the Information Society, The Hague". https://ecp.nl/wp-content/uploads/2019/01/Artificial-Intelligence-Impact-Assessment-English.pdf. Accessed 30 September 2024.
15. Liu C. 2019. Multiple social credit systems in China. Social Science Research Network, Rochester.
16. Finn RL, Wright D, Friedewald M. 2013. "Seven types of privacy". In: Gutwirth S, Leenes R, de Hert P, and Poullet Y (Eds) "European Data protection: coming of age". Springer, Dordrecht, pp 3–32.
17. Gal D. 2020. China's approach to AI ethics. In: Elliott H (Ed) The AI powered state: China's approach to public sector innovation. Nesta, London, pp 53–62.
18. Benartzi S, Besears J, Mlikman K et al. 2017. "Governments are trying to nudge us into better behavior. Is it working?" The Washington Post, 11 Aug. https://www.washingtonpost.com/news/wonk/wp/2017/08/11/governments-are-trying-to-nudge-us-into-better-behavior-is-it-working/. Accessed 29 September 2024.
19. Andersen R. 2020. "The panopticon is already here". The Atlantic, Sept. https://www.theatlantic.com/magazine/archive/2020/09/china-ai-surveillance/614197/. Accessed 29 September 2024.
20. Hartley N, Wood C. 2005. "Public participation in environmental impact assessment: implementing the Aarhus convention". Environ Impact Assess Rev 25:319–340. https://doi.org/10.1016/j.eiar.2004.12.002.
21. ICO. 2021. AI and data protection risk toolkit beta. Information Commissioner's Office, Wilmslow.
22. IEEE. 2020. 7010-2020: "IEEE recommended practice for assessing the impact of autonomous and intelligent systems on human well-being. IEEE Standards Association, Piscataway. https://doi.org/10.1109/IEEESTD.2020.9084219.
23. Ivanova Y 2020. "The data protection impact assessment as a tool to enforce non-discriminatory AI". In: Antunes L, Naldi M, Italiano GF et al (eds) Privacy technologies and policy. 8th annual privacy forum, APF 2020, Lisbon, Portugal, 22–23 Oct. Springer Nature, Switzerland, Cham, pp 3–24. https://doi.org/10.1007/978-3-030-55196-4_1.
24. Kaplan A, Haenlein M. 2019. "Siri, Siri, in my hand: who's the fairest in the land? On the interpretations, illustrations, and implications of artificial intelligence". Bus Horiz 62:15–25. https://doi.org/10.1016/j.bushor.2018.08.004.
25. Alrefaei AF, Hawsawi YM, Almaleki D et al. 2022. "Genetic data sharing and artificial intelligence in the era of personalized medicine based on a cross-sectional analysis of the Saudi human genome program". Sci Rep 12:1405. https://doi.org/10.1038/s41598-022-05296-7.
26. Moor JH. 2000. "Toward a theory of privacy in the information age". In: Baird RM, Ramsower RM, and Rosenbaum SE (Eds) "cyberethics: social and moral issues in the computer age". Prometheus, Amherst, pp 200–212.

27. Nissenbaum H. 2004. "Symposium: privacy as contextual integrity". Wash Law Rev 79:119–158.
28. Piccolo JJ. 2017. "Intrinsic values in nature: objective good or simply half of an unhelpful dichotomy?". J Nat Conserv 37:8–11. https://doi.org/10.1016/j.jnc.2017.02.007.
29. Reisman D, Schultz J, Crawford K, Whittaker M. 2018. Algorithmic impact assessments: a practical framework for public agency accountability. AI Now Institute, New York.
30. Severson RW. 1997. The principles for information ethics, 1st edn. Routledge, Armonk.
31. Hemalatha S, Mahalakshmi M, Vignesh V, Geethalakshmi M, Balasubramanian D, A JA., "Deep Learning Approaches for Intrusion Detection with Emerging Cybersecurity Challenges," 2023 International Conference on Sustainable Communication Networks and Application (ICSCNA), Theni, India, 2023, pp. 1522–1529.
32. Takashima K, Maru Y, Mori S et al. "Ethical concerns on sharing genomic data including patients' family members". BMC Med Ethics 19, 2018, 61. https://doi.org/10.1186/s12910-018-0310-5.
33. Rosenbaum E 2018. "5 biggest risks of sharing your DNA with consumer genetic-testing companies". CNBC, 16 June 2018.
34. Tavani H. *Privacy and security*. In: Langford D (Ed) Internet ethics, 2000th edn. Palgrave, Basingstoke, 2000, pp 65–95.
35. Warren SD, Brandeis LD. 1890. "The right to privacy". Harv Law Rev 4(5):193–220. https://doi.org/10.2307/1321160.
36. Wright D. 2011. "A framework for the ethical impact assessment of information technology". Ethics Inf. Technol 13:199–226. https://doi.org/10.1007/s10676-010-9242-6.
37. McCusker EA, Loy CT. 2017. "Huntington disease: the complexities of making and disclosing a clinical diagnosis after premanifest genetic testing". Tremor Other Hyperkinet Mov (NY) 7:467. https://doi.org/10.7916/D8PK0TDD.
38. Hill K 2020. "Another arrest, and jail time, due to a bad facial recognition match". The New York Times, 29 Dec. https://www.nytimes.com/2020/12/29/technology/facial-recognition-misidentify-jail.html. Accessed 29 September 2024.
39. Balli E 2021. "The ethical implications of facial recognition technology". ASU News, 17 Nov. https://news.asu.edu/20211117-solutions-ethical-implications-facial-recognition-technology. Accessed 29 September 2024.
40. Brown WS. 2000. "Ontological security, existential anxiety and workplace privacy". J Bus Ethics 23:61–65. https://doi.org/10.1023/A%3A1006223027879.
41. Giddens A. 1984. The constitution of society: outline of the theory of structuration. Polity, Cambridge.
42. Schep T. (n.d). "Social cooling". https://www.tijmenschep.com/socialcooling/. Accessed 29 September 2024.

18 AI-Powered Predictive Analysis for Proactive Cyber Defense

Pankaj Bhambri and Paula Bajdor

18.1 INTRODUCTION

This chapter delves into the core principles, technologies, and applications of AI-powered predictive analysis for cybersecurity. It provides a comprehensive exploration of key methodologies such as anomaly detection and threat intelligence while addressing challenges like data privacy and ethical concerns [1].

18.1.1 Context of Cybersecurity in the Digital Era

The rapid digitization of industries, driven by advancements in cloud computing, the Internet of Things (IoT), and digital communication, has revolutionized how businesses and individuals interact. However, with this digital transformation comes an explosion in cyber threats. Traditional cybersecurity defenses, which rely on static rules and manual intervention, have struggled to keep pace with the speed and complexity of modern attacks [2–5].

Artificial intelligence (AI) emerges as a powerful ally in this landscape, enabling a shift from reactive to proactive cybersecurity. By leveraging predictive analysis, AI can detect potential threats before they materialize, ensuring systems remain secure while minimizing disruption [6].

18.2 OBJECTIVES OF PREDICTIVE ANALYSIS IN CYBERSECURITY

Predictive analysis aims to identify vulnerabilities and potential attack vectors using historical data and real-time analytics. Its objectives include:

- *Early threat detection*: Identifying signs of an impending cyberattack before it causes damage.
- *Risk mitigation*: Reducing the impact of potential threats by pre-emptively addressing vulnerabilities.
- *Operational efficiency*: Automating processes to free up human analysts for strategic decision-making.

 DOI: 10.1201/9781003521020-18

18.2.1 Importance of a Proactive Approach

18.2.1.1 The Shift from Reactive to Proactive Defense

Traditional defenses work by responding after an incident has occurred, often resulting in significant downtime and data loss. In contrast, predictive analysis foresees issues, enabling organizations to take preventative measures [7].

18.2.1.2 Cost Implications

According to a report by IBM, the average cost of a data breach in 2023 was over $4 million, with costs escalating as breaches go undetected for longer periods. AI-driven predictive systems help reduce detection time, mitigating financial and reputational damages.

18.3 THE ROLE OF PREDICTIVE ANALYSIS IN CYBERSECURITY

18.3.1 Defining Predictive Analysis

Predictive analysis employs statistical techniques, data mining, and machine learning models to forecast future outcomes based on historical and current data [8].

18.3.1.1 Application in Cybersecurity

In cybersecurity, predictive analysis is pivotal in identifying malicious activities before they escalate. For example:

- *Phishing attack predictions*: By analyzing email headers and metadata, predictive systems can flag potential phishing attempts.
- *Behavioral anomalies*: Deviations in user behavior, such as accessing restricted files at odd hours, can signal insider threats.

18.3.2 Components of Predictive Cyber Defense

18.3.2.1 Anomaly Detection

Anomaly detection involves identifying unusual patterns that deviate from established baselines. AI models monitor:

- Network traffic: Spikes in data transfer rates or unusual IP connections.
- User behavior: Frequent failed login attempts or access to sensitive areas.

18.3.2.1.1 Real-world Example

A financial institution deployed AI to monitor transactions. The system flagged a series of small, unauthorized transactions, uncovering a larger fraud operation [9].

18.3.2.2 Behavior Modeling

Behavior modeling creates profiles of normal operations for users, devices, or systems. Over time, AI adapts these models to detect subtle shifts that may indicate a breach [10, 11].

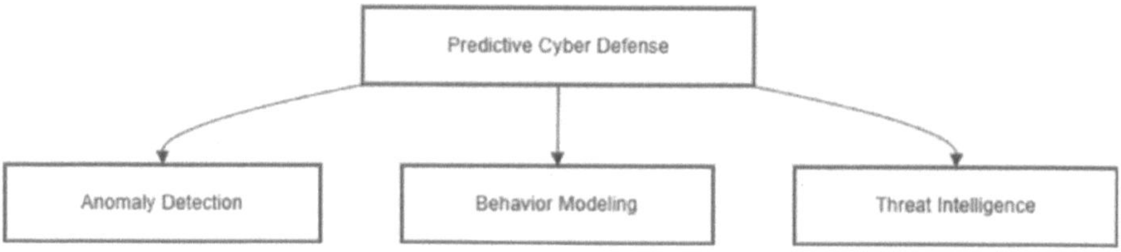

FIGURE 18.1 Components of predictive cyber defense.

18.3.2.3 Application

Behavior modeling is used in endpoint protection solutions like CrowdStrike, where each device is profiled to prevent malware execution.

18.3.2.4 Threat Intelligence Integration

Threat intelligence platforms aggregate data from various sources, including dark web forums, to predict emerging threats. AI systems process this information, ranking risks based on their relevance to an organization [12]. Figure 18.1 depicts the various components of predictive cyber defense system.

18.3.3 Advantages Over Traditional Methods

Predictive analysis outshines traditional methods due to its:

- *Scalability*: AI systems can handle vast amounts of data, making them suitable for large enterprises.
- *Speed:* Real-time analysis allows immediate responses to emerging threats.
- *Proactive defense*: By foreseeing potential risks, organizations can allocate resources more effectively.

18.4 CORE TECHNOLOGIES ENABLING AI-POWERED CYBER DEFENSE

AI-powered cybersecurity systems rely on a suite of advanced technologies to deliver predictive insights. These technologies include big data analytics, machine learning, anomaly detection systems, and threat intelligence platforms [13].

18.4.1 Big Data Analytics

Big data analytics forms the backbone of predictive cyber defense by processing and analyzing vast datasets collected from diverse sources, such as:

- *Network logs*: Capturing real-time network activity.
- *Endpoint telemetry*: Monitoring device-level behavior.
- *External threat feeds*: Aggregating known vulnerabilities and threat indicators.

FIGURE 18.2 Big data in cybersecurity.

18.4.1.1 Applications

1. *Detecting insider threats*: Analyzing employee activity for unauthorized access or data transfers.
2. *Preventing fraud*: Monitoring customer transactions for unusual patterns in banking and e-commerce.

Figure 18.2 shows the role of big data in cybersecurity.

18.4.2 Machine Learning Models

Machine learning (ML) powers the predictive capabilities of AI systems by enabling models to learn from historical data and make accurate predictions about potential cyber threats [14].

18.4.2.1 Types of Machine Learning Models in Cyber Defense

1. Supervised learning
   - Trains models using labeled datasets to identify known threats (e.g., phishing emails).
   - Example: Anti-spam filters using Naïve Bayes classification.
2. Unsupervised learning
   - Detects anomalies without prior knowledge of threats, using clustering algorithms like K-means.
   - Example: Flagging anomalous traffic patterns in network security.
3. Reinforcement learning
   - Learns optimal responses in dynamic environments by trial and error.
   - For example, autonomous systems for intrusion detection.

18.4.2.2 Machine Learning Lifecycle in Cyber Defense

1. Data collection
   - Sources include user activity logs, system telemetry, and external feeds.
2. Feature engineering
   - Extracting relevant attributes such as login times, IP addresses, and file access patterns.
3. Model training
   - Building and testing models to optimize accuracy and minimize false positives.
4. Deployment
   - Integrating models into real-time systems for continuous monitoring.

Figure 18.3 shows the machine learning workflow in cyber defense.

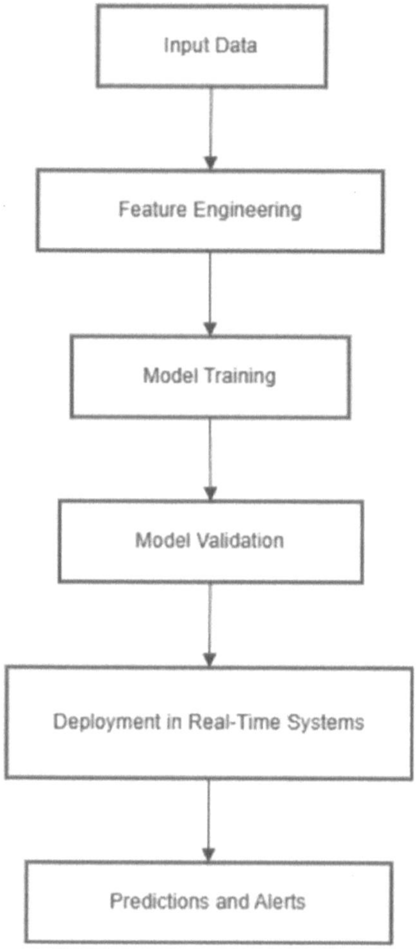

FIGURE 18.3 Machine learning workflow in cyber defense.

18.4.3 Advanced Anomaly Detection Systems

Anomaly detection is a critical function in predictive cyber defense, focusing on identifying patterns that deviate from established norms [15].

18.4.3.1 Techniques in Anomaly Detection

1. Statistical methods
   - Use thresholds and baselines for basic anomaly detection.
   - Limitations: Struggle with complex, evolving threats.
2. Machine learning models
   - Deep learning models, such as autoencoders, identify sophisticated anomalies.
   - Strengths: Handle high-dimensional data efficiently.

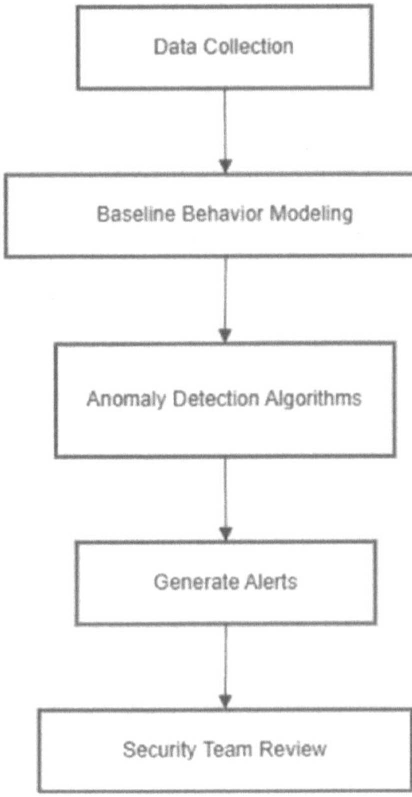

FIGURE 18.4 Anomaly detection workflow.

3. Hybrid systems
   - Combine statistical and machine learning approaches for improved accuracy.

Figure 18.4 shows the anomaly detection workflow.

18.4.4 Threat Intelligence Platforms

Threat intelligence platforms aggregate and analyze external threat data, such as indicators of compromise (IoCs), malware signatures, and exploits, to enrich predictive models [16].

18.4.4.1 Features of Threat Intelligence Platforms

1. *Data aggregation*: Collating feeds from multiple sources (e.g., dark web forums, threat databases).
2. *Correlation engines*: Linking data points to identify patterns and trends.
3. *Visualization tools*: Presenting actionable insights through dashboards and heatmaps.

18.4.4.2 Example Use Case

Organizations leverage platforms like Recorded Future to:

- Prioritize vulnerabilities based on real-world exploits.
- Identify zero-day threats affecting their systems.

18.4.5 Role of Automation

Automation amplifies the effectiveness of predictive cyber defense by:

1. *Accelerating response times*: Automatically blocking IPs associated with known threats.
2. *Reducing manual effort*: Handling routine tasks like log parsing.

18.4.5.1 Real-World Example: Automated Malware Defense

An AI-driven platform detects and isolates a malicious file within seconds of its introduction into a network, preventing lateral movement.

18.5 REAL-WORLD APPLICATIONS OF PREDICTIVE ANALYSIS IN CYBER DEFENSE

Predictive analysis plays a transformative role in several key industries where cybersecurity is paramount. We explore how this technology is applied in various sectors in following subsections.

18.5.1 Financial Institutions

Financial institutions are prime targets for cyberattacks due to the sensitive nature of the data they hold and the potential for significant financial loss.

18.5.1.1 Predictive Applications

1. *Fraud detection*: Machine learning algorithms analyze transaction patterns in real time to identify fraud.
2. *Threat intelligence integration*: Threat feeds provide insights into the latest tactics used in banking-related cybercrime.

18.5.1.2 Case Study: Bank of America's AI-Driven Cyber Defense

Bank of America implemented AI-powered tools for real-time fraud detection and user authentication, significantly reducing phishing and other fraudulent activities.

18.5.2 Healthcare Systems

Healthcare systems contain sensitive personal and health data, making them attractive targets for ransomware and data breaches [17].

18.5.2.1 Predictive Applications

1. *Ransomware prevention*: By monitoring network activity for signs of encryption or abnormal file access patterns, predictive analysis can halt ransomware before it spreads.
2. *Data privacy compliance*: AI helps ensure compliance by detecting unauthorized access to patient records.

18.5.2.2 Case Study: Predictive Defense in Hospitals

The National Health Service (NHS) in the UK uses predictive analysis to detect anomalies in patient data access, helping prevent insider threats and data breaches.

18.5.3 Critical Infrastructure

Critical infrastructure, including energy, water, and transportation systems, is essential for public safety and economic stability. These systems often use AI to safeguard against advanced persistent threats (APTs) [18].

18.5.3.1 Predictive Applications

1. *SCADA systems monitoring*: Supervisory control and data acquisition (SCADA) systems can be monitored for anomalies in critical infrastructure.
2. *Industrial control systems (ICS) security*: Predictive analysis identifies potential vulnerabilities in ICS protocols and communications.

18.5.3.2 Example: Predictive Analysis in Power Grid Protection

The US Department of Energy uses AI to detect anomalies in power grid operations, safeguarding against both physical and cyber threats.

Figure 18.5 shows the predictive analysis across industries.

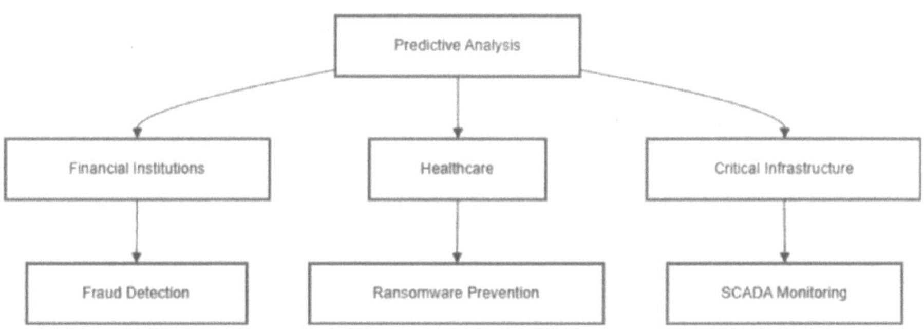

FIGURE 18.5 Predictive analysis across industries.

18.6 CHALLENGES AND LIMITATIONS

Despite the significant benefits, there are several challenges associated with implementing AI-powered predictive analysis in cybersecurity. These are discussed in subsequent subsections.

18.6.1 Data Privacy Concerns

Predictive analysis requires large volumes of data, raising concerns about data privacy and consent. Organizations must ensure compliance with regulations like GDPR, which limit data collection and sharing practices [19].

Solutions for these are:

- *Data anonymization*: Ensuring that sensitive data is anonymized before processing.
- *Privacy-preserving AI*: Using federated learning allows AI to train on data without leaving the device, maintaining user privacy.

18.6.2 Ethical Considerations in AI-Driven Security

AI in cybersecurity poses ethical challenges related to bias, transparency, and decision accountability.

1. *Algorithmic bias*: AI models may inadvertently prioritize certain users or behaviors as risky, leading to unfair outcomes.
2. *Transparency*: Ensuring that AI decisions can be explained to users and stakeholders.

The solutions for these changes are:

- Explainable AI: Developing models that provide human-readable justifications for decisions.
- Regular audits: Frequent assessments to mitigate potential biases and ensure ethical standards.

18.6.3 Continuous Learning for Evolving Threats

Cyber threats evolve rapidly, and static models become obsolete over time.

The solutions for these are:

- *Continuous model training*: Regular updates using the latest threat data.
- *Adversarial machine learning*: Training models to recognize and respond to adversarial techniques, such as data poisoning attacks.

18.7 CASE STUDIES: IMPLEMENTING AI-Powered PREDICTIVE ANALYSIS

This section presents in-depth case studies illustrating the effectiveness of AI in proactive cybersecurity [20].

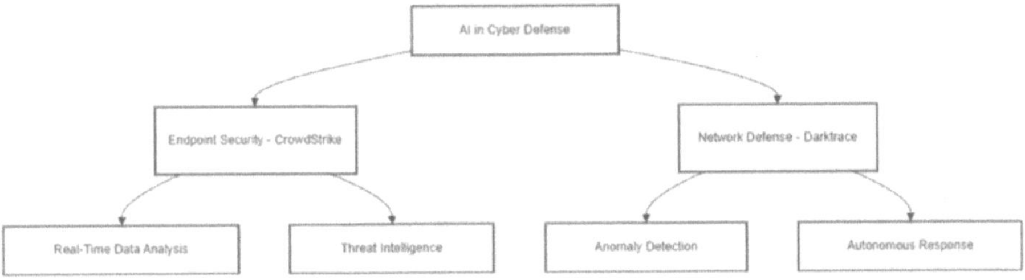

FIGURE 18.6 Key elements in AI case studies.

18.7.1 Case Study: AI in Endpoint Security – CrowdStrike Falcon

CrowdStrike Falcon uses AI-powered predictive models to identify endpoint threats. Key components include:

1. *Real-time data analysis*: Monitoring endpoint activity for signs of intrusion.
2. *Threat intelligence integration*: Incorporating global threat data for rapid response.

18.7.1.1 Outcome

CrowdStrike's predictive system blocked over 15,000 threats in a single year, providing early warnings for numerous potential breaches.

18.7.2 Case Study: AI in Network Defense – Darktrace

Darktrace's AI-driven system leverages machine learning for network threat detection.

1. *Anomaly detection*: Identifying unusual patterns in network traffic.
2. *Autonomous response*: Automatically responding to threats in real time.

18.7.2.1 Outcome

Darktrace prevented data breaches in a multinational organization by isolating infected devices before the threat could spread.

Figure 18.6 displays the key elements in AI case studies.

18.8 FUTURE DIRECTIONS AND EMERGING TRENDS

As AI technology advances, several emerging trends will shape the future of predictive cybersecurity.

18.8.1 AI-Augmented Human Intelligence

Human analysts and AI systems will work together, with AI handling routine tasks and humans focusing on complex strategic decisions [21, 22].

1. *Assisted decision-making*: AI provides recommendations, while humans make final security decisions.

2. *Cognitive security operations centers*: Integrating AI to enhance SOC capabilities.

18.8.2 AI and Quantum Computing in Cyber Defense

Quantum computing offers new possibilities for cybersecurity, as quantum algorithms can process large datasets quickly. However, it also poses risks, as adversaries could use quantum to break encryption [23–26].

18.8.2.1 Solutions

1. *Quantum-resistant algorithms*: Developing cryptographic methods resistant to quantum attacks.
2. *AI for quantum detection*: Predictive analysis can identify quantum-based attacks, helping secure sensitive data.

18.8.3 Increased Use of Blockchain for Data Security

Blockchain's decentralized structure can improve data integrity, making it harder for attackers to alter or delete data without detection [27–31].

18.8.3.1 Examples

1. *Smart contracts in cybersecurity*: Automatically executing security policies based on pre-set conditions.
2. *Immutable logs*: Using blockchain to store audit trails, ensuring they remain tamper-proof.

Figure 18.7 displays the future directions in AI-powered cybersecurity.

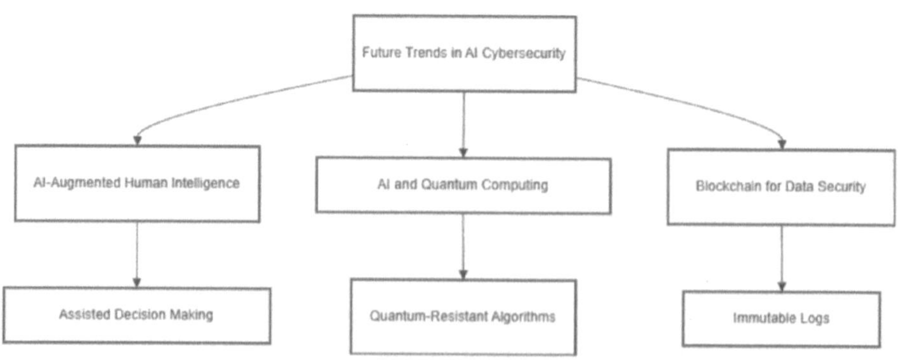

FIGURE 18.7 Future directions in AI-powered cybersecurity.

18.9 CONCLUSION

AI-powered predictive analysis offers a transformative shift in cybersecurity by enabling proactive defense mechanisms. By leveraging big data, machine learning, and anomaly detection, AI systems detect potential threats before they materialize. However, challenges such as ethical concerns, privacy, and the need for continuous learning must be addressed.

As AI advances, integrating it with human intelligence, quantum computing, and blockchain will further strengthen cybersecurity. Predictive analysis, thus, is poised to become a cornerstone in securing digital infrastructures.

REFERENCES

1. Chen, W., & Zhao, F. (2023). The role of artificial intelligence in cybersecurity threat prediction. Artificial Intelligence in Security, 30(3), 97–110.
2. Smith, J. (2023). AI-driven threat detection and prediction in modern cybersecurity practices. Cybersecurity Innovations, 45(3), 101–115.
3. Johnson, P., & Lee, D. (2024). Leveraging machine learning for proactive threat mitigation in cybersecurity. Journal of AI and Cyber Defense, 28(1), 59–74.
4. Kumar, R., & Patel, S. (2023). Enhancing cybersecurity with AI-based predictive analytics. International Journal of Cybersecurity, 19(4), 239–250.
5. Bhambri, P. (2025). Innovative Systems: Entertainment, Gaming, and the Metaverse. In R. C. Ho; B. L. Song; & P. K. Tee (Eds.), Managing Customer-Centric Strategies in the Digital Landscape (pp. 483–514). IGI Global. https://doi.org/10.4018/979-8-3693-5668-5.ch018
6. Zhang, H., & Wang, T. (2024). Proactive defense strategies in cybersecurity using AI and machine learning. AI for Cyber Defense, 33(2), 112–125.
7. Rana, R., & Bhambri, P. (2025). Generative AI in Web Application Development: Enhancing User Experience and Performance. In I. Shah; & N. Jhanjhi (Eds.), Generative AI for Web Engineering Models (pp. 471–486). IGI Global. https://doi.org/10.4018/979-8-3693-3703-5.ch021
8. Yi, R., & Liu, X. (2023). Machine learning techniques for proactive cyber defense. IEEE Transactions on Network and Service Management, 45(3), 312–324.
9. Davis, M., & Huang, J. (2024). AI-based anomaly detection for early cyber threat mitigation. Journal of Network Security, 23(1), 48–60.
10. Rana, R., & Bhambri, P. (2025). Generative AI-Driven Security Frameworks for Web Engineering: Innovations and Challenges. In I. Shah; & N. Jhanjhi (Eds.), Generative AI for Web Engineering Models (pp. 285–296). IGI Global. https://doi.org/10.4018/979-8-3693-3703-5.ch014
11. Williams, S., & Scott, R. (2024). Proactive security measures using predictive analytics in AI-based systems. Journal of AI and Digital Security, 12(2), 135–148.
12. White, G., & Zhang, B. (2023). Machine learning and its application in predictive cybersecurity. Security and Privacy Journal, 19(4), 143–157.
13. Bhambri, P., & Khang, A. (2025). Smart Universities and ICT Platforms. In M. L. Kolhe; P. Singh; S. Rani; & P. Kumar (Eds.), Planning of Sustainable Energy Systems in Urban Built Environments. CRC Press. https://www.appleacademicpress.com/planning-of-sustainable-energy-systems-in-urban-built-environments-/9781779640642

14. McCarter, D., & Harris, J. (2024). AI in proactive cybersecurity defense: Leveraging predictive analytics for early detection. Journal of Machine Learning and Security, 14(2), 120–134.
15. Clark, H., & Young, D. (2023). AI-enabled proactive defense for cybersecurity with predictive modeling. IEEE Transactions on AI, 17(7), 210–225.
16. Bhambri, P., & Kautish, S. K. (2024). Analytic Hierarchy Process and Business Value Creation. In S. Kautish (Ed.), Using Strategy Analytics for Business Value Creation and Competitive Advantage (pp. 54–77). IGI Global. https://doi.org/10.4018/979-8-3693-2823-1.ch003
17. Bennett, P., & Moore, A. (2023). Cyber threat prediction and prevention through AI-driven analytics. Cybersecurity Systems Journal, 27(3), 92–105.
18. Richards, M., & Chen, S. (2023). AI for predictive threat detection: Enhancing cybersecurity measures. Journal of AI and Machine Learning, 31(5), 73–88.
19. Rana, R., & Bhambri, P. (2024). Ethical Considerations in Artificial Intelligence for Environmental Solutions: Striking a Balance for Sustainable Innovation. In P. Bhambri; & P. Bajdor (Eds.), Handbook of Technological Sustainability: Innovation and Environmental Awareness (pp. 389–396). CRC Press. https://doi.org/10.1201/9781003475989-28
20. Murphy, T., & Singh, P. (2023). Proactive defense using AI-driven predictive analysis. AI and Digital Security Journal, 10(4), 56–70.
21. Ruby, S., Biju, T., & Bhambri, P. (2024). Catalysing Sustainable Progress: Empowering MSMEs Through Tech Innovation for a Bright Future. In P. Bhambri; & P. Bajdor (Eds.), Handbook of Technological Sustainability: Innovation and Environmental Awareness (pp. 374–388). CRC Press. https://doi.org/10.1201/9781003475989-27
22. Thirumalaiyammal, B., Steffi, P. F., & Bhambri, P. (2024). Green Horizons: Navigating Environmental Challenges Through Technological Innovation. In P. Bhambri; & P. Bajdor (Eds.), Handbook of Technological Sustainability: Innovation and Environmental Awareness (pp. 292–304). CRC Press. https://doi.org/10.1201/9781003475989-22
23. Bhambri, P., & Khang, A. (2024). AI-Integrated Biosensors and Bioelectronics for Healthcare. In A. Khang (Ed.), AI-Driven Innovations in Digital Healthcare: Emerging Trends, Challenges, and Applications (pp. 82–96). IGI Global. https://doi.org/10.4018/979-8-3693-3218-4.ch004
24. Bhambri, P., & Bakshi, P. (2022). An Energy Efficient Adhoc on-Demand Distance Vector Protocol for IoT (Vol. 1). Lap Lambert Academic Publishing. ISBN: 9786205488218.
25. Bhambri, P., Sharma, R., & Bedi, V. (2022). Energy Aware Bio Inspired Routing Technique for Mobile Adhoc Networks (Vol. 1). Lap Lambert Academic Publishing. ISBN: 9786204983202.
26. Bhambri, P., & Singh, S. (2013). Fundamentals of Information Technology: Introduction to Applications of IT (Vol. 1). Lap Lambert Academic Publishing. ISBN: 9783659330889.
27. Bhambri, P., & Khang, A. (2024). Machine Learning Advancements in E-Health: Transforming Digital Healthcare. In A. Khang (Ed.), Medical Robotics and AI-Assisted Diagnostics for a High-Tech Healthcare Industry (pp. 174–194). IGI Global. https://doi.org/10.4018/979-8-3693-2105-8.ch012
28. Bhambri, P., & Khang, A. (2024). Managing and Monitoring Patient's Healthcare Using AI and IoT Technologies. In A. Khang (Ed.), Driving Smart Medical Diagnosis Through AI-Powered Technologies and Applications (pp. 1–23). IGI Global. https://doi.org/10.4018/979-8-3693-3679-3.ch001
29. Bhambri, P., & Kaur, A. (2013). Novel Technique for Robust Image Segmentation: New Technique of Segmentation in Digital Image Processing (Vol. 1). Lap Lambert Academic Publishing. ISBN: 9783659331831.

30. Bhambri, P., & Kaur, P. (2015). Design and Implementation of Novel Algorithm Using Zero Watermarking: Digital Image Processing Technique for Text Documents (Vol. 1). Lap Lambert Academic Publishing. ISBN: 9783659796159.
31. Bhambri, P., & Kaur, J. (2020). Hybrid Classification Model for the Reverse Code Generation (Vol. 1). Lap Lambert Academic Publishing. ISBN: 9786202683432.

19 Deep Learning Techniques for Intrusion Detection in Critical Infrastructure

Pankaj Bhambri and Ilona Paweloszek

19.1 INTRODUCTION

Intrusion detection in critical infrastructure has emerged as a fundamental element of contemporary cybersecurity owing to the increasing complexity and prevalence of cyber threats. Critical infrastructure systems, including power grids, transportation networks, including water supply systems, are essential for societal operation but are increasingly targeted by malicious entities utilizing advanced persistent threats (APTs) and zero-day exploits. Conventional intrusion detection techniques, while somewhat effective, frequently encounter challenges related to scalability, heterogeneity, especially real-time demands of these systems. Deep learning techniques are establishing themselves as a revolutionary solution, providing unmatched proficiency in identifying intricate and nuanced attack patterns. Utilizing neural networks, including convolutional neural networks (CNNs) for spatial analysis and recurrent neural networks (RNNs) for temporal sequences, these approaches offer resilient, adaptive, and efficient solutions for intrusion detection [1, 2].

19.1.1 Importance of Critical Infrastructure Security

Critical infrastructure security is paramount as these systems form the backbone of modern society, encompassing essential sectors like energy, transportation, healthcare, and water supply. Any disruption or compromise in these systems can lead to cascading consequences, including economic losses, public safety hazards, and national security threats. The interconnected nature of critical infrastructure increases its vulnerability, as a single breach in one sector can ripple through others, magnifying the impact. Furthermore, the rising frequency and sophistication of cyberattacks, driven by APTs and state-sponsored actors, highlight the urgent need for robust security measures. Protecting critical infrastructure is not just about safeguarding physical assets but also ensuring the resilience and continuity of services that millions of people depend on daily. Thus, developing advanced detection and mitigation strategies, such as leveraging deep learning techniques, is essential to preempt and neutralize emerging threats effectively [3, 4].

 DOI: 10.1201/9781003521020-19

19.1.2 Overview of Cyber Threat Landscape

As digital transformation integrates advanced technologies into essential sectors like energy, transportation, and healthcare, these systems have become prime targets for cybercriminals, state-sponsored attackers, and hacktivist groups. Threat actors leverage sophisticated techniques, including APTs, ransomware, and zero-day exploits, to disrupt operations, steal sensitive data, and compromise system integrity. The increasing interconnectedness of critical infrastructure through Internet of Things (IoT) devices and cloud platforms amplifies vulnerabilities, making systems more susceptible to attacks like distributed denial-of-service (DDoS) and supply chain breaches. This dynamic and rapidly changing threat environment underscores the urgent need for proactive, intelligent cybersecurity measures, with deep learning emerging as a powerful tool for detecting and mitigating these complex and evolving cyber threats [5, 6].

19.1.3 Need for Advanced Intrusion Detection Systems

The escalating complexity and prevalence of cyberattacks on essential infrastructure underscore the pressing necessity for advanced intrusion detection systems (IDS). Conventional IDSs, which depend significantly on established rules and signature-based detection, find it challenging to address contemporary threats such as APTs, zero-day vulnerabilities, and covert attacks that change and develop over time. Critical infrastructure systems, such as power grids, transportation networks, and water supply chains, are diverse, highly interconnected, and require real-time responses and stringent security measures to avert catastrophic disruptions. Advanced IDSs, utilizing deep learning methodologies, possess the capability to scrutinize extensive data sets, identify anomalies with exceptional accuracy, and adjust to evolving threats. These systems address limitations of legacy methods by leveraging self-learning algorithms, making them indispensable for the dynamic and high-stakes environment of critical infrastructure security [7].

19.2 FOUNDATIONS OF INTRUSION DETECTION IN CRITICAL INFRASTRUCTURE

The foundations of intrusion detection in critical infrastructure are rooted in understanding the unique characteristics and challenges posed by these systems. Critical infrastructure encompasses a diverse range of physical and cyber-physical systems, including energy grids, transportation, water supply, and communication networks, all of which are integral to societal well-being. These systems operate in highly interconnected and heterogeneous environments, making them susceptible to complex cyber threats. Traditional IDS rely on signature-based or rule-based methods, which are often insufficient against APTs and zero-day exploits. The shift toward deep learning-based techniques addresses these limitations by enabling the analysis of high-dimensional, dynamic data streams in real time. Effective intrusion detection in critical infrastructure requires a balance between accuracy, scalability, and computational efficiency, alongside the ability to adapt to evolving threats, making the integration of deep learning a promising direction for robust cybersecurity [8, 9].

19.2.1 Characteristics of Critical Infrastructure Systems

Critical infrastructure systems are complex, interconnected networks that are essential for societal functions and economic stability. Their key characteristics include:

- *Heterogeneity*: They consist of diverse components, including physical assets, cyber-physical systems, and digital networks.
- *Interdependence*: Systems are interconnected, meaning disruptions in one can cascade to others.
- *High availability*: Continuous operation is crucial, with minimal tolerance for downtime.
- *Scalability challenges*: These systems must handle varying demands, often with legacy components.
- *Susceptibility to cyber threats*: Their increasing reliance on digital technologies makes them vulnerable to sophisticated cyberattacks.

Figure 19.1 displays the key characteristics of critical infrastructure, illustrating their dependencies and challenges [10, 11].

19.2.2 Challenges in Cybersecurity for Critical Infrastructure

Critical infrastructure systems face unique cybersecurity challenges due to their complexity, interdependence, and the critical services they provide [12]. Key challenges include:

- *Legacy systems and modernization*: Many critical infrastructure systems were designed decades ago with minimal or no cybersecurity considerations. Integrating modern security measures into these outdated systems is complex and costly.
- *Interconnectedness and interdependencies*: The integration of multiple systems across sectors increases the risk of cascading failures. A breach in one system can quickly propagate to others, amplifying the impact of an attack.

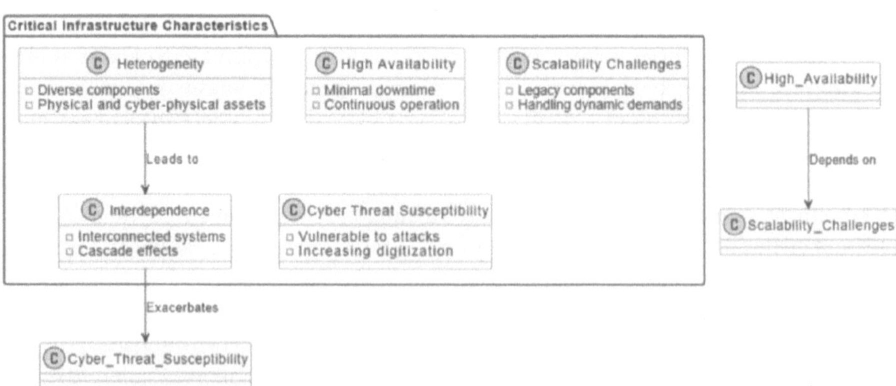

FIGURE 19.1 Key characteristics of critical infrastructure.

- *Real-time operational requirements*: Infrastructure like power grids and transportation networks require real-time responses, making traditional security solutions, which often involve latency, unsuitable.
- *Sophistication of cyber threats*: APTs and zero-day exploits pose significant risks, as attackers increasingly use sophisticated techniques to bypass detection.
- *Insufficient security awareness*: Operators and stakeholders may lack adequate training or awareness about evolving cyber threats, leading to gaps in security implementation and response readiness.
- *Data sensitivity and privacy*: Critical infrastructure often involves sensitive data, such as user information and operational metrics, making it a prime target for attackers aiming to compromise privacy or disrupt operations.
- *Regulatory and compliance challenges*: Variations in cybersecurity regulations across regions and industries can complicate the implementation of consistent and robust security measures.
- *Resource constraints*: Financial, human, and technological resources are often limited, especially for smaller entities within critical infrastructure sectors, hindering the adoption of advanced cybersecurity solutions.

19.2.3 Traditional vs. Deep Learning-Based Intrusion Detection

Traditional IDSs rely on predefined rules, signatures, and heuristics to detect known attack patterns [13]. These systems typically use two main approaches:

- *Signature-based detection*: Matches incoming data against a database of known attack signatures. It is effective for detecting known threats but struggles with zero-day attacks and unknown threats.
- *Anomaly-based detection*: Identifies deviations from a baseline of normal behavior. While it can detect unknown attacks, it is prone to generating false positives and requires ongoing tuning.

In contrast, deep learning-based intrusion detection uses advanced machine learning algorithms, particularly neural networks, to learn and adapt to new and evolving attack patterns. Key advantages include:

- *Improved accuracy*: Deep learning models can learn complex patterns from large datasets, improving detection accuracy, even for unknown attacks.
- *Automated feature extraction*: Unlike traditional methods, deep learning can automatically identify relevant features without needing manual intervention.
- *Adaptability*: Deep learning models can continuously improve their detection capabilities as they are exposed to new data, making them more robust against emerging threats.

Deep learning-based systems, though more powerful, often come with challenges such as higher computational costs and the need for large labeled datasets for

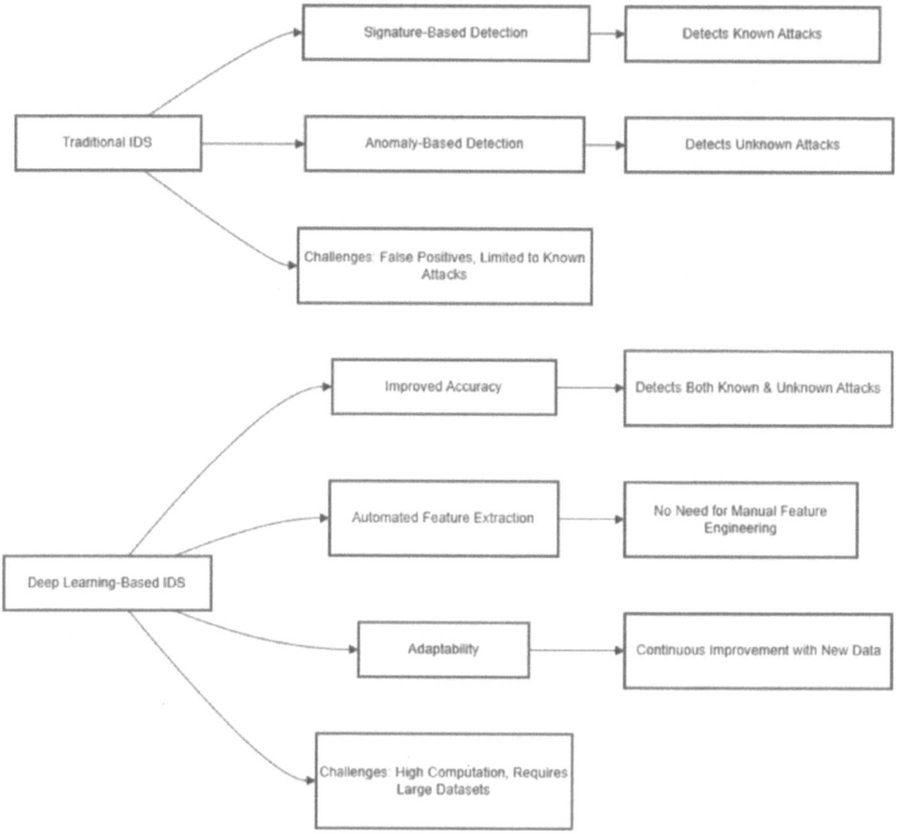

FIGURE 19.2 Flow from traditional methods with their limitations to the more advanced, adaptive capabilities of deep learning-based IDS [14–17].

training. Figure 19.2 shows the flow from traditional methods with their limitations to the more advanced, adaptive capabilities of deep learning-based IDS, while also highlighting the challenges faced by both approaches.

19.3 DEEP LEARNING TECHNIQUES FOR INTRUSION DETECTION

Deep learning methodologies for intrusion detection have transformed cybersecurity by improving the capacity to recognize intricate, evolving attack patterns in critical infrastructure systems. In contrast to conventional techniques, deep learning models such as CNNs, RNNs, and autoencoders can autonomously extract features from unprocessed data, thereby obviating the necessity for manual feature engineering. CNNs are proficient in recognizing spatial patterns in network traffic data, whereas RNNs excel at identifying temporal dependencies in time-series data, rendering them suitable for the analysis of event or activity sequences. Autoencoders are proficient in anomaly detection, as they can acquire a compressed representation of standard system behavior and identify deviations as potential intrusions. These methodologies provide enhanced precision, scalability, and flexibility relative to conventional IDSs, facilitating the identification of both recognized and unidentified threats, including APTs and zero-day exploits. Nonetheless, deep learning models encounter challenges including substantial computational expenses, the necessity for extensive labeled datasets, and interpretability concerns that necessitate further progress, such as the incorporation of XAI to enhance the transparency and comprehensibility of these systems [18].

19.3.1 CNNs for Pattern Recognition

CNNs are a category of deep learning models that excel in pattern recognition, particularly with image and spatial data. CNNs are engineered to autonomously identify and assimilate hierarchical features via a succession of convolutional layers, pooling layers, and fully connected layers. The convolutional layers utilize filters on input data (e.g., images or network traffic) to identify local patterns such as edges, textures, or more intricate structures in subsequent layers. The capacity to identify patterns across various levels of abstraction enables CNNs to excel in object recognition, image classification, and anomaly detection tasks. In cybersecurity, CNNs can facilitate pattern recognition in network traffic, detecting anomalous behavior or attack signatures, even amidst noise or minor data variations. Their strength resides in their ability to capture spatial dependencies and autonomously learn features from raw data, markedly enhancing performance compared to conventional manual feature extraction techniques. Nonetheless, CNNs necessitate extensive labeled datasets and considerable computational resources for effective training [19, 20].

19.3.2 RNNs for Temporal Data Analysis

RNNs are a category of deep learning models engineered to process sequential and temporal data by retaining a memory of prior inputs via feedback loops in their structure. In contrast to conventional feedforward neural networks, RNNs sequentially process data, modifying their internal state according to the current input and the preceding state, rendering them suitable for tasks requiring temporal information, including time series forecasting, natural language processing, and cybersecurity intrusion detection. RNNs excel at recognizing temporal patterns and dependencies in sequential data, such as network event sequences or system logs, which are essential for identifying attacks like APTs or other evolving zero-day exploits. Standard RNNs, however, encounter difficulties with long-term dependencies because of the vanishing gradient problem. To address this, more sophisticated architectures such as long short-term memory (LSTM) networks and gated recurrent units (GRUs) have been created, which more effectively capture long-range dependencies and improve the model's capacity to learn intricate temporal patterns. Notwithstanding their efficacy, RNNs necessitate considerable computational resources and extensive labeled data for training [21, 22].

19.3.3 Autoencoders for Anomaly Detection

Autoencoders are a category of neural networks utilized chiefly for unsupervised learning and anomaly detection. The system comprises two primary components: an encoder that compresses the input data into a lower-dimensional representation (latent space) and a decoder that reconstructs the data from this compressed format. The objective of training an autoencoder is to reduce the disparity between the input and the reconstructed output, thereby acquiring a concise representation of normal data. In the realm of anomaly detection, autoencoders are exceptionally proficient as they can discern the fundamental patterns of "normal" system behavior. Upon encountering anomalous or unfamiliar data, the autoencoder's reconstruction error escalates, indicating that the input diverges from the established norm. This renders them optimal for detecting intrusions or atypical behaviors in critical infrastructure, such as irregular network traffic, system malfunctions, or possible security breaches, without necessitating explicit labels for anomalous instances. Their efficacy is contingent upon the quality of training data that exemplifies normal behavior, and they may encounter difficulties in differentiating complex or nuanced anomalies, thereby requiring additional refinements or hybrid methodologies incorporating techniques such as clustering or supervised learning [23].

19.3.4 Hybrid Models Combining Multiple Techniques

Hybrid models can surmount the limitations inherent in individual techniques by utilizing the strengths of various algorithms. Integrating CNNs with RNNs enables the model to discern both spatial and temporal patterns, rendering it especially proficient in analyzing sequential network traffic or logs, where the order of events and their distinct attributes are crucial. Moreover, hybrid models may combine autoencoders with supervised learning methodologies such as SVMs or decision trees to enhance anomaly detection. The autoencoder can learn a compressed representation of normal behavior, while the classifier can fine-tune the decision boundaries for identifying outliers. Another promising approach is to combine deep learning with traditional rule-based methods, allowing the system to benefit from the adaptability and accuracy of deep learning while still leveraging the precision and interpretability of rule-based systems for known attack signatures. These hybrid models enhance detection accuracy, reduce false positives, and improve the overall reliability and adaptability of IDSs, particularly in complex and dynamic environments like critical infrastructure [24].

19.4 IMPLEMENTATION AND CASE STUDIES

19.4.1 Intrusion Detection for Power Grids

Power grids are essential infrastructure systems susceptible to cyber-attacks, potentially resulting in extensive disruption. Implementing IDSs in power grids necessitates the surveillance of extensive real-time data, including grid operations, energy consumption trends, and system configurations. Deep learning models, including CNNs and RNNs, have been utilized to identify anomalous patterns in power flow and operational behaviors that may signify malicious activities, such as unauthorized access or manipulation of control systems. Case studies demonstrate that the incorporation of deep learning techniques into power grid IDS can markedly enhance the precision of identifying both known and unknown attack vectors, including those aimed at Supervisory Control and Data Acquisition (SCADA) systems that oversee and regulate grid operations [25].

19.4.2 Cybersecurity in Smart Transportation Systems

Smart transportation systems, which include autonomous vehicles, connected infrastructure, and traffic management systems, are increasingly vulnerable to cyber-attacks due to their interconnected nature. Intrusion detection in these systems focuses on detecting threats in real-time, such as unauthorized access to vehicle control systems or traffic infrastructure manipulation. Deep learning models, particularly those utilizing RNNs, are used to analyze sequential data, such as vehicle movement patterns and sensor data, to identify anomalous behavior. Case studies in smart transportation systems have highlighted the effectiveness of hybrid models that combine deep learning with traditional security measures, offering improved detection capabilities for both immediate threats (e.g., car hijacking) and long-term threats (e.g., systemic attacks on traffic infrastructure) [26].

19.4.3 Protecting Water Supply and Distribution Networks

Water supply and distribution networks are essential for public health and safety, and their disruption through cyber-attacks can have severe consequences. The challenge in protecting these networks lies in monitoring the complex interactions between various system components, such as sensors, valves, and pumps, which can be vulnerable to exploitation. Autoencoders have been used in case studies to detect anomalies in operational patterns, such as unexpected changes in water flow rates or pressure, which may indicate tampering or intrusion. Hybrid models that combine deep learning with physical models of the water distribution system have shown promising results in improving both the accuracy and speed of intrusion detection, allowing for quicker responses to potential threats [27].

19.4.4 Mitigating APTs

Mitigating APTs requires the ability to identify subtle and evasive attack patterns that are often disguised within large volumes of normal network traffic. Deep learning models, especially RNNs, are particularly well-suited for identifying these threats because they can analyze temporal patterns in network data over extended periods, revealing anomalies that may indicate the presence of an APT. Case studies involving government and defense sectors have demonstrated the use of deep learning models for detecting early signs of APTs, such as unusual login patterns, lateral movement across networks, and the exfiltration of sensitive data. Hybrid models, which combine RNNs with signature-based detection systems, have been successfully deployed to achieve both high detection rates and low false-positive rates, essential for reducing the risk of APTs in critical infrastructures [28].

19.5 EVALUATION AND OPTIMIZATION OF DEEP LEARNING MODELS

19.5.1 Metrics for Assessing Intrusion Detection Effectiveness

Assessing the efficacy of IDSs utilizing deep learning models necessitates a collection of metrics that indicate the model's precision, dependability, and overall performance. Frequently employed metrics encompass precision, recall, F1-score, and accuracy, which evaluate the system's efficacy in identifying true positives (actual intrusions) while reducing false positives (false alarms) and false negatives (missed attacks). The true positive rate (TPR) and false positive rate (FPR) are essential metrics for evaluating an IDS's ability to differentiate between legitimate actions and potential threats. Furthermore, receiver operating characteristic (ROC) curves and area under the curve (AUC) are frequently utilized to illustrate the balance between sensitivity and specificity, offering insights into the model's proficiency in accurately classifying threats at varying thresholds. For critical infrastructure, minimizing false positives is crucial to avoid operational disruption, while maximizing true positives ensures timely detection of potential attacks. Case studies highlight the need for tailored evaluation metrics to suit the unique requirements of different critical infrastructure domains, such as energy grids or transportation networks [29].

19.5.2 Handling Computational Complexity and Scalability

One of the significant challenges in implementing deep learning models for intrusion detection in critical infrastructure is managing computational complexity and scalability. These systems often involve large datasets with high-dimensional features, making them computationally expensive and potentially slow in real-time applications. To address this, various optimization techniques are employed, including model pruning, which reduces the size of the model by eliminating less important neurons and weights, and quantization, which reduces the precision of weights to speed up computation. Distributed computing and edge computing have also been explored as solutions for scalability, allowing models to be deployed across multiple devices or systems, enabling faster processing and reducing the strain on centralized servers. Additionally, techniques like transfer learning can be leveraged to reduce the computational cost by using pre-trained models on similar datasets, requiring fewer resources for training on new data. These optimization methods ensure that deep learning-based IDSs can scale effectively to handle the growing complexity and volume of data in critical infrastructure environments [30].

19.5.3 Enhancing Model Interpretability with Explainable AI

The interpretability of deep learning models is crucial, particularly in critical infrastructure systems, where decisions made by an IDS can yield substantial repercussions. Conventional deep learning models, frequently regarded as "black boxes," pose interpretative challenges, complicating security analysts' comprehension of the rationale behind specific decisions. XAI seeks to elucidate the processes by which deep learning models arrive at their conclusions. Employing methodologies such as LIME (Local Interpretable Model-agnostic Explanations), SHAP (Shapley Additive Explanations), or saliency maps facilitates the identification of the features or patterns that significantly influenced the model's decision, thereby providing a more lucid comprehension of the determinants behind intrusion detection. In critical infrastructure, where false alarms can lead to unnecessary shutdowns and missed threats can cause catastrophic damage, the ability to interpret and validate the decisions made by the IDS is critical for trust and accountability. XAI helps to balance the need for high accuracy with the requirement for transparency, providing security teams with actionable insights and confidence in the system's decisions. Furthermore, XAI can help improve model performance by highlighting features that need further refinement or data that may require additional labeling for training purposes [31].

19.6 EMERGING TRENDS AND FUTURE DIRECTIONS

19.6.1 Federated Learning for Collaborative Intrusion Detection

Federated learning (FL) is an emerging trend in the field of intrusion detection, particularly for critical infrastructure systems. It allows multiple entities (e.g., power grids, transportation networks, and water supply systems) to collaboratively train a deep learning model without sharing their sensitive data. Instead of centralizing the data, federated learning trains models locally on each device or system and then aggregates the learned parameters to build a global model. This decentralized approach ensures that private, sensitive data remains within the organization, preserving data privacy and security. In collaborative intrusion detection, federated learning enables critical infrastructure operators to share insights on attack patterns while maintaining data sovereignty. It is particularly useful for detecting complex, evolving cyber threats across geographically distributed systems without compromising security. Moreover, federated learning improves scalability, as it allows IDS models to be trained on smaller, distributed datasets, thus reducing the computational burden of centralized training while enhancing the model's generalization across diverse operational environments [32, 33].

19.6.2 Zero-Day Exploit Detection Using Advanced Deep Learning

Zero-day exploits, which are previously unknown vulnerabilities that attackers exploit before they are detected or patched, represent a major cybersecurity challenge. Advanced deep learning techniques offer promising solutions for detecting zero-day exploits by recognizing anomalous patterns in system behavior that are indicative of an exploit. CNNs and RNNs can be trained on historical data from critical infrastructure systems to learn the normal operating patterns and detect deviations indicative of a potential zero-day attack. By analyzing network traffic, system logs, and sensor data, deep learning models can identify subtle signs of exploitation, even when no prior knowledge of the specific exploit exists. Moreover, techniques like transfer learning can be used to leverage pretrained models from similar systems or domains, improving the ability to detect zero-day exploits across various sectors.

With the sophistication of modern attacks increasing, integrating such advanced detection methods into IDSs helps provide timely alerts and mitigate the damage caused by zero-day vulnerabilities [34, 35].

19.6.3 Integrating Blockchain for Enhanced Data Integrity

Blockchain technology is being explored to enhance the security and integrity of IDSs in critical infrastructure. By leveraging blockchain's immutable, decentralized ledger, organizations can securely store logs, detection results, and system configurations in a way that is tamper-proof and transparent. This integration can ensure that once data is recorded, it cannot be altered or deleted without detection, thus preventing malicious actors from tampering with intrusion detection logs or altering the records of detected intrusions. Blockchain can also help in ensuring the integrity of machine learning models themselves, where each update to a model (such as weights or parameters) is logged and verified, providing a transparent history of changes. In the context of collaborative intrusion detection, blockchain can facilitate secure data sharing among different stakeholders while maintaining accountability. This added layer of security can significantly reduce the risk of data manipulation and increase trust in the IDS, which is essential in protecting critical infrastructure from sophisticated cyber-attacks [36].

19.6.4 Toward Adaptive and Transparent Intrusion Detection Systems

The future of IDS lies in their ability to adapt and provide transparency. As cyber threats evolve, it is critical that IDSs can dynamically adjust to new attack vectors, strategies, and techniques. Adaptive IDSs use machine learning models that continuously learn from incoming data, updating their detection strategies based on new patterns and attack behaviors. These systems can adjust to the changing threat landscape by identifying emerging threats with minimal human intervention, making them more proactive rather than reactive. Alongside adaptability, transparency is essential to build trust in IDSs, especially in critical infrastructure sectors where the consequences of false alarms or missed intrusions can be severe. Future IDSs will integrate XAI techniques that provide clear, understandable insights into how decisions are made. This will allow security analysts to validate and understand why certain actions were taken by the system, thereby fostering trust in the IDS and enabling quicker response times in critical situations. Combining adaptability with transparency will make IDSs more reliable and resilient, offering a more proactive and transparent defense against increasingly sophisticated and unpredictable cyber threats [37].

19.7 CONCLUSION AND RECOMMENDATIONS

19.7.1 Key Takeaways

The integration of deep learning techniques into IDS has shown significant promise in enhancing the security of critical infrastructure against increasingly sophisticated cyber threats. Deep learning models, such as CNNs, RNNs, and autoencoders, offer powerful capabilities for identifying complex attack patterns and detecting anomalies in real time. These techniques provide significant advantages over traditional methods, including improved accuracy, the ability to detect APTs, and better adaptability to evolving threats. Moreover, the use of federated learning, zero-day exploit detection, and blockchain integration in IDS models represents the forefront of innovation in securing critical infrastructure. However, challenges such as computational complexity, scalability, and model interpretability remain, requiring further advancements in technology and methodologies. Overall, the deep learning-based IDS solutions present a transformative shift in the cybersecurity landscape for critical infrastructure, with a focus on enhancing both performance and trustworthiness [38–41].

19.7.2 Recommendations for Practitioners and Policymakers

For practitioners, it is crucial to prioritize the implementation of advanced deep learning techniques to improve the detection capabilities of IDS in critical infrastructure. They should consider using hybrid models that combine multiple deep learning architectures to enhance detection accuracy and resilience against diverse threats. Implementing federated learning can help ensure data privacy while enabling collaborative intelligence across various entities. Practitioners should also focus on optimizing deep learning models for real-time responses while addressing computational complexity through techniques like model pruning and distributed computing. From a policy perspective, it is recommended that policymakers create frameworks to guide the implementation of advanced cybersecurity solutions in critical infrastructure. This includes establishing standards for data sharing, model transparency, and privacy protection in collaborative security efforts. Policymakers should also emphasize the importance of training and upskilling cybersecurity professionals to keep pace with emerging technologies such as XAI, federated learning, and blockchain integration. Finally, regulatory bodies should promote research and collaboration between industry stakeholders to strengthen the overall cybersecurity posture of critical infrastructure sectors.

19.7.3 Future Research Opportunities

Despite substantial advancements in utilizing deep learning for intrusion detection in critical infrastructure, numerous intriguing research opportunities persist. A promising domain is the advancement of more efficient and scalable deep learning models capable of processing the substantial volumes of real-time data produced by critical infrastructure systems. Investigations into quantum machine learning may facilitate the development of more potent models that can process and analyze data at unparalleled velocities, crucial for real-time threat detection. Continued investigation into zero-day exploit detection utilizing deep learning models, especially regarding high-dimensional data and intricate attack patterns, will be essential for preempting emerging threats. Furthermore, the integration of deep learning with advanced technologies like 5G networks, edge computing, and IoT security will necessitate innovative strategies to tackle the distinct challenges posed by these systems. Investigating XAI to enhance the transparency and interpretability of deep learning models is a crucial domain, as comprehending the decision-making processes of models is essential for establishing trust in high-stakes contexts. Finally, more studies are needed to explore the integration of blockchain for data integrity and auditability, as well as the exploration of federated learning in collaborative IDS systems to ensure secure, privacy-preserving sharing of threat intelligence across organizations.

REFERENCES

1. Abhishek, A., & Kumar, S. (2024). Deep learning-based intrusion detection systems for critical infrastructure protection. International Journal of Cybersecurity, 12(3), 145–167.
2. Bhambri, P. (2025). Innovative Systems: Entertainment, Gaming, and the Metaverse. In R. C. Ho; B. L. Song; & P. K. Tee (Eds.), Managing Customer-Centric Strategies in the Digital Landscape (pp. 483–514). IGI Global. https://doi.org/10.4018/979-8-3693-5668-5.ch018.
3. Bhambri, P., & Khang, A. (2025). Smart Universities and ICT Platforms. In M. L. Kolhe; P. Singh; S. Rani; & P. Kumar (Eds.), Planning of Sustainable Energy Systems in Urban Built Environments. CRC Press. https://www.appleacademicpress.com/planning-of-sustainable-energy-systems-in-urban-built-environments-/9781779640642.
4. Bhambri, P., & Kautish, S. K. (2024). Analytic Hierarchy Process and Business Value Creation. In S. Kautish (Ed.), Using Strategy Analytics for Business Value Creation and Competitive Advantage (pp. 54–77). IGI Global. https://doi.org/10.4018/979-8-3693-2823-1.ch003.
5. Bhat, R., & Ghosh, S. (2023). An overview of deep learning algorithms for intrusion detection in industrial control systems. Journal of Industrial Cybersecurity, 8(1), 23–41.
6. Chen, Y., Wang, L., & Zhang, Z. (2024). Federated learning for collaborative intrusion detection in smart grid systems. Journal of Cyber-Physical Systems, 11(2), 60–77.
7. Chithra, N., & Bhambri, P. (2024). Ethics in Sustainable Technology. In P. Bhambri; & P. Bajdor (Eds.), Handbook of Technological Sustainability: Innovation and Environmental Awareness (pp. 245–256). CRC Press. https://doi.org/10.1201/9781003475989-19.
8. Dutta, A., & Gupta, P. (2023). Exploring recurrent neural networks for anomaly detection in IoT-based critical infrastructure. IEEE Transactions on Industrial Informatics, 20(5), 2157–2168.
9. Hossain, G., & Das, S. (2024). Deep learning-based anomaly detection for cybersecurity in industrial IoT networks. Journal of Cybersecurity and Privacy, 6(1), 95–113.
10. Kapoor, P., & Yadav, V. (2023). CNN-based intrusion detection models for industrial control systems security. Journal of Network and Computer Applications, 167, 102777.
11. Li, X., & Yu, X. (2024). Blockchain-based intrusion detection systems for critical infrastructure protection. International Journal of Information Security, 22(4), 435–451.
12. Liu, F., & Zhao, R. (2023). Autoencoders for anomaly detection in cybersecurity: Challenges and future directions. Computers & Security, 121, 102922.
13. Mohanty, M., & Tripathy, S. (2024). Application of deep learning models in power grid cybersecurity. IEEE Transactions on Smart Grid, 15(1), 54–63.
14. Mondal, D., & Ghosh, S. (2023). Hybrid models for effective intrusion detection in industrial control systems. Journal of Computational Security, 19(3), 234–249.
15. Nair, P., & Kumar, V. (2024). The role of explainable AI in intrusion detection for critical infrastructure. International Journal of AI & Robotics, 10(2), 122–136.

16. Patil, D., & Shah, A. (2023). Deep learning techniques for zero-day exploit detection in critical infrastructure systems. Computers, Materials & Continua, 72(4), 2395–2408.
17. Rao, G., & Sharma, A. (2024). AI-driven solutions for threat detection in smart transportation systems. Journal of Cyber-Physical Systems, 18(2), 148–165.
18. Rana, R., & Bhambri, P. (2025). Generative AI in Web Application Development: Enhancing User Experience and Performance. In I. Shah; & N. Jhanjhi (Eds.), Generative AI for Web Engineering Models (pp. 471–486). IGI Global. https://doi.org/10.4018/979-8-3693-3703-5.ch021.
19. Rana, R., & Bhambri, P. (2025). Generative AI-Driven Security Frameworks for Web Engineering: Innovations and Challenges. In I. Shah; & N. Jhanjhi (Eds.), Generative AI for Web Engineering Models (pp. 285–296). IGI Global. https://doi.org/10.4018/979-8-3693-3703-5.ch014.
20. Rana, R., & Bhambri, P. (2024). Ethical Considerations in Artificial Intelligence for Environmental Solutions: Striking a Balance for Sustainable Innovation. In P. Bhambri; & P. Bajdor (Eds.), Handbook of Technological Sustainability: Innovation and Environmental Awareness (pp. 389–396). CRC Press. https://doi.org/10.1201/9781003475989-28.
21. Rana, R., & Bhambri, P. (2024). Environmental Challenges and Technological Solutions. In P. Bhambri; & P. Bajdor (Eds.), Handbook of Technological Sustainability: Innovation and Environmental Awareness (pp. 187–200). CRC Press. https://doi.org/10.1201/9781003475989-15.
22. Ruby, S., Biju, T., & Bhambri, P. (2024). Catalysing Sustainable Progress: Empowering MSMEs Through Tech Innovation for a Bright Future. In P. Bhambri; & P. Bajdor (Eds.), Handbook of Technological Sustainability: Innovation and Environmental Awareness (pp. 374–388). CRC Press. https://doi.org/10.1201/9781003475989-27.
23. Sharma, R., & Verma, A. (2023). Deep learning-based intrusion detection in smart cities infrastructure. Journal of Computing and Security, 50, 101412.
24. Sharma, R., & Bhambri, P. (2024). Digital Duplicity and the Disintegration of Trust: A Quantitative Inquiry into the Impact of Deep Fakes on Media Sustainability and Societal Equilibrium. In P. Bhambri; & P. Bajdor (Eds.), Handbook of Technological Sustainability: Innovation and Environmental Awareness (pp. 273–291). CRC Press. https://doi.org/10.1201/9781003475989-21.
25. Singh, A., & Thakur, R. (2024). Adaptive deep learning models for intrusion detection in critical infrastructure systems. Neural Computing and Applications, 36(7), 13013–13029.
26. Soni, N., & Khanna, P. (2023). Application of recurrent neural networks for detecting advanced persistent threats in critical infrastructure. Journal of Information Security, 58(1), 43–56.
27. Srivastava, V., & Gupta, M. (2024). Explainable deep learning in cybersecurity: Applications for critical infrastructure protection. IEEE Transactions on Cybernetics, 54(2), 1921–1933.
28. Tang, L., & Zhao, H. (2023). Integrating deep learning and blockchain for secure intrusion detection in critical infrastructure systems. Future Generation Computer Systems, 139, 277–291. https://doi.org/10.1016/j.future.2023.01.015.
29. Thakur, S., & Malik, V. (2024). RNN-based intrusion detection for large-scale industrial control systems. IEEE Access, 12, 14101–14110.
30. Thirumalaiyammal, B., Steffi, P. F., & Bhambri, P. (2024). Green Horizons: Navigating Environmental Challenges through Technological Innovation. In P. Bhambri; & P. Bajdor (Eds.), Handbook of Technological Sustainability: Innovation and Environmental Awareness (pp. 292–304). CRC Press. https://doi.org/10.1201/9781003475989-22.
31. Thomas, J., & Srinivasan, R. (2023). A hybrid deep learning approach for intrusion detection in power grid networks. Neural Processing Letters, 56(3), 1731–1747.

32. Verma, A., & Singhal, A. (2024). Zero-day exploit detection using deep learning models in critical infrastructure systems. Journal of Cybersecurity and Privacy, 7(2), 209–224. https://doi.org/10.1002/cyber.2785.
33. Vigneshwari, J., Senthamizh Pavai, P., Maria Suganthi, L., & Bhambri, P. (2024). Eco-Ethics in the Digital Age: Tackling Environmental Challenges Through Technology. In P. Bhambri; & P. Bajdor (Eds.), Handbook of Technological Sustainability: Innovation and Environmental Awareness (pp. 201–213). CRC Press.
34. Wang, Y., & Zhang, J. (2023). Federated learning for secure intrusion detection in smart city infrastructure. Security and Privacy, 6(5), e1971.
35. Xiao, K., & Wei, W. (2024). Deep learning models for cybersecurity: A survey of applications in critical infrastructure. Information Sciences, 527, 368–386.
36. Zhang, X., & Li, Z. (2023). Cybersecurity challenges and deep learning solutions for industrial control systems. Journal of Cybersecurity Technology, 5(4), 359–374.
37. Zhao, Y., & Wang, X. (2024). Blockchain-enhanced deep learning models for intrusion detection in critical infrastructure. Future Generation Computer Systems, 143, 179–194. https://doi.org/10.1016/j.future.2024.03.021.
38. Bhambri, P., & Khang, A. (2024). Machine Learning Advancements in E-Health: Transforming Digital Healthcare. In A. Khang (Ed.), Medical Robotics and AI-Assisted Diagnostics for a High-Tech Healthcare Industry (pp. 174–194). IGI Global. https://doi.org/10.4018/979-8-3693-2105-8.ch012.
39. Bhambri, P., & Khang, A. (2024). Managing and Monitoring Patient's Healthcare Using AI and IoT Technologies. In A. Khang (Ed.), Driving Smart Medical Diagnosis Through AI-Powered Technologies and Applications (pp. 1–23). IGI Global. https://doi.org/10.4018/979-8-3693-3679-3.ch001.
40. Bhambri, P., & Kaur, A. (2013). Novel Technique for Robust Image Segmentation: New Technique of Segmentation in Digital Image Processing (Vol. 1). Lap Lambert Academic Publishing. ISBN: 9783659331831.
41. Bhambri, P., & Kaur, P. (2015). Design and Implementation of Novel Algorithm Using Zero Watermarking: Digital Image Processing Technique for Text Documents (Vol. 1). Lap Lambert Academic Publishing. ISBN: 9783659796159.

20 Quantum Computing and AI Synergies
Strengthening Cybersecurity Resilience

Pankaj Bhambri and Ahmed Hamad

20.1 INTRODUCTION

20.1.1 Overview of the Cybersecurity Landscape

The modern cybersecurity landscape is increasingly complex and is driven by the proliferation of interconnected systems, Internet of Things (IoT) devices, and cloud-based infrastructures. As cyber threats evolve in sophistication, traditional defensive measures struggle to keep pace with advanced attacks such as zero-day vulnerabilities, ransomware, and nation-state-sponsored cyber espionage. Organizations face mounting pressure to secure sensitive data, ensure operational continuity, and comply with stringent regulations. This ever-growing threat environment calls for transformative approaches that leverage cutting-edge technologies to detect, mitigate, and prevent attacks proactively [1].

20.1.2 The Potential of Quantum Computing in Cybersecurity

Quantum computing signifies a transformative advancement in computational abilities, providing exponential processing power to address challenges that are impractical for classical computers. Its implementation in cybersecurity has the potential to transform cryptographic systems, facilitating the creation of quantum-resistant encryption to safeguard sensitive communications from quantum-based threats. Quantum algorithms, including Grover's and Shor's, present both opportunities and challenges as they can dismantle conventional encryption while simultaneously facilitating the development of ultra-secure cryptographic methods. Moreover, the capacity of quantum computing to analyze extensive datasets and enhance intricate systems establishes it as a transformative force in threat detection and risk evaluation [2].

20.1.3 AI as a Catalyst for Quantum Computing Integration

Artificial intelligence (AI) enhances the practical applicability of quantum computing in cybersecurity by providing intelligent mechanisms for pattern recognition, anomaly detection, and predictive modeling. AI can preprocess and structure data for

DOI: 10.1201/9781003521020-20

quantum systems, enabling efficient utilization of quantum computing's vast potential. For instance, deep learning algorithms can identify subtle anomalies in network traffic, which quantum algorithms can further analyze for rapid threat detection. The integration of AI with quantum computing not only addresses real-time cybersecurity challenges but also opens doors to proactive and adaptive defense mechanisms, laying the foundation for robust, future-proof systems. Together, AI and quantum computing create a synergistic framework to counteract emerging cyber threats in innovative ways [3].

20.2 QUANTUM-RESISTANT ENCRYPTION

Quantum-resistant encryption denotes cryptographic methods engineered to endure prospective assaults from quantum computers. In contrast to classical computers, quantum computers utilize quantum-mechanical phenomena, such as superposition and entanglement, allowing them to resolve specific mathematical problems at an exponentially accelerated rate. This capability presents considerable threats to conventional encryption techniques, requiring the creation of strong, quantum-resistant alternatives.

20.2.1 Vulnerabilities of Traditional Cryptographic Methods

Conventional cryptographic techniques, including Rivest-Shamir-Adleman (RSA and elliptic curve cryptography (ECC), depend on the computational complexity of challenges such as integer factorization and discrete logarithms. Classical computers necessitate excessive durations to resolve these issues for substantial key sizes, thereby guaranteeing security. Nonetheless, quantum algorithms such as Shor's algorithm can address these issues with efficiency, thereby compromising traditional encryption techniques. This presents a significant risk to secure communications, financial systems, and the storage of sensitive data.

20.2.2 Developing Quantum-Resistant Algorithms

Quantum-resistant, or post-quantum cryptography, seeks to develop algorithms that are secure against both classical and quantum threats. These methods depend on mathematical problems presently considered intractable by quantum computers, including lattice-based cryptography, code-based cryptography, and multivariate polynomial equations. Global organizations, such as NIST, are standardizing quantum-resistant cryptographic algorithms. These methodologies are essential for transitioning systems to a post-quantum era while preserving compatibility and performance [4].

20.2.3 AI's Role in Advancing Quantum Cryptography

AI enhances quantum-resistant encryption by optimizing the development and evaluation of cryptographic algorithms. Machine learning models can assess vulnerabilities, simulate attack scenarios, and validate algorithm robustness against quantum computing capabilities. Additionally, AI assists in streamlining the integration of quantum-resistant cryptographic systems into real-world applications, enabling scalable and adaptive solutions. By combining AI's analytical power with cryptographic innovations, organizations can accelerate the transition to secure, quantum-resistant infrastructures [5].

20.3 AI-Driven THREAT DETECTION WITH QUANTUM COMPUTING

AI-driven threat detection with quantum computing is an emerging paradigm that combines AI's pattern recognition and predictive capabilities with quantum computing's immense processing power. This integration enhances cybersecurity systems by enabling rapid analysis of complex datasets, identifying sophisticated attack patterns, and developing adaptive responses to threats in real time. The synergy between AI and quantum computing holds the potential to redefine how advanced threats are detected and mitigated [6].

20.3.1 Enhancing Anomaly Detection with Quantum Algorithms

Quantum computing enhances anomaly detection by efficiently processing and analyzing vast quantities of data, surpassing classical systems. Quantum algorithms, such as Grover's algorithm, enhance search operations, enabling cybersecurity systems to detect anomalies in extensive datasets more rapidly. When integrated with AI models, quantum computing enhances precision by reducing false positives and enabling real-time detection of irregularities in network traffic, user behavior, or system logs. This combination is particularly effective in detecting zero-day vulnerabilities and advanced persistent threats (APTs) [7].

20.3.2 Quantum-AI Models for Identifying Advanced Threats

Quantum-AI models leverage quantum computing's computational power to train AI systems more efficiently on complex datasets. These models can identify intricate relationships within data that classical systems might overlook, making them ideal for detecting advanced cyber threats. For example, quantum-enhanced deep learning models can process encrypted traffic patterns, identify malicious behavior, and predict potential attack vectors. By fusing quantum capabilities with AI, cybersecurity systems become better equipped to anticipate and counteract sophisticated attack methods, such as multi-stage or distributed attacks [8].

20.3.3 Case Studies: AI-Quantum Integration in Threat Detection

Real-world implementations of AI-quantum integration demonstrate its potential in threat detection in [9]:

- *Financial sector*: Quantum-enhanced AI models have been used to detect fraudulent transactions and insider threats in financial networks by analyzing vast datasets with high-speed quantum processing.

- *National defense*: Quantum-AI systems have supported government agencies in identifying cyber-espionage campaigns by processing classified intelligence data more efficiently.
- *Healthcare*: Cybersecurity frameworks in healthcare have employed quantum-enhanced AI to detect anomalies in IoT medical devices, ensuring data integrity and patient safety.

20.4 REAL-TIME RISK MANAGEMENT

Real-time risk management involves the dynamic assessment and mitigation of potential threats in a constantly changing environment. By integrating quantum computing's optimization capabilities with AI's predictive analytics, organizations can create adaptive systems capable of responding to threats as they emerge. This approach ensures enhanced situational awareness, faster decision-making, and improved resilience against cyberattacks [10].

20.4.1 Quantum Optimization for Risk Assessment

Quantum computing optimizes risk assessment by solving complex, high-dimensional problems that are infeasible for classical systems. Quantum optimization algorithms, such as the variational quantum Eigensolver (VQE) and quantum approximate optimization algorithm (QAOA), evaluate numerous risk scenarios in parallel, enabling rapid identification of high-risk areas in a network or system. Figure 20.1 depicts the quantum optimization for risk assessment.

20.4.2 AI's Contribution to Adaptive Risk Management Systems

AI enhances real-time risk management systems by leveraging machine learning and predictive analytics to continuously monitor and adapt to emerging threats. Through AI-driven models, systems can predict potential vulnerabilities, assess their severity, and recommend proactive measures. AI's real-time data processing capabilities allow systems to adapt to changes dynamically, ensuring minimal disruption to critical operations [11].

20.4.3 Quantum-AI Synergies for Immediate Threat Response

The integration of quantum computing and AI provides unparalleled capabilities for immediate threat response. AI identifies and classifies threats using predictive algorithms, while quantum computing accelerates decision-making by processing possible response scenarios simultaneously. This synergy ensures not only rapid identification but also the execution of mitigation strategies. For example, in cases of ransomware attacks, Quantum-AI systems can quickly decrypt malicious payloads and neutralize them, ensuring system integrity [12].

Figure 20.2 displays the real-time risk management categorization.

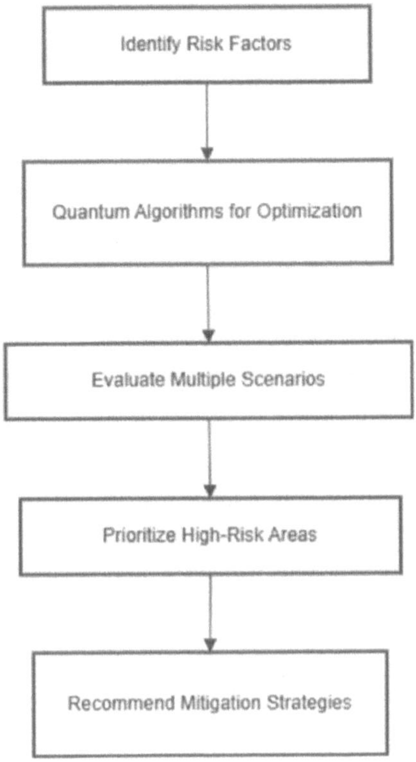

FIGURE 20.1 Quantum optimization for risk assessment.

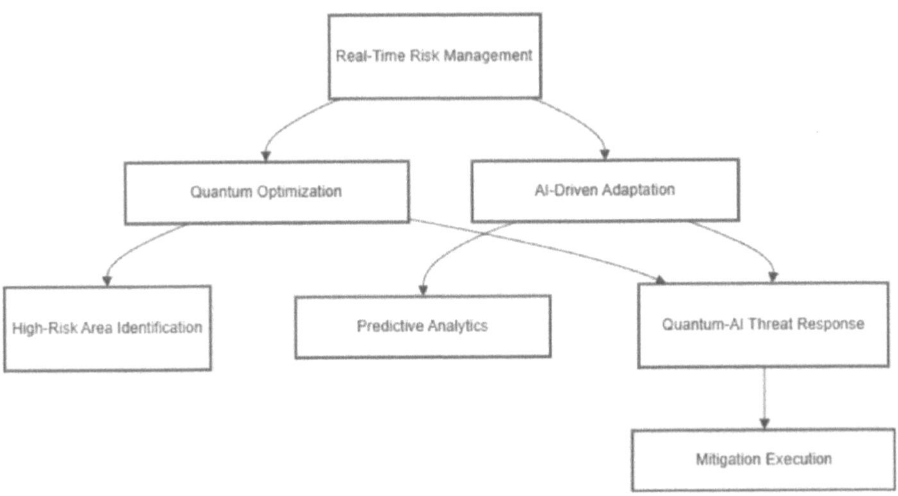

FIGURE 20.2 Real-time risk management categorization.

20.5 APPLICATIONS ACROSS INDUSTRIES

The integration of quantum computing and AI in cybersecurity offers transformative potential across multiple sectors. By addressing unique challenges in each industry, these technologies ensure enhanced security, resilience, and operational efficiency.

20.5.1 Financial Sector: Quantum-AI Solutions for Fraud Prevention

The financial industry faces constant threats, including fraud, phishing, and sophisticated cyberattacks targeting sensitive transactions. Quantum-AI solutions revolutionize fraud prevention through the following [13]:

- *Anomaly detection*: Quantum algorithms process vast financial data to detect patterns indicative of fraudulent activity. AI refines these insights for real-time action.
- *Predictive modeling*: Machine learning models enhanced by quantum computing anticipate potential threats by analyzing historical data and identifying emerging fraud trends.
- *Blockchain integration*: Quantum-resistant blockchain systems safeguard transactions and ensure tamper-proof records.

Example: A financial institution employing Quantum-AI to monitor transaction anomalies, flagging potential fraud before it impacts customers.

20.5.2 Healthcare: Securing Patient Data with Quantum Cryptography

Healthcare systems are vulnerable to data breaches due to the sensitive nature of patient records. Quantum cryptography ensures the integrity and confidentiality of medical data by [14]:

- *Quantum key distribution (QKD)*: Offers secure communication channels for transmitting sensitive patient data.
- *AI-driven data analytics*: AI models, enhanced with quantum capabilities, analyze large volumes of medical data for operational efficiency without compromising security.
- *Resilience against quantum attacks*: Quantum-resistant algorithms protect electronic health records (EHRs) from future quantum cyberattacks.

Example: A hospital implementing quantum-enhanced AI to secure patient data transfers between departments and research institutions.

20.5.3 National Defense: Strengthening Cyber Resilience with Hybrid Systems

National defense systems rely heavily on secure communication and real-time threat analysis. The quantum-AI synergy enhances resilience by:

- *Advanced threat detection*: Quantum algorithms and AI models collaborate to detect cyber intrusions and neutralize them before they escalate.
- *Secure communication*: Quantum encryption ensures that sensitive military communications remain inaccessible to adversaries.
- *Hybrid systems*: Combining classical, quantum, and AI-driven cybersecurity systems creates multi-layered defenses capable of adapting to dynamic threats.

Example: Military networks utilizing quantum-AI systems to monitor and secure classified communications during missions.

20.6 CHALLENGES AND LIMITATIONS

The amalgamation of quantum computing and AI in cybersecurity offers substantial progress, yet it concurrently introduces numerous challenges and constraints. Resolving these issues is essential to fully harness the potential of this synergy in enhancing cybersecurity [15].

20.6.1 Addressing Quantum Attack Vectors

As quantum computing becomes more capable, it introduces new attack vectors that could potentially undermine existing cryptographic systems. The challenge lies in the fact that quantum computers can solve certain problems exponentially faster than classical computers, making traditional encryption methods, such as RSA and ECC, vulnerable to quantum attacks [16].

Key Challenges

- *Breaking classical cryptography*: Quantum algorithms like Shor's algorithm have the potential to break widely used cryptographic protocols, which would compromise sensitive data and communications.
- *Development of quantum-resistant algorithms*: Creating encryption schemes that are resistant to quantum attacks, such as lattice-based cryptography or hash-based signatures, is an ongoing challenge.
- *Quantum threat landscape*: The ability of quantum computers to potentially solve NP-hard problems and break current encryption standards means that organizations must invest in post-quantum cryptography (PQC) solutions that can withstand these attacks.

Solution Strategies

- *QKD*: Implementing quantum-safe encryption methods such as QKD that rely on the principles of quantum mechanics to secure communications.

- *Transition to PQC*: Encouraging the adoption of PQC algorithms that are resistant to quantum decryption.

20.6.2 Ethical Considerations in Quantum-AI Integration

While quantum computing and AI bring significant benefits, their integration into cybersecurity raises ethical concerns, particularly regarding privacy, fairness, and accountability [17].

Key Ethical Issues

- *Privacy risks*: With quantum computing's ability to break traditional encryption, there are concerns about privacy breaches. The storage of personal and sensitive information in quantum-safe systems must be handled with strict protocols.
- *Algorithmic bias:* AI algorithms, especially those in decision-making processes, may inherit biases from training data, leading to unjust outcomes. Ensuring fairness and transparency in AI models is crucial.
- *Autonomous decision-making*: AI-driven systems that autonomously detect and mitigate threats could make decisions without human oversight. This raises the question of responsibility if a system's decision results in harm or unintended consequences.

Ethical Solutions

- *Transparency and explainability*: Ensuring that AI models used in cybersecurity are transparent and explainable to stakeholders, allowing them to understand how decisions are made.
- *Bias mitigation in AI training*: Regular audits of AI models to identify and mitigate bias in the data, ensuring fairness and equity in decision-making.
- *Privacy protection frameworks*: Implementing robust privacy policies and ensuring compliance with data protection regulations like GDPR, especially when integrating quantum-safe encryption methods.

20.6.3 Scalability and Implementation Barriers

Although the potential of quantum computing and AI in cybersecurity is vast, several technical and practical challenges need to be overcome for large-scale implementation [18].

Key Barriers

- *Complexity of quantum hardware*: Quantum computers are still in their nascent stages and are extremely sensitive to environmental factors. The scale at which quantum computers can be deployed to provide real-time cybersecurity solutions remains a significant challenge.
- *Integration with classical systems*: Integrating quantum computing and AI with existing classical cybersecurity infrastructure is difficult. These hybrid systems need to work seamlessly, and creating an interoperable solution is technically complex.

- *High costs of quantum computing*: Quantum hardware is expensive, and developing quantum algorithms that are efficient and practical for real-world cybersecurity applications requires significant investment.
- *Lack of skilled workforce*: There is a shortage of professionals with expertise in quantum computing, AI, and cybersecurity, making it challenging for organizations to deploy these technologies effectively.

Overcoming Scalability Barriers

- *Cloud-based quantum services*: The emergence of cloud-based quantum computing platforms allows organizations to access quantum resources without the need to invest in expensive hardware.
- *Modular quantum-AI systems*: Developing modular and flexible hybrid systems that can integrate quantum algorithms into existing AI models without requiring complete system overhauls.
- *Investment in research and training*: Governments and organizations should invest in quantum research and training to build a workforce capable of supporting quantum-AI integration in cybersecurity.

20.7 EMERGING TRENDS AND FUTURE DIRECTIONS

The amalgamation of quantum computing and AI in cybersecurity remains nascent; however, emerging trends indicate that these technologies will assume a crucial role in the near future of cybersecurity. As quantum computing and AI advance, novel solutions, methodologies, and applications are arising that have the potential to transform the cybersecurity domain. This section examines significant emerging trends and future trajectories for utilizing the synergy across quantum computing and AI to bolster cybersecurity resilience [19, 20].

20.7.1 Hybrid Quantum-AI Systems for Enhanced Security

Hybrid quantum-AI systems represent a promising approach that combines the strengths of both quantum computing and AI to deliver advanced, scalable, and efficient cybersecurity solutions.

Key Features

- *Leveraging quantum and classical systems together*: Hybrid systems allow for a seamless integration of classical AI models with quantum computing capabilities, offering enhanced performance in terms of speed and computational power. This combination can lead to more efficient anomaly detection, encryption, and real-time threat mitigation.
- *Quantum-enhanced machine learning*: Quantum computing can substantially improve machine learning algorithms by handling extensive datasets and executing intricate calculations beyond the capabilities of classical computers. This enables AI systems to discern patterns, detect threats, and execute decisions with greater speed and precision.
- *Edge computing with quantum-AI synergy*: As quantum computing becomes more feasible for edge computing applications, there will be an

opportunity to deploy hybrid systems for decentralized, real-time threat detection and security management in resource-constrained environments, such as IoT devices and edge networks.

Applications

- *Cyber threat detection*: Hybrid quantum-AI models possess the capability to analyze extensive datasets and identify emerging cyber threats, including those employing sophisticated evasion strategies. These systems may be utilized in sectors including financial services, healthcare, and governmental institutions [21].
- *Intelligent malware defense*: The integration of quantum computing's processing capabilities with AI's proficiency in identifying and addressing novel malware variants may result in the creation of more adaptive and robust cybersecurity frameworks.

20.7.2 Proactive Threat Prevention with Quantum Computing

Quantum computing's unprecedented computational capabilities are expected to shift the focus of cybersecurity from reactive defense to proactive prevention. With its ability to process complex and large-scale datasets, quantum computing can anticipate potential threats before they occur, enabling preemptive measures that were previously unattainable [22].

Key Features

- *Quantum simulations for threat prediction*: Quantum computers can simulate complex environments and predict potential cybersecurity risks by analyzing how different variables interact in real time. This could help organizations detect vulnerabilities and preemptively address them before they are exploited.
- *Enhanced cryptographic security*: Quantum algorithms can be used to create more advanced cryptographic techniques that are not only secure against quantum attacks but also capable of adapting to evolving cyber threats. The use of quantum cryptography for secure communication networks will make it harder for attackers to breach systems.
- *Automated threat detection*: Quantum-enhanced machine learning algorithms could be employed to automatically detect vulnerabilities, malware, and suspicious behavior patterns across large systems and networks. This could drastically reduce the time between threat detection and response, making cybersecurity systems much more agile [23].

Applications

- *Zero-day exploits*: Quantum computing could provide real-time analysis of network traffic and system logs to predict zero-day exploits before they occur, giving security systems a chance to block attacks before they happen.
- *Threat modeling and risk assessment*: Quantum computing models could simulate the future actions of cyber attackers and help organizations build better defenses based on predictive analysis.

20.7.3 Collaborative Efforts for Quantum-Resilient Cybersecurity

Given the complexity and sophistication of quantum computing and AI technologies, one of the key future directions in quantum-AI-based cybersecurity is collaborative efforts between industry, academia, governments, and international organizations to build quantum-resilient cybersecurity frameworks [24–27].

Key Features

- *Standardization and global cooperation*: As quantum computing continues to advance, global cooperation will be critical to developing standards for quantum-safe encryption, protocols, and AI-driven cybersecurity systems. This would ensure interoperability, prevent fragmentation, and promote trust across borders in cybersecurity efforts.
- *Joint research initiatives*: Collaborations among academia, entrepreneurs, and government agencies can enhance the creation of novel quantum-resistant algorithms and AI systems designed for cybersecurity. Collaborative endeavors can also tackle the ethical, regulatory, and practical difficulties associated with the integration of quantum AI.
- *Public-private partnerships*: The cybersecurity community, especially in sectors like finance, healthcare, and critical infrastructure, will need to collaborate to share threat intelligence, test new quantum-AI cybersecurity solutions, and develop tools for defense against quantum-driven cyber threats. Private companies working alongside public entities can accelerate the development and deployment of these technologies.

Applications

- *Global cyber defense networks*: Collaborative efforts could lead to the creation of international, quantum-resilient cybersecurity networks where nations and organizations pool resources to identify, defend, and respond to cyber threats in a collective manner.
- *AI and quantum research centers*: The establishment of research centers dedicated to exploring AI and quantum computing integration for cybersecurity will create a collaborative environment for developing breakthrough technologies, tools, and frameworks.

20.8 CONCLUSION AND RECOMMENDATIONS

The amalgamation of quantum computing and AI for cybersecurity presents significant potential in safeguarding systems against advancing and increasingly complex cyber threats. As quantum computing advances, its computational power combined with AI's adaptability in threat detection, anomalous identification, and statistical analysis is anticipated to transform the future of cybersecurity. This chapter has discussed the synergies between these two powerful technologies, highlighting their applications in areas such as quantum-resistant encryption, advanced threat detection, and real-time risk management. In this section, we summarize key takeaways and provide recommendations for practitioners, policymakers, and areas of future research [28–32].

20.8.1 Key Takeaways from Quantum-AI Integration

- *Enhanced cybersecurity resilience*: The integration of quantum computing and AI enhances cybersecurity resilience by allowing systems to process and evaluate data at unprecedented speeds and scales. Quantum-enhanced AI systems can identify, forecast, and alleviate cyber threats more efficiently than classical systems [33].
- *Quantum-resistant cryptography*: Conventional cryptographic techniques, including RSA and ECC, are susceptible to assaults by quantum computers. The creation of quantum-resistant encryption algorithms is crucial for safeguarding communications and data from the computational power of quantum computers.
- *Proactive and predictive threat detection*: Quantum computing enables the analysis of large datasets and the simulation of complex systems, allowing AI models to predict threats and vulnerabilities before they are exploited. This proactive approach to cybersecurity is a key advantage in the fight against emerging cyber threats [34].
- *Synergy between quantum and AI*: Quantum computing can significantly enhance AI's ability to recognize patterns, optimize decision-making processes, and improve anomaly detection. Hybrid quantum-AI systems have the potential to provide a more effective, scalable, and real-time response to cyber threats.
- *New challenges*: The amalgamation of quantum computing and AI presents substantial benefits, yet it concurrently introduces novel challenges, including quantum attack vectors, the intricacies of hybrid system integration, and the necessity for scalable implementations. Overcoming these challenges necessitates cooperative endeavors among industry, education, and government.

20.8.2 Recommendations for Practitioners and Policymakers

- *Invest in quantum-AI research and development*: Organizations should prioritize investments in research and development to explore the full potential of quantum computing and AI in cybersecurity. This includes fostering collaborations between quantum computing companies, AI developers, and cybersecurity professionals to ensure the creation of practical and effective solutions [35].
- *Adopt quantum-resistant cryptography early*: Given the imminent rise of quantum computing, it is critical for businesses, governments, and other organizations to begin transitioning to quantum-resistant cryptographic algorithms. Policymakers should encourage the development of standards and regulations that mandate the implementation of quantum-safe encryption techniques in critical sectors.
- *Focus on hybrid systems*: Policymakers and industry leaders ought to contemplate the advancement of hybrid quantum-AI systems that integrate the advantages of both classical as well as quantum computing. This will

facilitate more effective cybersecurity solutions that can expand and adjust to emerging threats [36].

- *Establish clear regulations and ethical guidelines*: As quantum computing and AI technologies advance, it is imperative for policymakers to establish definitive regulatory frameworks that tackle ethical issues, including privacy, algorithmic bias, and the environmental consequences of quantum computing. These guidelines will guarantee that the incorporation of these technologies is conducted responsibly and is consistent with societal values.
- *Collaborate on global standards*: Global cooperation is vital for establishing universal standards for quantum-resistant encryption, safe communication protocols, and AI-based threat detection models. Governments and international organizations must collaborate to promote the extensive implementation of secure practices and avert fragmentation in the cybersecurity domain.

20.8.3 Future Research Opportunities in Quantum Cybersecurity

- *Quantum machine learning for cybersecurity applications*: The convergence of quantum computing with machine learning offers substantial unexplored possibilities. Researchers ought to concentrate on creating quantum machine learning methods tailored for cybersecurity, including enhanced anomaly detection, improved encryption techniques, and the improvement of risk management frameworks [37].
- *Quantum-AI hybrid security models*: Future research should explore hybrid models that combine quantum computing, classical computing, and AI in an integrated manner to solve complex cybersecurity challenges. These systems could evolve to handle high-risk, high-complexity environments such as critical infrastructure, IoT, and national defense.
- *Quantum cyber threat simulations*: Developing quantum-enhanced simulations for cyber threats could help predict attack vectors, model the impact of cyberattacks, and optimize defensive strategies. Research in this area could lead to the creation of AI-driven systems that simulate real-world cybersecurity scenarios in quantum environments, offering preemptive threat responses.
- *Quantum cryptography and blockchain integration*: Quantum computing presents new challenges and opportunities for cryptography and blockchain technology. Research should explore how quantum-resistant encryption can be integrated into blockchain systems to ensure secure, tamper-proof digital ledgers in the face of quantum threats [38].
- *Scalability and practical implementation of quantum-AI systems*: One of the key challenges for the integration of quantum computing in cybersecurity is the scalability and practical implementation of quantum-AI systems. Future research should focus on building quantum-AI systems that are both scalable and cost-effective for widespread adoption, especially for small and medium-sized enterprises (SMEs) with limited resources [39, 40].

- *Ethical and regulatory frameworks for quantum-AI integration*: As quantum computing and AI technologies converge, it is important to develop ethical and regulatory frameworks to ensure that these technologies are used responsibly. Research should focus on the development of standards and regulations that address privacy concerns, security threats, and potential misuse of these technologies [41–43].

REFERENCES

1. Liu, N. (2024). 2024 quantum predictions in computing, AI, and cybersecurity. SDxCentral. Retrieved from https://www.sdxcentral.com.
2. Bansal, S., & Jain, R. (2024). Quantum-AI hybrid systems in cybersecurity: Future outlook and challenges. Journal of Quantum Computing, 10(1), 15–28.
3. Zhang, X., & Patel, S. (2024). AI and quantum computing for cybersecurity: Synergies and applications. IEEE Access, 12, 14462–14475.
4. Smith, J., & Green, M. (2023). The future of quantum computing in cryptography and AI for cybersecurity. International Journal of Quantum Security, 5(3), 202–213.
5. Dwyer, D., & Brown, A. (2023). Post-quantum cryptography: Emerging standards and cybersecurity resilience. Cybersecurity Trends, 11(2), 98–112.
6. Kim, Y., & Lee, H. (2023). Integrating quantum computing with AI-driven threat detection systems. IEEE Transactions on Cybersecurity, 8(4), 234–248.
7. Carter, S., & Johnson, K. (2023). Real-time risk management with hybrid quantum-AI models. Quantum Technology Review, 16(1), 56–72.
8. Bhambri, P. (2025). Innovative Systems: Entertainment, Gaming, and the Metaverse. In R. C. Ho; B. L. Song; & P. K. Tee (Eds.), Managing Customer-Centric Strategies in the Digital Landscape (pp. 483–514). IGI Global. https://doi.org/10.4018/979-8-3693-5668-5.ch018.
9. Chen, L., & Xu, F. (2023). Quantum algorithms for AI-powered anomaly detection in cybersecurity. Journal of Quantum Computing in Security, 7(2), 37–50.
10. Thompson, D., & Baker, T. (2023). Exploring the role of AI in enhancing quantum cryptography for cybersecurity. Journal of Cryptography and Quantum Security, 12(1), 15–33.
11. Bhambri, P., & Kumar, S. (2024). Cloud and IoT Integration for Smart Healthcare. In P. Bhambri; R. Soni; & T. A. Tran (Eds.), Smart Healthcare Systems: AI and IoT Perspectives (pp. 69–84). CRC Press. https://doi.org/10.1201/9781032698519-6
12. Smith, L., & Jackson, R. (2024). Challenges in scaling quantum-AI hybrid systems for cybersecurity applications. AI. & Quantum Journal, 3(1), 42–58.
13. Wright, J., & Turner, P. (2024). Quantum computing and AI for advanced threat detection systems. Advances in Cybersecurity, 22(2), 102–117.
14. Evans, B., & Thompson, J. (2023). The convergence of quantum computing and AI in cybersecurity: Opportunities and challenges. Quantum Computing Journal, 9(3), 130–146.
15. Rana, R., & Bhambri, P. (2025). Generative AI in Web Application Development: Enhancing User Experience and Performance. In I. Shah; & N. Jhanjhi (Eds.), Generative AI for Web Engineering Models (pp. 471–486). IGI Global. https://doi.org/10.4018/979-8-3693-3703-5.ch021
16. Patel, R., & Rao, S. (2023). Quantum machine learning models for enhanced cybersecurity threat detection. International Journal of Machine Learning and Cybersecurity, 14(2), 89–101.
17. Ritu, Bhambri, P., & Tripathy, B. (2024). Role of AI and IoT Based Medical Diagnostics Smart Healthcare System for Post-Covid-19 World. In P. Bhambri; R. Soni; & T. A. Tran (Eds.), Smart Healthcare Systems: AI and IoT Perspectives (pp. 135–145). CRC Press. https://doi.org/10.1201/9781032698519-10

18. Johnson, M., & Williams, E. (2023). Hybrid quantum-AI models in the financial sector for fraud detection. Journal of Financial Cybersecurity, 6(1), 72–85.
19. Bhambri, P., & Khang, A. (2024). Computational Intelligence in Manufacturing Technologies. In S. Mehta; S. K. Gupta; A. A. Aljohani; & M. Khayyat (Eds.), Impact and Potential of Machine Learning in the Metaverse (pp. 327–356). IGI Global. https://doi.org/10.4018/979-8-3693-5762-0.ch013
20. Zhang, Y., & Zheng, L. (2023). AI-enhanced quantum cryptography for securing healthcare data. Quantum Health Security, 4(2), 31–45.
21. Rana, R., & Bhambri, P. (2025). Generative AI-Driven Security Frameworks for Web Engineering: Innovations and Challenges. In I. Shah; & N. Jhanjhi (Eds.), Generative AI for Web Engineering Models (pp. 285–296). IGI Global. https://doi.org/10.4018/979-8-3693-3703-5.ch014
22. Zhou, F., & Huang, W. (2023). Real-time quantum computing solutions for national defense cybersecurity. Journal of Cyber Defense, 11(4), 205–218.
23. Kaur, A., Bhambri, P., & Singla, S. K. (2024). Sentimental analysis using RNN, CNN, and LSTM: A comparative study of accuracy and computational efficiency. Library Progress International, 44(3), 4424–4431.
24. Chen, T., & Zhang, K. (2024). Quantum AI-driven threat detection models in cloud computing security. Journal of Cloud Security and Quantum Computing, 8(1), 23–37.
25. Bhambri, P. (2024). Artificial Intelligence Enabled Internet of Medical Things for Enhanced Healthcare Systems. In P. Bhambri; R. Soni; & T. A. Tran (Eds.), Smart Healthcare Systems: AI and IoT Perspectives (pp. 1–17). CRC Press. https://doi.org/10.1201/9781032698519-1
26. Sanchez, L., & Garcia, R. (2024). Quantum-AI applications for national defense cybersecurity. Defense Cybersecurity Insights, 7(1), 59–74.
27. P. Bhambri; R. Soni; & T. A. Tran (Eds.). (2024). Smart Healthcare Systems: AI and IoT Perspectives (1st ed., pp. 374). CRC Press. https://doi.org/10.1201/9781032698519
28. Patel, S., & Gupta, R. (2024). Quantum resilience in AI cybersecurity applications. IEEE Quantum Computing Journal, 2(1), 12–26.
29. Li, Z., & Wang, J. (2024). Hybrid quantum-AI systems for fraud prevention in financial institutions. Journal of Quantum Finance, 9(3), 148–160.
30. P. Bhambri; & P. Bajdor (Eds.). (2024). Handbook of Technological Sustainability: Innovation and Environmental Awareness (1st ed., pp. 412). CRC Press. https://doi.org/10.1201/9781003475989
31. Garcia, J., & Singh, P. (2023). The role of quantum computing in enhancing AI-driven cybersecurity resilience. Journal of Computational Security, 18(4), 223–235.
32. Liu, Y., & Cheng, W. (2023). Post-quantum cryptography for AI systems in cybersecurity. Cybersecurity Innovations, 12(1), 18–32.
33. Geetha, K., Vigneshwari, J., Bhambri, P., & Thangam, A. (2024). Sustainable Solutions for Global Waste Challenges: Integrating Technology in Disposal and Treatment Methods. In P. Bhambri; & P. Bajdor (Eds.), Handbook of Technological Sustainability: Innovation and Environmental Awareness (pp. 46–56). CRC Press. https://doi.org/10.1201/9781003475989-5
34. Brown, H., & Patel, S. (2024). Advancements in quantum cryptographic protocols for AI-enhanced cybersecurity. International Journal of Cybersecurity and Quantum Technologies, 2(2), 71–85.
35. Liu, W., & Wang, Z. (2024). AI-quantum integration for real-time cybersecurity threat mitigation. Journal of AI in Cyber Defense, 10(1), 56–72.
36. Zhang, X., & Lee, J. (2023). Leveraging quantum computing for AI-based real-time risk management. Journal of Real-Time Cybersecurity, 8(3), 143–157.
37. Rana, R., & Bhambri, P. (2024). Blockchain for Transparent, Privacy Protected and Secure Health Data Management. In P. Bhambri; R. Soni; & T. A. Tran (Eds.), Smart

Healthcare Systems: AI and IoT Perspectives (pp. 33–43). CRC Press. https://doi.org/10.1201/9781032698519-3
38. Bajdor, P., & Bhambri, P. (2024). The Principles of Sustainability. In P. Bhambri; & P. Bajdor (Eds.), Handbook of Technological Sustainability: Innovation and Environmental Awareness (pp. 3–17). CRC Press. https://doi.org/10.1201/9781003475989-2
39. Bhambri, P., & Khang, A. (2024). Machine Learning Advancements in E-Health: Transforming Digital Healthcare. In A. Khang (Ed.), Medical Robotics and AI-Assisted Diagnostics for a High-Tech Healthcare Industry (pp. 174–194). IGI Global. https://doi.org/10.4018/979-8-3693-2105-8.ch012
40. Bhambri, P., & Khang, A. (2024). Managing and Monitoring Patient's Healthcare Using AI and IoT Technologies. In A. Khang (Ed.), Driving Smart Medical Diagnosis Through AI-Powered Technologies and Applications (pp. 1–23). IGI Global. https://doi.org/10.4018/979-8-3693-3679-3.ch001
41. Bhambri, P., & Kaur, A. (2013). Novel Technique for Robust Image Segmentation: New Technique of Segmentation in Digital Image Processing (Vol. 1). Lap Lambert Academic Publishing. ISBN: 9783659331831.
42. Bhambri, P., & Kaur, P. (2015). Design and Implementation of Novel Algorithm Using Zero Watermarking: Digital Image Processing Technique for Text Documents (Vol. 1). Lap Lambert Academic Publishing. ISBN: 9783659796159.
43. Bhambri, P., & Kaur, J. (2020). Hybrid Classification Model for the Reverse Code Generation (Vol. 1). Lap Lambert Academic Publishing. ISBN: 9786202683432.

21 Integrating AI with Blockchain for Decentralized Security and Threat Prevention

Pankaj Bhambri and Marta Starostka-Patyk

21.1 INTRODUCTION

In today's rapidly evolving digital landscape, cybersecurity challenges are becoming more complex, especially in decentralized networks. As organizations and industries increasingly rely on distributed systems, they are exposed to a growing array of cyber threats. Decentralized networks, while offering greater flexibility and scalability, introduce unique security vulnerabilities that traditional centralized security mechanisms struggle to address. In this context, artificial intelligence (AI) and blockchain technologies have emerged as powerful tools for enhancing security. This chapter introduces the critical role of AI and blockchain integration in fortifying decentralized networks against cyber threats and explores the synergy between the two technologies in providing a comprehensive, adaptive, and scalable security solution [1].

21.1.1 Overview of Cybersecurity Challenges in Decentralized Networks

Decentralized networks, such as those used in blockchain-based applications, distributed ledgers, and peer-to-peer systems, are particularly vulnerable to various cybersecurity threats due to their inherent characteristics. In a decentralized environment, trust is distributed among participants rather than being centralized in a single authority. While this promotes transparency and resistance to single points of failure, it also opens the door to a range of potential security risks, including data breaches, malicious attacks, fraud, and identity theft. In addition, these networks often suffer from scalability challenges, which render traditional security solutions less effective. These challenges are exacerbated by the increasing sophistication of cyber-attacks, such as advanced persistent threats (APTs), which can evade conventional detection systems. In this environment, traditional security models that rely on perimeter defense or centralized control are insufficient [2].

21.1.2 The Need for Advanced Security Solutions

Due to the changing landscape of cyber threats, there is an immediate necessity for sophisticated security solutions that can mitigate the unique vulnerabilities present

DOI: 10.1201/9781003521020-21

in decentralized networks. Conventional security measures, including firewalls, intrusion detection systems (IDS), and antivirus software, are tailored for centralized networks and frequently inadequately safeguard decentralized or distributed environments. The primary limitations of these conventional methods are their inadequate scalability, dependence on a singular point of failure, and challenges in identifying complex, multi-faceted attacks. Furthermore, as the complexity of cyber threats continues to increase, the demand for more proactive, adaptive, and automated security mechanisms has grown. AI and blockchain provide complementary solutions that can overcome these limitations and offer enhanced protection [3].

21.1.3 Synergy between AI and Blockchain for Threat Prevention

The integration of AI and blockchain establishes a robust, synergistic framework for addressing cybersecurity issues in decentralized networks. AI, with its sophisticated abilities in data analysis, anomaly detection, and decision-making, can be utilized to recognize and address emerging threats in real time. Machine learning (ML) algorithms can perpetually learn from network traffic patterns and adjust to novel attack types, whereas deep learning (DL) models can improve threat detection by scrutinizing intricate data structures and uncovering concealed patterns indicative of malicious behavior. Conversely, blockchain offers a decentralized, immutable ledger that guarantees data integrity, transparency, and tamper resistance. Integrating AI with blockchain enables organizations to establish a more resilient and transparent security framework. For example, blockchain's tamper-proof nature can be used to securely log AI-driven threat detection results, while AI models can enhance the decision-making process for smart contract-based security automations. This integration provides a scalable, real-time, and adaptive approach to detecting, preventing, and mitigating cyber threats across decentralized networks [4, 5].

21.2 AI ALGORITHMS FOR CYBERSECURITY

AI algorithms are essential in improving cybersecurity systems, especially in detecting, analyzing, and alleviating emerging cyber threats. Utilizing ML and DL methodologies, AI can analyze extensive datasets, identify anomalies, and discern intricate patterns that may signify potential cyberattacks. These algorithms can improve cybersecurity systems' capacity to respond to threats in real time while also anticipating and averting attacks by identifying early warning indicators. This section examines three fundamental AI techniques employed in cybersecurity: ML for threat detection, DL models in cybersecurity, and anomaly detection and pattern recognition utilizing AI [6].

21.2.1 Machine Learning for Threat Detection

ML has emerged as a fundamental element of contemporary cybersecurity methodologies owing to its capacity to autonomously learn from data and generate predictions or decisions without direct programming. In threat detection, ML models examine historical data, including network traffic, user behavior, and system logs, to discern patterns and develop models capable of predicting and detecting anomalous or suspicious activities. Prevalent ML algorithms in threat detection comprise classification techniques such as decision trees, support vector machines (SVM), and k-nearest neighbors (KNN), which categorize network traffic or system events as benign or malicious based on historical data. Moreover, unsupervised learning methods, such as clustering, can discern previously unrecognized attack patterns by aggregating analogous data points, thereby enabling the detection of novel or zero-day attacks that may not have been previously encountered. The capacity of ML algorithms to enhance their performance with the influx of new data is a considerable benefit in responding to emerging and evolving threats [7].

21.2.2 Deep Learning Models in Cybersecurity

DL, a subset of ML, has gained significant prominence in cybersecurity due to its robust capacity to analyze complex and extensive datasets. DL models, especially neural networks, are engineered to autonomously recognize complex patterns in data that conventional ML techniques may find challenging to discern. Convolutional neural networks (CNNs) and recurrent neural networks (RNNs) are two prevalent DL architectures utilized in cybersecurity. CNNs excel in the analysis of spatial data, including images or logs, to identify patterns indicative of potential security breaches, such as intrusion attempts or malware [8]. RNNs are adept at analyzing sequential data, including network traffic or event logs over time, enabling the detection of anomalies or attacks that develop gradually, such as distributed denial-of-service (DDoS) attacks or APTs. Utilizing DL, cybersecurity systems can achieve enhanced detection accuracy, diminished false positives, and more adaptive threat response mechanisms that evolve with increased data exposure [9].

21.2.3 Anomaly Detection and Pattern Recognition Using AI

Anomaly detection and pattern recognition are essential components of AI-powered cybersecurity, as they enable systems to identify unusual behavior or deviations from established baselines that could indicate malicious activity. AI-based anomaly detection systems work by learning the normal behavior of network traffic, user actions, or system operations through training on historical data [10]. Once trained, these systems can flag any behavior that deviates from this normal pattern, signaling potential threats such as data breaches, insider threats, or malicious software activity. Pattern recognition, closely related to anomaly detection, involves the identification of recurrent and recognizable attack patterns within vast datasets. Techniques like clustering, decision trees, and neural networks are used to classify new data into known categories of malicious activity. This helps cybersecurity systems not only detect specific types of attacks but also predict future attacks based on recognized patterns. Furthermore, AI algorithms can integrate multiple detection techniques, enabling more comprehensive threat identification that covers a broader range of attack vectors [11].

21.3 BLOCKCHAIN'S ROLE IN SECURITY

Blockchain technology has emerged as a transformative instrument in the cybersecurity domain, chiefly owing to its decentralized, immutable, and transparent characteristics. In contrast to conventional centralized systems, blockchain functions on a distributed ledger, wherein each participant retains a copy of the data, thereby preventing any singular entity from exerting control over the entire system. This decentralized architecture, coupled with cryptographic methods, provides substantial benefits for improving security across multiple domains, especially in contexts where trust, data integrity, and transparency are essential. This section examines the function of blockchain in security, emphasizing decentralized trust and tamper-resistant systems, blockchain's contribution to data integrity and transparency, and the significance of consensus mechanisms in maintaining system security [12, 13].

21.3.1 Decentralized Trust and Tamper-Proof Systems

A significant advantage of blockchain technology is its capacity to foster trust independently of a central authority. In conventional centralized systems, trust is vested in intermediaries or authorities to oversee and authenticate data transactions. This centralized approach creates vulnerabilities, including single points of failure, which can be exploited by malicious entities. Blockchain obviates the necessity for trusted intermediaries by decentralizing the responsibility for transaction validation among a network of participants, commonly known as nodes. Every transaction is cryptographically authenticated and documented in blocks that are interconnected in a chain, guaranteeing that once information is inscribed in the blockchain, it cannot be modified or interfered with without altering all subsequent blocks, thereby rendering fraud and data manipulation exceedingly challenging. The tamper-proof characteristic of blockchain renders it an optimal technology for safeguarding sensitive data in sectors like finance, healthcare, and critical infrastructure, where data integrity is essential [14].

21.3.2 Blockchain for Data Integrity and Transparency

Blockchain offers a formidable solution for guaranteeing data integrity and transparency. In a blockchain network, each transaction or data entry is chronologically stamped and documented in a secure and immutable fashion. Upon verification and addition to the blockchain, a transaction is irrevocably recorded, preventing any party from altering or deleting the data. This feature is especially significant in contexts where data authenticity is paramount, such as supply chain management, contract execution, and healthcare records. The transparency provided by blockchain allows all participants to access the transaction history, facilitating the verification of data authenticity and ensuring accountability. The decentralized characteristic of blockchain allows all participants to access identical data in real time, fostering transparency and eradicating information asymmetry that could be exploited by nefarious entities.

Blockchain facilitates the creation of verifiable systems, as each modification or transaction is documented sequentially and can be traced to its source. This traceability is particularly advantageous in sectors like finance, where fraud detection and regulatory compliance are critical. Efficient and secure data auditing is essential in mitigating cyberattacks, including data tampering and unauthorized access to sensitive information.

21.3.3 The Role of Consensus Mechanisms in Security

Consensus mechanisms are essential elements of blockchain networks that guarantee agreement among distributed nodes regarding the validity of transactions prior to their inclusion in the blockchain. These mechanisms establish the basis for security and trust in decentralized systems. By necessitating agreement among numerous participants, blockchain reduces the risks linked to centralized control and malevolent entities. The predominant consensus mechanisms encompass Proof of Work (PoW), Proof of Stake (PoS), as well as more sophisticated mechanisms such as Delegated Proof of Stake (DPoS) and Practical Byzantine Fault Tolerance (PBFT) [15].

- Proof of Work (PoW), utilized in Bitcoin and other cryptocurrencies, entails resolving intricate mathematical problems to authenticate transactions, guaranteeing that only those who allocate computational resources can append blocks to the blockchain. This mechanism, although energy-intensive, offers robust security assurances by rendering it computationally impractical for adversaries to modify the blockchain.
- Proof of Stake (PoS) is a more energy-efficient mechanism in which participants are selected to validate transactions based on the quantity of cryptocurrency they possess and are prepared to "stake" as collateral. This mechanism motivates validators to behave ethically, as dishonest actions may lead to the forfeiture of their staked assets.

Alternative consensus mechanisms, including PBFT and DPoS, offer varying trade-offs among security, scalability, and decentralization, yet all aim to prevent double-spending, guarantee transaction validity, and uphold the integrity of the blockchain. The efficacy of a consensus mechanism is determined by its capacity to thwart attacks such as 51% attacks, wherein a nefarious entity acquires control of over half of the network's computational power (in PoW) or stake (in PoS), thereby jeopardizing the network's integrity. Consensus mechanisms create a decentralized, transparent, and secure environment by ensuring that most participants concur on the validity of transactions, thereby significantly mitigating the risks of fraud, data manipulation, and system compromise [16].

21.4 INTEGRATION OF AI AND BLOCKCHAIN FOR THREAT PREVENTION

The amalgamation of AI and blockchain technology represents a revolutionary strategy in cybersecurity. By integrating AI's capacity to analyze extensive datasets, identify patterns, and generate predictions with blockchain's decentralized, transparent, and tamper-proof framework, organizations can improve threat detection, automate responses, and safeguard data transmission in decentralized networks. This collaboration offers a holistic security solution that adjusts to evolving threats instantaneously while maintaining data integrity and transparency. This section examines the integration of AI and blockchain to enhance cybersecurity, specifically in threat detection, smart contract automation, and the protection of data communication in decentralized settings [17].

21.4.1 Enhancing Threat Detection with AI and Blockchain

AI-powered threat detection systems, such as ML and DL models, are highly effective in identifying patterns and anomalies that indicate potential security breaches. However, AI models often face challenges such as data manipulation, adversarial attacks, and lack of trustworthiness, which can compromise the integrity of threat detection systems. Blockchain's decentralized, immutable nature addresses these challenges by ensuring that data used in AI models cannot be tampered with once recorded, thus enhancing the overall accuracy and reliability of AI-based threat detection [18].

The integration of AI with blockchain can substantially enhance the threat detection process. Blockchain guarantees the integrity of data input into AI models, while AI models persistently scrutinize this data to identify potential anomalies, including unauthorized access, network intrusions, or malware. The integration of AI's real-time analytics with blockchain's transparent record-keeping can yield more precise predictions, allowing systems to detect advanced threats such as APTs and zero-day vulnerabilities. AI can perpetually enhance its detection abilities by assimilating data from the blockchain. As an increasing number of cybersecurity incidents are documented on the blockchain, the AI system can leverage this data to refine its algorithms, thereby augmenting its capacity to identify and address emerging threats. This integration facilitates the development of a resilient, adaptive defense system that can respond to the dynamic nature of cyberattacks [19].

21.4.2 Automating Threat Response via Smart Contracts

Smart contracts, self-executing agreements with terms encoded in lines of code, serve as a potent instrument in the amalgamation of AI and blockchain for cybersecurity. These contracts operate on blockchain networks and autonomously execute the stipulated actions upon the fulfillment of predetermined conditions. In cybersecurity, smart contracts can automate real-time threat responses, minimizing human intervention and facilitating quicker, more efficient reactions to attacks.

Upon detection of a threat by AI-driven systems, a smart contract can autonomously implement predetermined actions, including isolating compromised devices, blocking malicious IP addresses, or deploying countermeasures to inhibit the proliferation of malware. For instance, if an AI system identifies an intrusion attempt, a smart contract could promptly execute a sequence of actions, including issuing an alert, limiting access to sensitive information, and commencing incident response protocols. This automation diminishes response time, guarantees uniformity in threat mitigation, and facilitates the swift containment of security breaches [20].

Smart contracts enhance security by ensuring transparency and immutability in the response process. Upon the execution of an action by a smart contract, the entire process is documented on the blockchain, establishing an auditable trail for the assessment of responses to cyber threats. This transparency guarantees that actions executed during a cybersecurity incident are authentic, traceable, and impervious to alteration by malicious entities.

21.4.3 Securing Data Flow and Communication in Decentralized Networks

Decentralized networks, characterized by data distribution across numerous nodes, present considerable security challenges, such as data interception, unauthorized access, and manipulation. Blockchain offers a robust method for safeguarding data transmission and communication within these networks. Organizations can guarantee that data exchanged between decentralized nodes is secure, tamper-proof, and verifiable by employing blockchain's cryptographic methods and distributed ledger technology. AI and blockchain can collaborate to safeguard data transmission in decentralized networks by utilizing AI for oversight and blockchain for maintaining data integrity. AI algorithms can incessantly surveil the network for anomalous activities, including unauthorized data access, man-in-the-middle attacks, or attempts at data exfiltration. Upon identification of a potential threat, AI can initiate a response via blockchain's secure communication channels, guaranteeing that any data pertaining to the threat remains safeguarded and unmodified [21].

The decentralized architecture of blockchain reduces the vulnerabilities linked to centralization in conventional networks, where data is more susceptible to attacks. In blockchain-based systems, data is disseminated across numerous nodes, complicating efforts by malicious actors to undermine the entire system. This distribution, combined with AI's continuous monitoring capabilities, ensures that data flows within the network are secure, and any communication between nodes is encrypted and validated through blockchain's consensus mechanisms. In addition to securing data flow, blockchain's tamper-proof nature ensures that the communication logs between nodes remain intact and auditable. Any malicious attempt to alter data transmitted within the network would require altering the entire chain, which is computationally infeasible. This makes it highly difficult for attackers to manipulate or intercept data flows, providing a robust solution for securing communication in decentralized networks [22].

21.5 USE CASES AND APPLICATIONS

The integration of AI and blockchain technologies has a transformative impact across multiple industries, enabling more secure, efficient, and adaptive cybersecurity solutions. This section discusses several key sectors where this integration is particularly beneficial: the financial sector, healthcare, critical infrastructure, and the Internet of Things (IoT) in smart cities. Each of these sectors faces unique challenges that AI and blockchain can address, making them prime candidates for advanced, decentralized security and threat prevention strategies [23].

21.5.1 Financial Sector: Protecting Transactions and Assets

The financial sector is a principal target for cyberattacks owing to the substantial value of digital transactions and assets. Cybercriminals often seek to exploit weaknesses in payment systems, banking services, and digital wallets. The integration of AI and blockchain can substantially improve security in this sector by offering a dual layer of protection. AI-powered algorithms can oversee transactions in real time, identifying anomalies or suspicious patterns that may signify fraud or unauthorized activity. ML and DL models can perpetually learn from transactional data, enhancing their capacity to detect fraud, money laundering, and other financial offenses. These models can examine extensive transactional data and identify any anomalous behavior, such as sudden fluctuations in transaction volume or irregular spending patterns. Blockchain is essential for maintaining the integrity of financial transactions. The decentralized ledger technology of blockchain guarantees that once a transaction is documented, it remains immutable and secure from alteration. This immutability ensures that all transactions are transparent and auditable, offering a secure and verifiable record. Furthermore, blockchain's cryptographic techniques guarantee that only authorized individuals can access or validate transactions, thereby mitigating the risk of unauthorized actions. The integration of real-time AI surveillance and blockchain's immutable record-keeping establishes a formidable security framework for safeguarding digital assets and financial transactions against cyber threats [24].

21.5.2 Healthcare: Securing Patient Data and Medical Records

The healthcare sector encounters escalating cybersecurity challenges as patient information becomes increasingly digital and interconnected. Safeguarding sensitive patient information, including medical records and personal health data, is essential for preserving privacy and averting identity theft. The integration of AI with blockchain offers a robust solution for safeguarding healthcare data. AI algorithms can identify anomalous access patterns in healthcare systems, including unauthorized attempts to access medical records or data breaches. These AI models can be trained to identify behaviors that signify a security threat, such as the utilization of compromised credentials or irregular network traffic. Through the continuous observation of these patterns, AI can promptly detect and alleviate potential threats prior to their escalation [25].

The function of blockchain in healthcare is its capacity to offer a decentralized and unalterable record of patient information. Medical records for each patient can be securely stored in a blockchain system that guarantees data integrity and restricts access to authorized individuals only. The transparency and auditability of blockchain enable healthcare providers to uphold an accurate and verifiable record of patient interactions, thereby deterring fraudulent activities like the unauthorized alteration of medical histories or prescriptions. Blockchain facilitates secure data sharing among healthcare providers, enabling the seamless and secure exchange of patient records between authorized medical institutions while safeguarding data privacy and consent. The integration of AI for anomaly detection and blockchain for data integrity and privacy constitutes an optimal solution for the protection of patient data [26].

21.5.3 Critical Infrastructure: Safeguarding Energy and Utility Systems

Critical infrastructure systems, including energy grids, water supply networks, and transportation systems, are essential for societal functioning and are increasingly targeted by cyberattacks. Safeguarding these systems from cyber threats is paramount to guarantee national security and public safety. The integration of AI and blockchain provides a novel method to augment the security of these intricate, frequently decentralized networks. AI models can be utilized to oversee critical infrastructure systems in real time, identifying anomalies such as irregular energy consumption patterns, abrupt system failures, or breaches in control systems. These systems can also forecast future threats by examining historical data to pinpoint vulnerabilities or potential attack vectors. ML algorithms can be trained to identify indicators of cyberattacks or operational problems and autonomously activate alarms or mitigation protocols [27].

Blockchain fortifies the security of essential infrastructure by offering a transparent and immutable record of system events, communications, and transactions. Blockchain can secure communications among components of an energy grid or water distribution network, guaranteeing that data exchanged between systems is immutable and resistant to manipulation. Moreover, the decentralized characteristic of blockchain diminishes the likelihood of a singular point of failure, thereby complicating efforts for attackers to undermine the entire system. Integrating AI with blockchain enables energy and utility systems to achieve real-time monitoring alongside a secure, transparent record of operations. This hybrid methodology facilitates the identification and mitigation of threats while preserving the integrity of essential infrastructure systems.

21.5.4 IoT and Smart Cities: Enhancing Security in Connected Environments

The emergence of the Internet of Things (IoT) and the advancement of smart cities have presented novel challenges in cybersecurity. As millions of interconnected devices communicate across extensive networks, the security of IoT systems and the safeguarding of user privacy in smart cities have emerged as critical issues. The integration of AI and blockchain can effectively tackle these challenges by offering scalable and adaptive solutions for IoT security. AI algorithms can monitor IoT devices for anomalous behavior, including unexpected activity, data breaches, or communication with unauthorized devices. These AI models can incessantly analyze data from IoT sensors, identifying patterns indicative of a cyberattack or device failure. AI can assist in identifying DDoS attacks on IoT networks by recognizing anomalous traffic patterns prior to causing substantial disruptions. The function of blockchain in IoT security is its capacity to offer decentralized governance and authentication of device interactions. Every IoT device can be registered on the blockchain, guaranteeing that only authorized devices may communicate within the network. Blockchain can serve to maintain an immutable record of all device activities, offering a transparent and verifiable log of interactions. This transparency guarantees that any efforts to modify data or undermine the integrity of the IoT system can be identified and tracked.

In the realm of smart cities, blockchain enables secure data exchange among diverse municipal services, including traffic management, waste management, and public safety, whereas AI enhances these services by analyzing extensive volumes of real-time data. The amalgamation of AI and blockchain establishes a formidable framework for augmenting the security and efficiency of smart cities, guaranteeing that interconnected environments remain secure, resilient, and responsive to emerging threats.

21.6 EMERGING TRENDS AND FUTURE DIRECTIONS

The integration of AI and blockchain technologies is rapidly evolving, with new advancements continuously shaping the landscape of cybersecurity and threat prevention. As cyber threats grow increasingly sophisticated, the need for more dynamic, decentralized, and adaptive security systems has driven the emergence of several key trends. This section explores the rising significance of decentralized AI models, blockchain-based AI for proactive threat prevention, smart contracts for automated security, and future research opportunities in AI–blockchain integration [28].

21.6.1 The Rise of Decentralized AI Models

The conventional centralized methodology for AI models frequently encounters issues including data privacy concerns, the potential for single points of failure, and dependence on centralized servers, which may be susceptible to cyberattacks. In light of these constraints, there is an increasing transition towards decentralized AI models. Through the application of blockchain technology, AI algorithms can be disseminated across a network of nodes, guaranteeing that data and processing power are collectively managed rather than monopolized by a singular entity.

Decentralized AI models facilitate the secure, transparent, and trustless implementation of ML algorithms, with each participating node contributing to both training and inference activities. This decentralized methodology not only reduces security vulnerabilities but also improves the scalability and robustness of AI systems. In cybersecurity, decentralized AI can facilitate the detection and response to threats in real time, independent of a central authority. Moreover, decentralized AI can preserve user privacy, as data does not require centralized storage or processing, thereby reducing susceptibility to breaches and attacks [29].

21.6.2 Blockchain-Based AI for Proactive Threat Prevention

The immutable, transparent, and decentralized characteristics of blockchain provide a robust framework for developing AI-driven cybersecurity systems that proactively mitigate threats. Rather than simply responding to attacks, blockchain-based AI systems can consistently analyze network traffic, user behavior, and system activities to anticipate and avert potential threats prior to their manifestation.

Utilizing blockchain, AI models can securely store and authenticate threat intelligence, guaranteeing that the data employed for training predictive models is impervious to tampering and reliable. AI algorithms can examine this data to detect emerging attack patterns, anomalous activities, and vulnerabilities. Moreover, blockchain's consensus mechanisms guarantee that the information disseminated among network nodes is consistent and precise, facilitating a collective, decentralized reaction to potential threats. AI-driven blockchain networks could independently obstruct dubious IP addresses or identify compromised systems, proactively mitigating attacks before they affect critical infrastructure [30].

The proactive characteristics of blockchain-based AI offer an adaptive security layer, diminishing the necessity for reactive strategies like patching or incident response. It represents a progression from conventional threat detection systems to more proactive and preventive cybersecurity frameworks.

21.6.3 Smart Contracts for Automated Security Solutions

Smart contracts—self-executing agreements with terms encoded in programming language—are progressively utilized to automate security procedures in decentralized settings. Within the framework of AI–blockchain integration, smart contracts can autonomously initiate security measures based on predetermined criteria, eliminating the necessity for manual intervention or centralized oversight. For example, smart contracts can autonomously prevent transactions that display dubious patterns or are linked to recognized attack vectors. Upon detection of a breach, they can initiate automatic responses, including disabling compromised accounts, isolating affected systems, or alerting security personnel. This automation markedly improves the velocity and efficacy of threat detection and mitigation, guaranteeing the prompt implementation of security measures upon the identification of a potential risk [31].

Furthermore, smart contracts can significantly contribute to the integrity and transparency of security procedures. Due to their operation on blockchain technology, all actions executed by smart contracts are inscribed in an immutable ledger, guaranteeing that every security event is documented and subject to audit. This degree of transparency cultivates trust among stakeholders, enabling them to confirm that security measures are executed as intended.

21.6.4 Future Research Opportunities in AI–Blockchain Integration

The integration of AI and blockchain is still an emerging field with vast potential for innovation. As cybersecurity threats become more complex, there are numerous avenues for research to explore how these two technologies can work together more effectively. Some key future research opportunities in AI–blockchain integration include [32–35]:

- *AI and blockchain for multilayered security*: Exploring how AI can enhance blockchain's existing security features, such as encryption, with advanced algorithms for intrusion detection, fraud prevention, and real-time threat analysis.

- *AI-powered blockchain consensus mechanisms*: Investigating how AI could optimize or enhance blockchain consensus protocols (e.g., PoW, PoS) to improve efficiency, security, and scalability.
- *AI for blockchain performance optimization*: Researching how AI can help optimize the performance of blockchain networks by predicting bottlenecks, optimizing transaction processing, and enhancing scalability without compromising security.
- *Decentralized autonomous security systems*: Exploring the concept of fully decentralized security systems powered by AI and blockchain, where AI models autonomously detect, prevent, and respond to threats without any human intervention.
- *Interoperability between blockchain platforms*: Investigating how AI can facilitate interoperability between different blockchain platforms, enabling seamless integration and data sharing across multiple decentralized networks while maintaining high-security standards.
- *Ethical and governance issues*: As AI and blockchain intersect, addressing the ethical and governance challenges of using these technologies, especially in terms of data privacy, algorithmic bias, and transparency, will be a critical area for future research.

21.7 CHALLENGES AND LIMITATIONS

The integration of AI and blockchain offers significant potential for improving cybersecurity, yet it also introduces various challenges and limitations that must be resolved for the effective implementation of these technologies in practical security solutions. This section examines key challenges, such as scalability and performance issues, privacy and data protection concerns, and the regulatory and ethical implications of AI-blockchain security solutions [36].

21.7.1 Scalability and Performance Issues

A primary challenge in the integration of AI with blockchain is scalability. Blockchain technology, especially public blockchains, is characterized by restricted transaction throughput and elevated latency resulting from the consensus mechanisms (e.g., PoW or PoS) necessary for transaction verification within a distributed network. With the escalation of users and transactions, these systems may become sluggish and ineffective. When combined with AI, which necessitates extensive datasets and considerable computational resources, scalability emerges as an increasingly vital concern. ML and DL models, essential for AI-based threat detection and prevention, must analyze vast quantities of data in real time. The decentralized and distributed characteristics of blockchain can intensify latency issues, complicating the prompt delivery of threat responses and updates throughout an extensive network. The dependence of blockchain on each node to validate transactions and store data can escalate storage demands and burden system performance as data accumulates, particularly when AI models necessitate regular updates or large datasets for training. Consequently, identifying efficient solutions to scale blockchain technology, while ensuring the adequate performance of AI-driven threat detection and prevention models, presents a substantial challenge [37].

To address scalability challenges, researchers and developers are investigating solutions including off-chain processing, hybrid consensus mechanisms, and layer-2 scaling solutions. Nonetheless, these technologies remain in the developmental phase and require additional refinement to be effective in extensive implementations.

21.7.2 Privacy and Data Protection Concerns

A notable challenge of AI–blockchain integration is guaranteeing privacy and data protection. Although blockchain offers a secure and transparent framework for data storage and sharing, its intrinsic transparency may jeopardize privacy, particularly in sectors such as healthcare and finance that handle sensitive personal information. The immutability of blockchain implies that once data is inscribed, it cannot be altered or deleted, potentially conflicting with data protection regulations like the General Data Protection Regulation (GDPR) in the European Union, which encompasses the "right to be forgotten."

Conversely, AI systems, especially ML algorithms, necessitate substantial data access to train models efficiently. The utilization of sensitive personal or corporate data in AI model training may result in unauthorized access or data leakage, particularly within a decentralized system where data control is dispersed among various entities. This heightens the risk of data exploitation, necessitating the training and deployment of AI models in compliance with privacy regulations to safeguard sensitive information [38].

To mitigate these privacy concerns, techniques such as differential privacy, federated learning, and encryption are being investigated. Federated learning facilitates the training of ML models across decentralized nodes while preserving raw data confidentiality, thereby mitigating the risk of disclosing sensitive information. Moreover, integrating AI with blockchain's encryption and pseudonymization capabilities could offer a means to protect data while preserving the integrity and transparency of the blockchain.

21.7.3 Regulatory and Ethical Implications of AI-Blockchain Security Solutions

The integration of AI and blockchain for cybersecurity also raises important regulatory and ethical issues. Regulatory bodies across the globe are still grappling with how to properly govern the use of these technologies, particularly when they are applied to sensitive domains such as financial transactions, healthcare, and critical infrastructure. The decentralized nature of blockchain, combined with the autonomous decision-making capabilities of AI, makes it difficult to establish clear accountability for actions taken by these systems. If a security breach occurs, or if AI models make erroneous or harmful decisions, determining who is responsible can be a complex legal issue. Moreover, as AI models become more advanced and capable of autonomously detecting and responding to threats, ethical concerns surrounding their use in security contexts become more pronounced. For example,

AI systems in cybersecurity might be used to deploy automated responses without human intervention, potentially leading to privacy violations, overreach, or unintended consequences. In addition, ethical questions arise around the potential biases in AI models, which could lead to discriminatory security measures or the wrongful targeting of specific individuals or groups based on flawed data [39].

Regulatory frameworks that govern the use of AI and blockchain are still in the early stages, and there is a lack of consensus on how to regulate these technologies in the context of cybersecurity. Policymakers must consider how to balance the benefits of decentralized, autonomous security systems with the need for accountability, transparency, and protection of individual rights. Ethical considerations also include ensuring that AI systems are transparent and explainable, particularly when they are used to make critical decisions in security, such as blocking transactions or identifying threats. Without clear insights into how these AI models operate, it becomes difficult for stakeholders to trust and accept their decisions [40, 41].

Overall, regulatory and ethical frameworks need to evolve in parallel with technological advancements to ensure that AI-blockchain security solutions are deployed in a responsible, accountable, and fair manner. Addressing these concerns will be essential to gaining public trust and ensuring that AI-blockchain security systems are sustainable and legally compliant.

21.8 EMERGING TRENDS AND FUTURE DIRECTIONS

The integration of AI and blockchain technology has emerged as a transformative approach to enhancing cybersecurity, particularly in decentralized and complex systems. This section summarizes the key insights from the chapter and provides recommendations for practitioners and policymakers while exploring the future potential of decentralized security systems.

21.8.1 Key Takeaways from AI and Blockchain Integration

The integration of AI and blockchain technology constitutes a formidable convergence that bolsters cybersecurity via improved automation, transparency, and data integrity. The capability of AI to analyze extensive datasets and identify patterns is enhanced by blockchain's decentralized, immutable, and transparent framework, which guarantees data integrity and fosters trust in a distributed setting. Collectively, these technologies provide substantial enhancements in threat detection, response automation, and overall system security.

A primary advantage of merging AI with blockchain is the capacity to establish decentralized, tamper-resistant security systems capable of identifying and addressing cyber threats in real time. AI-driven algorithms, including ML and DL models, facilitate the detection of anomalous behaviors and potential threats through the analysis of extensive data across numerous network nodes. Conversely, blockchain offers a secure and transparent method for storing and disseminating information, rendering it impervious to tampering and guaranteeing accountability. Moreover, smart contracts on blockchain networks facilitate automated threat response, diminishing the necessity for human involvement and accelerating the mitigation of threats.

The collaboration between AI and blockchain facilitates real-time threat monitoring and improves data security in sectors including finance, healthcare, critical infrastructure, and IoT networks.

21.8.2 Recommendations for Practitioners and Policymakers

Practitioners must adopt a comprehensive strategy that encompasses both technical and operational aspects when integrating AI and blockchain into cybersecurity practices. Practitioners should prioritize the adoption of scalable AI models capable of efficiently processing extensive datasets with minimal latency. They should investigate privacy-preserving AI methodologies, including federated learning and differential privacy, to ensure data confidentiality and adhere to regulations such as GDPR. Moreover, investing in blockchain infrastructure that facilitates high transaction throughput and minimal latency will be essential to fulfilling the performance requirements of AI systems in practical applications.

Policymakers must create explicit regulatory frameworks to govern the application of AI and blockchain in cybersecurity. These frameworks must prioritize privacy, accountability, and transparency in AI-driven decision-making processes. Policymakers must collaborate with industry stakeholders to establish standards that address the ethical ramifications of AI and blockchain integration, including data protection, bias in AI models, and the alignment of automated decisions with human oversight. Policymakers should promote research and development in AI and blockchain technologies to overcome their current limitations, including scalability, performance, and interoperability. Enhancing cooperation between the public and private sectors can expedite the implementation of secure, decentralized AI-blockchain solutions while maintaining adherence to privacy and security regulations.

21.8.3 The Future of Decentralized Security Systems

The future of decentralized security systems driven by AI and blockchain appears promising, with substantial advancements anticipated in the forthcoming years. As AI models advance in sophistication and their ability to comprehend intricate security landscapes, their amalgamation with blockchain will yield more potent instruments for countering emerging threats. As decentralized AI models become more prevalent, we can anticipate an increase in autonomous systems that can detect, analyze, and respond to threats in real time, thereby diminishing dependence on centralized control and reducing human error. Furthermore, the advancing capabilities of blockchain, including the implementation of more scalable consensus mechanisms and the utilization of private or permissioned blockchains, will significantly improve the performance and efficiency of AI-blockchain systems. The emergence of quantum-resistant blockchain and AI algorithms may provide enhanced security against the escalating threat posed by quantum computing.

The integration of AI and blockchain in cybersecurity will be essential for managing the escalating complexity and interdependence of smart cities, IoT, and other connected systems. The integration of blockchain's decentralized characteristics with AI's adaptive learning capabilities will facilitate the development of security systems that autonomously address threats, thereby guaranteeing scalability and resilience in rapidly changing environments.

The future of decentralized security systems utilizing AI and blockchain is expected to experience heightened adoption across multiple sectors, propelled by technological innovations and regulatory progress. As these systems evolve, they will offer more resilient, efficient, and transparent security solutions, effectively mitigating the expanding array of cyber threats in today's interconnected environment.

REFERENCES

1. Ali, M., & Zohra, K. (2023). Blockchain-based AI-driven cybersecurity: A novel approach for securing decentralized networks. International Journal of Security and Networks, 18(3), 238–254.
2. Chen, L., Zhang, Y., & Guo, X. (2024). Combining AI and blockchain for cybersecurity: A survey and future directions. Computers & Security, 117, 102728.
3. Kumar, R., & Singh, P. (2023). Leveraging machine learning for blockchain security in smart contracts. Journal of Blockchain Research, 5(2), 93–110.
4. Zhao, Y., & Li, F. (2024). Decentralized AI for blockchain-enabled cybersecurity: A case study on IoT networks. Future Generation Computer Systems, 126, 283–299. https://doi.org/10.1016/j.future.2024.01.013
5. Wang, S., & Liu, T. (2023). Blockchain-based distributed AI for enhancing security in decentralized networks. IEEE Transactions on Network and Service Management, 20(1), 1–14.
6. Bhambri, P., & Bajdor, P. (2024). Technological Sustainability Unveiled: A Comprehensive Examination of Economic, Social, and Environmental Dimensions. In P. Bhambri; & P. Bajdor (Eds.), Handbook of Technological Sustainability: Innovation and Environmental Awareness (pp. 80–98). CRC Press. https://doi.org/10.1201/9781003475989-8
7. Zhang, Q., Wang, X., & Gu, Q. (2024). The integration of AI and blockchain for enhanced threat prevention in cloud computing environments. Cloud Computing and Security, 22(4), 412–427.
8. Wu, Z., & Yang, F. (2024). Blockchain-enabled AI techniques for detecting advanced persistent threats. International Journal of Information Security, 31(1), 97–113.
9. Nguyen, H., & Tran, T. (2023). Smart contracts and AI in blockchain-based cybersecurity systems. Blockchain Technology Journal, 8(3), 158–171.
10. Li, M., & Zhao, X. (2023). A hybrid AI-blockchain model for real-time threat detection in decentralized networks. IEEE Transactions on Dependable and Secure Computing, 20(5), 1357–1370.
11. Bhambri, P., & Kautish, S. K. (2024). Analytic Hierarchy Process and Business Value Creation. In S. Kautish (Ed.), Using Strategy Analytics for Business Value Creation and Competitive Advantage (pp. 54–77). IGI Global. https://doi.org/10.4018/979-8-3693-2823-1.ch003
12. Sharma, A., & Yadav, P. (2024). Blockchain and AI-driven cybersecurity for next-generation autonomous vehicles. Journal of AI in Cybersecurity, 10(1), 35–51.
13. Gupta, M., & Verma, S. (2023). AI and blockchain for securing digital healthcare systems. Health Informatics Journal, 29(2), 103–119.
14. Zhang, Y., & Wu, X. (2023). Machine learning and blockchain for real-time intrusion detection in decentralized networks. Journal of Computing and Security, 28(3), 212–227.

15. Yang, H., & Zhang, L. (2024). Blockchain and AI for privacy-preserving cybersecurity solutions. Journal of Privacy and Confidentiality, 10(3), 112–130.
16. Bhambri, P., & Khang, A. (2025). Smart Universities and ICT Platforms. In M. L. Kolhe; P. Singh; S. Rani; & P. Kumar (Eds.), Planning of Sustainable Energy Systems in Urban Built Environments. CRC Press. https://www.appleacademicpress.com/planning-of-sustainable-energy-systems-in-urban-built-environments-/9781779640642
17. Chen, X., & Liu, Q. (2023). Decentralized security systems using AI and blockchain: Concepts and case studies. Cybersecurity Review, 17(2), 85–102.
18. Singh, A., & Bhatt, M. (2024). Securing blockchain-based IoT networks with AI algorithms. IoT and Blockchain Journal, 6(2), 44–59.
19. Zhao, J., & Wang, Y. (2023). Blockchain and AI in the cybersecurity landscape: A survey of current trends. Journal of Cybersecurity, 21(1), 55–70.
20. Liu, P., & Zhou, Z. (2024). Decentralized AI for blockchain-enabled intrusion detection systems. Computational Intelligence and Cybersecurity, 12(3), 177–194.
21. Rana, R., & Bhambri, P. (2025). Generative AI-Driven Security Frameworks for Web Engineering: Innovations and Challenges. In I. Shah; & N. Jhanjhi (Eds.), Generative AI for Web Engineering Models (pp. 285–296). IGI Global. https://doi.org/10.4018/979-8-3693-3703-5.ch014
22. Zhao, W., & Wei, Y. (2023). Combining AI and blockchain for autonomous cybersecurity defense in smart cities. Journal of Smart Cities, 9(4), 124–139.
23. Wu, Y., & Zhang, X. (2024). Blockchain-powered AI for securing decentralized financial systems. IEEE Transactions on Blockchain, 5(1), 19–34.
24. Vigneshwari, J., Senthamizh Pavai, P., Maria Suganthi, L., & Bhambri, P. (2024). Eco-Ethics in the Digital Age: Tackling Environmental Challenges Through Technology. In P. Bhambri; & P. Bajdor (Eds.), Handbook of Technological Sustainability: Innovation and Environmental Awareness (pp. 201–213). CRC Press.
25. Mohana Sundari, V., Ganeshkumar, M., & Bhambri, P. (2024). Environmental Stewardship in the Digital Age: A Technological Blueprint. In P. Bhambri; & P. Bajdor (Eds.), Handbook of Technological Sustainability: Innovation and Environmental Awareness (pp. 57–67). CRC Press. https://doi.org/10.1201/9781003475989-6
26. Hong, J., & Zhang, H. (2023). Integrating AI with blockchain to secure data sharing in healthcare systems. Journal of Medical Cybersecurity, 14(2), 76–92.
27. Kim, S., & Park, J. (2024). Blockchain and AI for real-time cybersecurity in smart manufacturing. Journal of Industrial Internet Security, 11(1), 33–49.
28. Patel, D., & Thakur, P. (2024). Leveraging AI and blockchain for next-generation digital asset security. Digital Security Journal, 13(4), 189–202.
29. Bhambri, P. (2025). Innovative Systems: Entertainment, Gaming, and the Metaverse. In R. C. Ho; B. L. Song; & P. K. Tee (Eds.), Managing Customer-Centric Strategies in the Digital Landscape (pp. 483–514). IGI Global. https://doi.org/10.4018/979-8-3693-5668-5.ch018
30. Li, X., & Sun, L. (2023). Blockchain with AI for tamper-proof digital signatures in secure communication systems. Journal of Cryptography and Blockchain Technology, 17(1), 44–58.
31. He, R., & Luo, Q. (2024). Blockchain and AI for securing critical infrastructure: A comprehensive review. Critical Systems Security, 10(2), 211–227.
32. Patel, R., & Mehta, A. (2023). AI-based anomaly detection in blockchain-powered financial systems. Journal of Financial Technology, 15(3), 79–93.
33. Rana, R., & Bhambri, P. (2025). Generative AI in Web Application Development: Enhancing User Experience and Performance. In I. Shah; & N. Jhanjhi (Eds.), Generative AI for Web Engineering Models (pp. 471–486). IGI Global. https://doi.org/10.4018/979-8-3693-3703-5.ch021

34. Shen, J., & Zhao, T. (2024). The role of AI and blockchain in securing supply chains and logistics. Journal of Blockchain Applications, 11(1), 120–136.
35. Verma, S., & Aggarwal, M. (2023). Blockchain and AI for decentralized identity management in digital systems. IEEE Access, 11, 32345–32358.
36. Wang, Q., & Lin, X. (2024). Blockchain and AI for advanced persistent threat detection: A new paradigm. Journal of Cybersecurity Technologies, 17(2), 85–100.
37. Yang, Z., & Liu, G. (2023). AI-driven blockchain technology for decentralized risk management. Journal of Risk Management and Security, 6(4), 112–125.
38. Rana, R., & Bhambri, P. (2024). Environmental Challenges and Technological Solutions. In P. Bhambri; & P. Bajdor (Eds.), Handbook of Technological Sustainability: Innovation and Environmental Awareness (pp. 187–200). CRC Press. https://doi.org/10.1201/9781003475989-15
39. Shah, S., & Kumar, V. (2024). AI-enhanced blockchain protocols for securing decentralized applications. Journal of Software Security, 10(1), 45–58. https://doi.org/10.1016/j.jos.2024.02.003
40. Bhambri, P., & Kaur, P. (2015). Design and Implementation of Novel Algorithm Using Zero Watermarking: Digital Image Processing Technique for Text Documents (Vol. 1). Lap Lambert Academic Publishing. ISBN: 9783659796159.
41. Bhambri, P., & Kaur, J. (2020). Hybrid Classification Model for the Reverse Code Generation (Vol. 1). Lap Lambert Academic Publishing. ISBN: 9786202683432.

Index

Q

R

S

T

U

V

Z